Algorithmic reflections:

Selected works

Richard Crandall

PSIpress

Perfectly Scientific Press
www.perfscipress.com

Perfectly Scientific Press
3754 SE Knight St.
Portland, OR 97202

First Perfectly Scientific Press paperback edition: March 2012.
Perfectly Scientific Press paperback ISBN: 978-1-935638-19-3

Cover design by Julia Canright and Emily Buford.

Cover graphic by Ellen Crandall.

Visit our website at www.perfscipress.com

Printed in the United States of America.
9 8 7 6 5 4 3 2 1 0

Preface

In my college years I was admittedly envious of peers who could program machines and get answers thereby. I had thought that an algorist was something like the baseball player I could never be, that algorithms were works of art that my rougher hands—trained in physics as they were—could not create with sufficient care & delicacy. But in later years, I discovered for myself that algorithms are not just the "recipes," or "step-by-step" procedures of the dry dictionary world. Indeed, I now think of an algorithm in an heuristic sense: A pathway that wanders toward the truth. Even though "algorithm" is a noun, I have come to think of it in some sense as a verb. Or more precisely, the word algorithm carries for me an aura of actions & results. Just one example: A paper herein focuses essentially on an algorithm for diagnosing retinal (eye) disease. It is hard to write such out as step-by-step; yet, one now sees clearly how such diagnosis can be implemented.

Algorithmic reflections has, in common with a previous collection,[1] a distinct interdisciplinary flavor. The difference is that *Algorithmic reflections* is geared more toward actual problem-solving; or, should I say, new ways to look at very hard problems.

The opening of this collection, namely Part I: Number theory, has some older algorithms, but such were selected because they either a) stand as "models" for current work by other researchers, or b) as in the case of newer papers such as "On the googol-th bit of the Erdős–Borwein constant," employ algorithms developed by pioneers in experimental mathematics to illuminate modern problems. So this book part has something old, something new.

One might wonder why a paper such as "Effective Laguerre asymptotics"—essentially a foray into the deeper properties of Laguerre functions of large index—has to do with algorithms per se. In that case, the answer is: Classical contour integration, usually thought of as a fine abstraction, can actually be used for *numerical* evaluations. The contour algorithm, if you will, then leads to a new way—and yes, a companion algorithm—for developing an intricate asymptotic series with effective bounds. That paper is placed in Part II: Analytical algorithms, along with other treatments in which algorithms and analysis live in the same apartment building.

Part III focuses upon some problems of physics and biology/medicine. In the case of the physics paper "The potential within a crystal lattice," there is a meld of intuitive physics and algorithmic formulae—meaning relations that are programmer-friendly, what with their fast convergence and straightforward layout. Then a subcollection of papers enjoying biological themes finished the collection. Probably the very last paper is most representative of the theme of this whole compendium; namely, the fractal character of brain-synapse distributions is studied from an algorithmic point-of-view. Mathematics, leading to programmable methods in this instance, actually enables us to ascertain various truths in regard to brain structure.

R. E. Crandall
Portland Oregon
March 2012

---

[1] R. Crandall, *Scientific reflections: Selected multidisciplinary works*, PSIpress (2011).

Contents

Part I

Number theory

"If numbers aren't beautiful, I don't know what is."

—Erdős Pál

1. On the $3x + 1$ problem

Discussion

Perhaps the most salient thing to observe about the "$3x + 1$" problem—even before defining said problem—is that the eminent Paul Erdős supposedly once said "Mathematics is not yet ripe for such problems."

The problem is one of those special beauties for which it can be said a child could understand its posing, but nobody, of any age or station, has solved it.

As for the problem, also called the Collatz problem (amongst many other names), it starts thus: If a number n is odd, it becomes $3n + 1$, else if n is even, it becomes $n/2$. Plain and simple. But the open question is, does every starting number n eventually reach 1? At 1, the orbit just goes $1 \to 4 \to 2 \to 1 \to$ ad infinitum. Starting with 27, the orbit goes $27 \to 82 \to 41 \to 124 \to \cdots \to 1$ but involving 111 arrows!

It was over 30 years ago that my fascination with this problem began, and the enclosed paper was my first publication. Not profound, I think, but one can see the basic style here of an aspiring young interdisciplinarian. (Imagine a young physicist trying to grapple with what is more like a computer-science problem...in the 1970s...) My main result on this 1978 pass was that the number of starting n not exceeding x that do end up at 1 is greater than x^c, i.e. a positive power of x. Since 1978, the positive power c has gone up, and a great many other accomplishments by enthusiastic researchers have been posited. (Even on a very hard problem, several decades can bring at least partial progress.)

Good summaries of the last few decades of effort are available.[1]

Challenges

One challenge is to try to significantly advance the known bound on successful starting numbers. Evidently, it is known (by 2011) that all $n \leq 2^{60}$ do end up at 1.

Source

Crandall, R.E. "On the $3x + 1$ problem," *Math. Comp.*, **32**, 1281–1292 (1978).

---

[1] A classic compendium is:
J. C. Lagarias, "The 3x + 1 problem and its generalizations."' American Mathematical Monthly 92 (1): 3Ð23 (1985).
while a modern statement of records is found at:
`http://www.ericr.nl/wondrous/`
while an exceptionally good brief with useful bibliography is at:
`http://mathworld.wolfram.com/CollatzProblem.html`

MATHEMATICS OF COMPUTATION, VOLUME 32, NUMBER 144
OCTOBER 1978, PAGES 1281–1292

On the "$3x+1$" Problem

By R. E. Crandall

Abstract. It is an open conjecture that for any positive odd integer m the function

$$C(m) = (3m + 1)/2^{e(m)},$$

where $e(m)$ is chosen so that $C(m)$ is again an odd integer, satisfies $C^h(m) = 1$ for some h. Here we show that the number of $m \leqslant x$ which satisfy the conjecture is at least x^c for a positive constant c. A connection between the validity of the conjecture and the diophantine equation $2^x - 3^y = p$ is established. It is shown that if the conjecture fails due to an occurrence $m = C^k(m)$, then k is greater than 17985. Finally, an analogous "$qx + r$" problem is settled for certain pairs $(q, r) \neq (3, 1)$.

1. Introduction. The "$3x + 1$" problem enjoys that appealing property, often attributed to celebrated number-theoretic questions, of being quite easy to state but difficult to answer. The problem can be expressed as follows: define, for odd positive integers x, the function:

$$(1.1) \qquad\qquad C(x) = (3x + 1)/2^{e(x)},$$

where $2^{e(x)}$ is the highest power of two dividing $3x + 1$. Since $C(x)$ is again an odd integer, one can iterate the C function any number of times. The problem: for any initial odd positive x is some iterate $C^k(x)$ equal to one?

This "$3x + 1$" problem has found a certain niche in modern mathematical folklore, without, however, having been extensively discussed in the literature. The function C defined in equation (1.1) is essentially Collatz' function [1] but the true origin of the problem seems obscure. The algorithm defined by successive iteration of C has been called the "Syracuse algorithm" [2]. Some authors [3], [8] have defined functions equivalent to C and proclaimed the conjecture, that $C^k(x) = 1$ for some $k(x)$, a long-standing one. Some partial results concerning the conjecture are known, but for the most part the behavior of the C function remains shrouded in mystery. It is hoped that the partial results contained in the next sections will shed some light on the problem.

2. Preliminary Observations. Let Z^+ denote the positive integers and let D^+ denote the odd elements of Z^+. Some elementary properties of the function C; $D^+ \rightarrow D^+$ as defined in (1.1) will now be discussed.

DEFINITION. For $m \in D^+$ the *trajectory* of m is the sequence $T_m = \{C(m), C^2(m), \dots \}$, where it is understood that the sequence terminates upon the first occurrence of $C^k(m) = 1$, $k \in Z^+$. If there is no such k, then T_m is an infinite sequence.

Received March 28, 1977; revised March 20, 1978.
AMS (MOS) subject classifications (1970). Primary 10A25.
Key words and phrases. Algorithm, diophantine equation.

DEFINITION. For $m \in D^+$ the *height* of m, denoted $h(m)$, is the cardinality of the trajectory T_m. In the case that T_m is a finite sequence, $h(m)$ will be the least number of iterations of C required to reach 1.

DEFINITION. For $m \in D^+$, we denote by $\inf T_m$ the least positive integer in the sequence T_m. Further, if T_m is bounded, we denote by $\sup T_m$ the greatest integer in the sequence T_m. If T_m is unbounded, we say that $\sup T_m$ is infinite.

The following table should serve as an example for the previous notation:

m	T_m	$h(m)$	$\sup T_m$
1	$\{1\}$	1	1
7	$\{11, 17, 13, 5, 1\}$	5	17
27	$\{41, \ldots, 1\}$	41	3077
$2^{1000} - 1$	$\{?\}$	4316	$> 10^{476}$
$2^{1000} + 1$	$\{?\}$	2417	$< 10^{301}$
$2^{4096} - 1$	$\{?\}$	19794	?

It is partly the erratic behavior of the height function h that gives interest to the "$3x + 1$" problem. The main conjecture is:

CONJECTURE (2.1). *For every $m \in D^+$, $h(m)$ is finite.*

This unsolved conjecture has been verified for $m < 10^9$ [1], [5]. One of the few partial results concerning the problem is that of Everett [3]:

THEOREM (2.1) (EVERETT). *For almost all $m \in D^+$, $\inf T_m < m$.*

This theorem is interesting because if $\inf T_m < m$ for all $1 < m \in D^+$, then the main conjecture (2.1) is clearly true. It should be noted, however, that while Theorem (2.1) reveals a definite tendency for trajectories T_m to descend below their generating integers m, the theorem has little to say concerning Conjecture (2.1). In fact, it is not even known whether a positive density of odd integers m satisfy the conjecture.

The next observation gives further indication of computational difficulties encountered in the "$3x + 1$" problem.

THEOREM (2.2). *As $m \in D^+$ increases, $(\sup T_m)/m$ is unbounded.*

Proof. Define, for $k \in Z^+$, the number $m_k = 2^k - 1$. It is readily verified that $C(m_k) = 3 \cdot 2^{k-1} - 1$ if $k > 1$, and in general $C^j(m_k) = 3^j 2^{k-j} - 1$ if $k > j$. Thus, for $k > 1$ the number

$$C^{k-1}(m_k) = 3^{k-1} 2 - 1$$

is a member of the trajectory of m_k. Therefore, for $k > 1$:

$$\frac{\sup T_{m_k}}{m_k} \geqslant \frac{3^{k-1} 2 - 1}{2^k - 1} > (3/2)^{k-1}$$

and the right-hand side grows without bound as k runs through the positive integers.

It is evident that numerical calculations of the heights of various numbers by computer methods must necessarily involve the storage of trajectory elements which are, in proportion to the starting numbers m, arbitrarily large; unless, of course, some theoretical method is discovered to simplify such computations. It is natural to ask whether $(\sup T_m)/m^a$ is unbounded for various powers a. The set of integers used in Theorem (2.2) is sufficient to show unboundedness only for $a < \log_2 3$. The irrational number $t = \log_2 3 = \log 3/\log 2$ will arise in a natural way in later results.

3. A Random-Walk Argument. An heuristic argument that lends credibility to the main conjecture (2.1) can be stated as follows. Assume that for odd integers m sufficiently large the real number $\log(C(m)/m)$ is a "random variable" with a distribution determined by the behavior of the function $e(m)$ as defined in Eq. (1.1). One "expects" that $e(m) = k$ with probability 2^{-k}. Thus, $\log(C(m)/m)$ is approximately $\log 3 - k \log 2$ with probability 2^{-k}. Therefore, one "expects" the number $\log(C(m)/m)$ to be:

$$\sum_{k \in Z^+} 2^{-k} \log(3/2^k) = -\log(4/3),$$

indicating a tendency for $C(m)$ to be less than m. If one then imagines that iteration of the C function induces a random walk beginning at $\log m$ on the real number line, an heuristic estimate for the height function $h(m)$ might be:

$$(3.1) \qquad h(m) \sim \frac{\log m}{\log(4/3)}.$$

What is notable about this argument, aside from its lack of precision, is that estimate (3.1) is not too far from the mark in a certain sense. Since we expect the h function to behave erratically, we define a smoother function called the average order of h:

$$(3.2) \qquad H(x) = \frac{2}{x} \sum_{\substack{m \leqslant x \\ m \in D^+}} h(m).$$

The heuristic estimate (3.1) translates into an estimate for $H(x)$:

$$(3.3) \qquad H(x) \sim 2(\log(16/9))^{-1} \log x = (3.476 \ldots) \log x.$$

This simple statistical argument seems to be partially supported by the following data:

x	$H(x)/\log x$
11	1.440 ...
101	2.546 ...
1001	3.206 ...
10001	3.330 ...
100001	3.298 ...

It would be of interest to compute $H(x)/\log x$ for some much larger value of x, say

10^{10}. In the absence of such knowledge, we simply conjecture:

CONJECTURE (3.1). $H(x) \sim 2 \log x / \log(16/9)$.

This conjecture is stronger than the main conjecture (2.1) in the sense that if there be even one $m \in D^+$ with infinite height, then (3.1) is false.

4. Uniqueness Theorem. We shall presently establish a certain uniqueness theorem which is useful in obtaining partial results concerning the main conjecture (2.1). Define, for $a \in Z^+$ and n rational the function:

$$(4.1) \qquad\qquad B_a(n) = (2^a n - 1)/3 .$$

In general, $B_a(n)$ is not an integer. Further define, for (backwards-ordered) sequences

$$\{a_i | i = k, \, k - 1, \ldots, 2, 1; a_i, \, k \in Z^+ \},$$

the functions

$$(4.2) \qquad B_{a_k \cdots a_1}(n) = B_{a_k}(B_{a_{k-1} \cdots a_1}(n)) = (2^{a_k} B_{a_{k-1} \cdots a_1}(n) - 1)/3.$$

Every function $B_{\{a_i\}}(n)$ is thus rational by construction for any sequence $\{a_j, a_{j-1}, \ldots, a_2, a_1\} = \{a_i\}$.

LEMMA (4.1).   *If* $n \in D^+$ *and* $B_{a_j \cdots a_1}(n)$ *is an integer, then for* $1 \leqslant i < j$ *all numbers* $B_{a_i \cdots a_1}(n)$ *are odd integers; and further*

$$C(B_{a_{i+1} \cdots a_1}(n)) = B_{a_i \cdots a_1}(n), \qquad C(B_{a_1}(n)) = n.$$

Proof. Assume for some k with $1 < k \leqslant j$ that $B_{a_k \cdots a_1}(n) \in Z^+$. Then from (4.2) we conclude $2^{a_k} B_{a_{k-1} \cdots a_1}(n)$ is also an integer. But by construction, $B_{a_{k-1} \cdots a_1}(n) = y/3^{k-1}$ for some integer y since n is an integer. Since 2 and 3 are coprime it follows that $B_{a_{k-1} \cdots a_1}(n)$ is an integer itself. Thus, as $B_{a_j \cdots a_1}(n)$ is an integer it follows by induction that all $B_{a_i \cdots a_1}(n)$ for $1 \leqslant i < j$ are integers. Further, from (4.2) we conclude that $3 B_{a_i \cdots a_1}(n) + 1$ is even so each $B_{a_i \cdots a_1}(n)$ must be odd. Finally,

$$C(B_{a_{i+1} \cdots a_1}(n)) = B_{a_i \cdots a_1}(n) 2^{a_{i+1} - e} = B_{a_i \cdots a_1}(n)$$

from (4.2), (1.1), and the fact that application of the C function on an odd integer yields another odd integer. That $C(B_{a_1}(n)) = n$ follows from (4.1).

LEMMA (4.2).   *If an integer* $m = B_{a_j \cdots a_1}(1)$ *and* $a_1 > 2$, *then the trajectory of* m *is*

$$T_m = \{B_{a_{j-1} \cdots a_1}(1), B_{a_{j-2} \cdots a_1}(1), \ldots, B_{a_1}(1), 1\}.$$

Proof. From Lemma (4.1) the assumption $m = B_{a_j \cdots a_1}(1)$ is an integer yields the correct iterations for the trajectory. It remains to show that none of the $B_{a_i \cdots a_1}(1)$ is equal to 1. This follows from the fact that $C(1) = 1$, so that we only need show $B_{a_1}(1) \neq 1$. But this follows from the assumption $a_1 > 2$.

DEFINITION.   Denote by G the set of finite sequences $\{a_j, a_{j-1}, \ldots, a_1\}$, where each $a_i \in Z^+$; $a_1 > 2$; and the following congruences are satisfied:

$$2^{a_1} \equiv 1 \pmod 3$$
$$\not\equiv 1 \pmod 9,$$
$$2^{a_i} B_{a_{i-1} \cdots a_1}(1) \equiv 1 \pmod 3 \qquad \text{for } 2 \leqslant i \leqslant j-1,$$
$$\not\equiv 1 \pmod 9$$
$$2^{a_j} B_{a_{j-1} \cdots a_1}(1) \equiv 1 \pmod 3.$$

LEMMA (4.3). *Let* $\{a_i\} = \{a_j, a_{j-1}, \ldots, a_1\}$. *Then* $B_{\{a_i\}}(1)$ *is an integer of height* j *if and only if* $\{a_i\} \in G$.

Proof. Assume $B = B_{a_j \cdots a_1}(1)$ is an integer of height j. Then from Lemma (4.2) each $B_{a_j \cdots a_1}(1)$ is in the trajectory of B, and thus $B_{a_1}(1) \neq 1$, so $a_1 > 2$. The congruences (mod 3) follow from Eq. (4.2) and the congruences (mod 9) follow from (4.2) applied twice. Thus, the sequence $\{a_i | i = j, j-1, \ldots, 1\}$ is in G.

Now assume $\{a_i\}$ is in G. Then the last congruence implies $B_{a_j \cdots a_1}(1)$ is an integer, and Lemma (4.2) shows this integer has height j.

The previous lemmas and definition of the set of sequences G enable us to prove the uniqueness theorem:

THEOREM (4.1). *Consider the set of integers which satisfy the main conjecture* (2.1):

$$M = \{ m \in D^+, \, m > 1 \, | \, h(m) \text{ finite} \}.$$

Then, for each $m \in M$ *there is a unique sequence* $\{a_i\} \in G$ *such that*

$$m = B_{a_{h(m)} \cdots a_1}(1),$$

and conversely, for each $\{a_i\} \in G$, $B_{\{a_i\}}(1) \in M$.

Proof. If $m \in M$, then define the integers

$$a_i = e(C^{h(m)-i}(m)) \quad \text{for } 1 \leqslant i \leqslant h(m),$$

where the e function is as defined in (1.1). Then by inspection we have $m = B_{a_{h(m)} \cdots a_1}(1)$ and by Lemma (4.3) we have $\{a_i\} \in G$. To show uniqueness, assume for some $\{b_i | i = k, k-1, \ldots, 1\}$ we have $m = B_{b_k \cdots b_1}(1)$. Then from Lemma (4.2) $k = h(m)$ and, since the C function is well defined, the exponents b_i must agree termwise with the a_i so the sequences $\{a_i\}$ and $\{b_i\}$ are equal. For the converse, we note that $\{a_i\} \in G$ implies $B_{\{a_i\}}(1)$ has finite height by Lemma (4.3)

5. Numbers with Given Height. The set M of integers $m > 1$, $m \in D^+$ of finite height is naturally partitioned by the height function h. There are infinitely-many numbers m with $h(m) = 1$, in fact these m are just numbers of the form $(4^k - 1)/3$. It is natural to ask whether there are always numbers of any given height. This question can be answered in the affirmative. We shall give an estimate of the relative density of the odd integers with given height.

LEMMA (5.1). $B_{a_j \cdots a_1}(1) < 2^{a_1 + a_2 + \cdots + a_j}/3^j.$

Proof. Since $B_{a_1}(1) = (2^{a_1} - 1)/3$, the assertion is true for $j = 1$. From Eq. (4.2) the result follows by induction.

We shall also need a lower bound for the number of solutions to the congruences that define the set G:

LEMMA (5.2). *For a real number $z > 0$ the number of sequences of length j in the set G with $a_1 + \cdots + a_j \leqslant z$ is greater than or equal to $(2[(z-2)/6j])^j$.*

Proof. Solutions can be restricted by the inequalities

$$a_1 - 2 \leqslant (z-2)/j; \quad a_i \leqslant (z-2)/j \quad \text{for } 1 < i \leqslant j,$$

which together force the sum $a_2 + \cdots + a_j$ to be $\leqslant z - a_1$. The number of ways of choosing the quantity $a_1 - 2 \leqslant (z-2)/j$ is at least $2[(z-2)/6j]$; since only $2^{a_1} \equiv 4$ or $7 \pmod 9$ is required, and out of every six consecutive integers at least two must satisfy the congruence. Similarly, once a_1 is chosen there must be at least $2[(z-2)/6j]$ choices for a_2 since the congruence $2^{a_2}B_{a_1}(1) \equiv 4$ or $7 \pmod 9$ has at least two solutions a_2 out of every set of six consecutive integers. In this way the number of solutions to $a_1 + \cdots + a_j \leqslant z$ is seen to be at least the jth power of the quantity $2 \cdot [(z-2)/6j]$.

These lemmas can be used to estimate the number of $m \in D^+$, $m \leqslant x$ which have a given height h.

THEOREM (5.1). *Let $\pi_h(x)$ be the number of $m \in M$ with $h(m) = h$ and $m \leqslant x$. Then there exist real positive constants r, x_0 independent of h such that for $x > \max(x_0, 2^{h/r})$,*

$$\pi_h(x) > (\log_2^h(x^r))/h!.$$

Remark. This theorem shows that for any positive integer h there are infinitely many $m \in D^+$ with height h.

Proof. From the uniqueness theorem (4.1) and Lemma (5.1) it follows that $\pi_h(x)$ is at least as large as the number of sequences $\{a_i\} \in G$, of length h, with $a_1 + \cdots + a_h \leqslant \log_2(3^h x)$. By Lemma (5.2) this implies

$$\pi_h(x) \geqslant \left(\frac{\log_2(3^h x/4)}{3h} - 2\right)^h,$$

obtained by bounding the square-brackets from below. Now we choose x_0 such that $x > x_0$ implies $x/4 > x^{1/2}$; and choose a real number r, $0 < r < 1$, such that

$$r\log_2(3/64) + 1/2 > 3er > 0.$$

Then, noting from Stirling's formula that $h!e^h > h^h$, we can transform our last inequality for $\pi_h(x)$, in the case that x is greater than both x_0 and $2^{h/r}$, to

$$\pi_h(x) \geqslant h^{-h}\left(\frac{\log_2 x^{r\log_2(3/64)+1/2}}{3}\right)^h > \frac{\log_2^h(x^r)}{h!}.$$

6. An Estimate for $\pi(x)$. Let $\pi(x)$ be the number of integers $m \leqslant x$ which belong to the set M of Theorem (4.1). Conjecture (2.1) is equivalent to the statement that $\pi(x)$ is precisely the number of odd integers greater than one but not greater than x. We establish here a lower bound for the function $\pi(x)$.

ON THE "$3x + 1$" PROBLEM   **1287**

LEMMA (6.1).

$$\lim_{t \to \infty} e^{-t} \sum_{\substack{u \in Z^+ \\ u \leqslant [t]}} \frac{t^u}{u!} = \frac{1}{2}.$$

Remark. This lemma can be proved using asymptotic expansions of certain error functions. An exposition of various formulas pertaining to the lemma can be found in [4].

Using the results of the last section, we now prove

THEOREM (6.1). *There exists a positive constant c such that for sufficiently large x*

$$\pi(x) > x^c.$$

Proof. Using Theorem (5.1), we can find $r > 0$ and $x_0 > 0$ such that $x > x_0$ implies

$$\pi(x) = \sum_{h \in Z^+} \pi_h(x) \geqslant \sum_{h=1}^{[r \log_2 x]} \frac{\log_2^h(x^r)}{h!}.$$

From Lemma (6.1) we deduce that for any $d > 0$ there is an $x_1 > x_0$ such that $x > x_1$ implies

$$\pi(x) > (1/2 - d)e^{r \log_2 x},$$

which is enough to obtain the lower bound x^c.

7. Cycles. Assume that all trajectories T_m for $m \in D^+$ are bounded, and assume that no $m > 1$ appears in its own trajectory. Then the main conjecture (2.1) is true, for under the two assumptions, the iterates $C(m)$, $C^2(m), \ldots, C^{\sup T_m}(m)$ are distinct except for possible ones; and since each iterate is less than $\sup T_m$, the number 1 must in fact appear in the list of iterates. It is not known whether either of the two assumptions is true. We shall show that numbers $m > 1$ which appear in their own trajectories, if they exist at all, are necessarily difficult to uncover. More precisely, we shall establish a lower bound for the period of an infinite cyclic trajectory in terms of its smallest member.

Let $m = B_{b_k \ldots b_1}(n)$ for some positive integer sequence $\{b_i\}$. Define

(7.1)
$$A_i = \sum_{j=k-i+1}^{k} b_j \quad \text{for } 1 \leqslant i \leqslant k;$$

$$A_0 = 0.$$

Then the B function can be expanded to give the identity:

(7.2)
$$2^{A_k} n - 3^k m = \sum_{j=0}^{k-1} 2^{A_j} 3^{k-1-j}.$$

The identity always follows from the statement that $n \in T_m$, for if n appears at the kth position of T_m then the assignment $b_j = e(C^{k-j}(m))$ gives $m = B_{b_k \ldots b_1}(n)$ and finally (7.2). Conversely, if (7.2) holds for some monotone sequence of increasing integers $0 = A_0 < A_1 < \cdots < A_k$, then the assignment $d_i = A_{k-i+1} - A_{k-1}$ gives $m = B_{d_k \cdots d_1}(n)$, so by Lemma (4.1) $C^k(m) = n$, implying either $n = 1$ or n appears in the kth position of T_m. In either case $n \in T_m$ and we have demonstrated

THEOREM (7.1). *Let* $m, n \in D^+$. *Then* $n \in T_m$ *if and only if there exists a sequence of integers*

$$0 = A_0 < A_1 < A_2 < \cdots < A_k$$

such that Eq. (7.2) holds. Further, if such a sequence $\{A_i\}$ *exists, then* $n = 1$ *or the* kth *element of* T_m *is* n.

COROLLARY (7.1). *If* $1 < m \in D^+$ *and* $m \in T_m$ *and the period of* T_m *is* k, *then there exists a sequence of integers* $0 = A_0 < A_1 < \cdots < A_k$ *such that*

$$(7.3) \qquad m(2^{A_k} - 3^k) = \sum_{j=0}^{k-1} 2^{A_j} 3^{k-1-j}.$$

Conversely, if Eq. (7.3) holds for such a monotone sequence, then $m \in T_m$ *and either* $m = 1$ *or* m *is the* kth *element of* T_m.

It is of interest that if m appears in its own trajectory, then in Eq. (7.3) the factor $2^{A_k} - 3^k$ must divide the right-hand side. If Conjecture (2.1) is true, the diophantine equation (7.3) must have no solutions for $m > 1$, subject to the constraint of monotonicity on the $\{A_i\}$. It is known that the equation $2^x - 3^y = z$ has only finitely-many solutions in integers x, y for each value of z [9] and, in fact, can have at most one solution for sufficiently large z [10]. But sharp results for special prime values of z can be obtained on the assumption that the main conjecture (2.1) is true.

If Conjecture (2.1) is true, then the impossibility of solutions to the diophantine equation (7.3) can be shown to imply that if 2 is a primitive root of a prime p, any solution in integers x, y to $2^x - 3^y = p$ must have $y < p/(t - 1)$, where $t = \log_2 3$.

It is of interest that if a pair of integers (a, b) can be found such that $b > 1$ and

$$0 < 2^a - 3^b \quad \text{divides} \quad 2^{a-b} - 1,$$

then the main conjecture (2.1) is false. Indeed, if we write

$$2^{a-b} - 1 = d(2^a - 3^b),$$

then the number $m = 2^b d - 1$ satisfies:

$$m(2^a - 3^b) = 3^b - 2^b = \sum_{j=0}^{b-1} 2^j 3^{b-j-1},$$

and since $m > 1$, Corollary (7.1) implies $h(m)$ is infinite.

Corollary (7.1) shows that if the main conjecture is true then powers of two and three tend to be poor approximations of each other. The number $t = \log_2 3$ must be accordingly difficult to approximate with rational numbers. This notion will be made precise shortly. For the moment we display the first 50 elements of the continued fraction for $t = \log_2 3$

ON THE "$3x + 1$" PROBLEM   1289

(7.5)
$$t = [1, 1, 1, 2, 2, 3, 1, 5, 2, 23, 2, 2, 1, 1, 55, 1, 4, 3, 1, 1, 15,$$
$$1, 9, 2, 5, 7, 1, 1, 4, 8, 1, 11, 1, 20, 2, 1, 10, 1, 4, 1, 1, 1$$
$$1, 1, 37, 4, 55, 1, 1, 49, \ldots]$$
$$= [a_0, a_1, a_2, \ldots, a_{49}].$$

The convergents to this fraction are determined by recurrences [6], [7]

(7.6)
$$\begin{aligned}
p_0 &= a_0; & p_{-1} &= 1; \\
q_0 &= 1; & q_{-1} &= 0; \\
p_n &= a_n p_{n-1} + p_{n-2} & &\text{for } n \in Z^+; \\
q_n &= a_n q_{n-1} + q_{n-2} & &\text{for } n \in Z^+.
\end{aligned}$$

The ratio p_n/q_n is called the nth convergent to t.

The next three lemmas stem from the theory of rational approximation. Relevant material can be found in [6] and [7].

LEMMA (7.1). *Let p_n/q_n denote the nth convergent to $t = \log_2 3$. Then for any pair of integers (x, y) with $y < q_n$,*

$$|p_n - q_n t| < |x - yt|.$$

LEMMA (7.2). *For p_n/q_n convergents to t,*

$$|p_n - q_n t| > (q_n + q_{n+1})^{-1}.$$

LEMMA (7.3). *For p_n/q_n convergents to t, let $y < q_m$. Then*
$$|2^x - 3^y| > 3^y \log 2 |p_m - q_m t|.$$

Proof.

$$\begin{aligned}
|2^x - 3^y| &= 3^y |\exp(x \log 2 - y \log 3) - 1| \\
&> 3^y \log 2 |x - yt|.
\end{aligned}$$

When $y < q_m$, Lemma (7.1) implies the desired inequality.

We now focus our attention on numbers $m > 1$ for which $m \in T_m$. It is clear that every infinite cyclic trajectory contains such an m.

LEMMA (7.4). *If $1 < m = \inf T_m$, then for the A_j as defined in Eq. (7.1),*

$$2^{A_j} \leqslant (3 + 1/m)^j.$$

Proof. From $C(x) = (3x + 1)/2^{e(x)}$ we infer, since $m \leqslant C^j(m)$ for each j, that

$$C^{j+1}(m) \leqslant C^j(m)(3 + 1/m)/2^{e(C^j(m))}.$$

But this implies

$$m \leqslant C^j(m) \leqslant m(3 + 1/m)^j/2^{A_j}$$

giving the desired inequality.

LEMMA (7.5). *Let $1 < m = \inf T_m$ and let k be the period of the trajectory T_m. Then*

$$m < k(3 + 1/m)^{k-1}/(2^{Ak} - 3^k).$$

Proof. From Corollary (7.1) and Lemma (7.4) we have

$$m(2^{4}k - 3^{k}) < \sum_{j=0}^{k-1} (3 + 1/m)^{k-1}$$

and the desired inequality follows from direct evaluation of the sum.

These lemmas enable us to establish a connection between the size of the period of a possible cyclic trajectory and the continued fraction for $t = \log_2 3$.

THEOREM (7.2). *Let p_n/q_n be convergents to t. Let $1 < m = \inf T_m$ and let k be the period of T_m. Then for $n \geq 4$:*

$$k > \min(q_n, 2m/(q_n + q_{n+1})).$$

Proof. If $k > q_n$, the result is trivial so assume $k \leq q_n$. Then from Lemmas (7.3) and (7.5) we have

$$m < \frac{k(3 + 1/m)^{k-1}}{3^k \log 2|p_n - q_n t|},$$

from which we see that

$$k > m(\log 2)(3 + 1/m)|p_n - q_n t|(1 - k/3m).$$

But $n \geq 4$ implies $q_n + q_{n+1} > 20$ using Eq. (7.5). Thus, if $k \geq m/10$, the result of the theorem is trivially true. So we assume $k < m/10$. But this implies $(1 - k/3m) > 29/30$ and from Lemma (7.2),

$$k > \frac{(3m + 1) \log 2}{q_n + q_{n+1}} (29/30) > \frac{2m}{q_n + q_{n+1}}.$$

COROLLARY (7.2). *For any given k there are finitely many cyclic trajectories with period k.*

Proof. This corollary follows directly from the observation that $\min(q_n, (2 \inf T_m)/(q_n + q_{n+1}))$ is unbounded for any infinite set of trajectories as n increases.

The known result that no $1 < m < 10^9$ appears in its own trajectory can be used to establish a lower bound for any period k.

THEOREM (7.3). *Assume $m > 1$ and $C^k(m) = m$. Then $k > 17985$.*

Proof. For $1 < m = C^k(m)$ the trajectory T_m is infinite cyclic, and there is thus an $m_0 = \inf T_m$ with $m_0 > 10^9$. Since $m_0 = C^k(m_0)$, the number k is greater than or equal to the period of T_{m_0}. Therefore, from Theorem (7.2) we infer that

$$k > \min(q_n, 2 \cdot 10^9/(q_n + q_{n+1}))$$

for any $n \geq 4$. From Eq. (7.5) we compute:

$$q_{10} = 31867; \quad q_{11} = 79335;$$

so either $k > 31867$ or $k > 2 \cdot 10^9/111202$. The latter bound is greater than 17985.

ON THE "3x + 1" PROBLEM   **1291**

8. The "$qx + r$" Problem. Define for $q, r \in D^+$; $q > 1$ the function

(8.1) $$C_{qr}(m) = (qm + r)/2^{e_{qr}(m)}$$

valid for $m \in D^+$ with $e_{qr}(m)$ always chosen to force $C_{qr}(m) \in D^+$. The generalized "$qx + r$" problem can be stated thus: given (q, r), for every $m \in D^+$ is there a k such that $C_{qr}^k(m) = 1$? What is remarkable about this problem is that strong evidence points to the following rather curious conjecture:

CONJECTURE (8.1). *In the "$qx + r$" problem, with $q, r \in D^+$ and $q > 1$, some $m \in D^+$ fails to satisfy an equation $C_{qr}^k(m) = 1$; except in the case $(q, r) = (3, 1)$.*

This conjecture says in a word that unless $q = 3$ and $r = 1$, the generalized height function $h_{qr}(m)$ will be infinite for some $m \in D^+$. We shall presently attempt to argue the plausibility of the conjecture.

It is easy to prove Conjecture (8.1) in the special case $r > 1$. Indeed for such an r choose any $m \equiv 0 \pmod r$. Then $C_{qr}(m) \cdot 2^{e_{qr}(m)} = qm + r \equiv 0 \pmod r$ and, as r is odd, $C_{qr}(m) \equiv 0 \pmod r$. But this means all iterates of m are destined to be divisible by r, hence $C_{qr}^k(m) = 1$ is impossible for $r > 1$ and $m \equiv 0 \pmod r$.

The status of Conjecture (8.1) is unsettled for most of the remaining "$qx + 1$" problems. It is known that the conjecture is true for $q = 5, 181$, and 1093; but all other cases remain elusive.

The case $q = 5$ is settled by the observation that $m = 13$ gives rise to a cyclic trajectory $\{33, 83, 13, \dots\}$, whose occurrence can be traced back to the diophantine equation $2^7 - 5^3 = 3$; and an equation similar to (7.3) exists with 3 replaced by 5 and k set equal to 3.

The case $q = 181$ is resolved by the observation that $m = 27$ gives rise to a cyclic trajectory $\{611, 27, \dots\}$; arising from the diophantine equation $2^{15} - 181^2 = 7$.

The third solvable case, $q = 1093$, is rather peculiar. In this problem, a number m has height one if and only if it is of the form

$$m = (2^{364p} - 1)/1093;$$

but, as it turns out, all other m have infinite height. This follows from the fact that in the "$1093x + 1$" problem, no m can have height 2, since Eq. (7.2), with 3 replaced by 1093, $n = 1$, and k set equal to two, is easily seen to be impossible. This case is the only one for which it is known that almost all m have infinite height.

The examples $q = 5, 181, 1093$ being the only known q for which Conjecture (8.1) is true (with $r = 1$), why is the conjecture plausible? The answer is, simply, it is likely that for any $q > 3$, some m has a diverging trajectory in the "$qx + 1$" problem. In fact, the heuristic arguments of Section 3 indicate that $\log(C_{q1}(m)/m)$ should be about $\log(q/4)$, which is positive for odd $q > 3$, so that numbers should tend to be "pushed upward" by application of the C_{q1} function.

In spite of the above considerations, it is unknown whether even a single m in a single "$qx + 1$" problem gives rise to an unbounded trajectory. The outstanding unsolved case is the "$7x + 1$" problem, for which there may be no infinite cyclic

1292 R. E. CRANDALL

trajectories. What makes the "$7x + 1$" problem all the more interesting is the empirical observation that the number $m = 3$ gives rise to a trajectory reaching to 10^{2000} and beyond, with no apparent tendency to return.

Acknowledgements. The author wishes to acknowledge the hospitality of the administration of Reed College, at whose Computer Center the calculations for this paper were performed. The author also wishes to thank Joe Roberts and J. P. Buhler for their valuable suggestions concerning the manuscript.

2234 S. E. 20th Street
Portland, Oregon 97214

1. J. H. CONWAY, "Unpredictable iterations," *Proc.* 1972 *Number Theory Conference*, Univ. of Colorado, Boulder, Colorado, 1972, pp. 45–52.

2. I. N. HERSTEIN & I. KAPLANSKY, *Matters Mathematical*, 1974.

3. C. J. EVERETT, "Iteration of the number-theoretic function $f(2n) = n$, $f(2n + 1) = 3n + 2$," *Advances in Math.*, v. 25, 1977, pp.42–45.

4. M. ABRAMOWITZ & I. STEGUN (Editors), *Handbook of Mathematical Functions*, 9th printing, Dover, New York, 1965.

5. *Zentralblatt für Mathematik*, Band 233, p. 10041.

6. J. ROBERTS, *Elementary Number Theory– A Problem Oriented Approach*, M.I.T. Press, Cambridge, Mass., 1977.

7. Y. A. KHINCHIN, *Continued Fractions*, Univ. of Chicago, Chicago, Ill., 1964.

8. *Pi Mu Epsilon Journal*, v. 5, 1972, pp. 338, 463.

9. S. S. PILLAI, *J. Indian Math. Soc.*, v. 19, 1931, pp. 1–11.

10. A. HERSCHEFELD, *Bull. Amer. Math. Soc.*, v. 42, 1936, pp. 231–234.

2. On a conjecture of Crandall concerning the $qn + 1$ problem

Discussion

This paper is by two good colleagues; I include it partly out of pride, but also to show that a rather minor thought—in this case, concerning orbits in the "$qn + 1$" problem—can turn into nontrivial number-theoretical ruminations. I was completely surprised to find out thus from Franco and Pomerance that almost all numbers have what would seem on the face of it a rare property.

Challenges

Various open problems are stated within this next paper. The previous paper by myself, on its p. 1291, has the basics of the $qn + 1$ problem, and an earlier set of open questions.

Source

Z. Franco and C. Pomerance, "On a conjecture of Crandall concerning the $qn + 1$ problem," *Math. Comp.*, **64**, 1333–1336 (1995).

MATHEMATICS OF COMPUTATION
VOLUME 64, NUMBER 211
JULY 1995, PAGES 1333–1336

ON A CONJECTURE OF CRANDALL CONCERNING
THE $qx + 1$ PROBLEM

ZACHARY FRANCO AND CARL POMERANCE

ABSTRACT. R. E. Crandall has conjectured that for any odd integer $q > 3$, there is a positive integer m whose orbit in the "$qx + 1$ problem" does not contain 1. We show that this is true for almost all odd numbers q, in the sense of asymptotic density.

For positive odd numbers q and m, let $C_q(m)$ denote the largest odd factor of $qm + 1$. For $q = 3$, the sequence of iterates m, $C_3(m)$, $C_3(C_3(m))$, ... consists of the odd numbers in the orbit of m under the $3x + 1$ function. The (unproved) $3x + 1$ conjecture asserts that such an orbit always contains the number 1, cf. [3]. When $q > 3$, one might still consider the sequence of iterates m, $C_q(m)$, $C_q(C_q(m))$, ... Let $C_q^j(m)$ denote the jth iterate of C_q applied to m.

Definition 1. We say a positive odd integer q is a *Crandall number* if there is some positive odd integer m such that $C_q^j(m) \neq 1$ for every positive integer j.

Let $\mathscr{C}$ denote the set of Crandall numbers. In [1], Crandall conjectured that every odd number $q > 3$ is in $\mathscr{C}$ and showed that $\mathscr{C}$ contains 5, 181 and 1093. The proofs for 5 and 181 involve cycles: for $q = 5$, we have the cycle 13, 33, 83, 13, and for $q = 181$, we have the cycle 27, 611, 27. The proof for 1093 depends on the fact that it is a "Wieferich prime," that is, a prime p with $2^{p-1} - 1$ divisible by p^2. For an odd number q, let $\ell(q)$ denote the order of 2 mod q in the group $(\mathbb{Z}/(q\mathbb{Z}))^*$.

Definition 2. We say a positive odd integer q is a *Wieferich number* if q and $(2^{\ell(q)} - 1)/q$ are not relatively prime.

Note that Wieferich primes are Wieferich numbers and that prime Wieferich numbers are Wieferich primes, so there should be no confusion with the two definitions.

Wieferich primes are historically connected with Fermat's last theorem: if a prime p is *not* a Wieferich prime, all integer solutions to $x^p + y^p = z^p$ have $p | xyz$. It is conjectured that very few primes are Wieferich primes, but

---

Received by the editor February 1, 1994.

1991 *Mathematics Subject Classification.* Primary 11A99.

Key words and phrases. $3x + 1$ problem, $qx + 1$ problem, Wieferich prime.

This paper is based in part on the first author's Ph.D. dissertation at the University of California, Berkeley, 1990. The second author is supported in part by an NSF grant.

　　　　　　　ZACHARY FRANCO AND CARL POMERANCE

that there are infinitely many. The only other Wieferich prime known is 3511, though they have been searched for up to $6 \cdot 10^9$ (see [5]).

As it turns out, there are quite a few Wieferich numbers. In this paper we shall first show, following Crandall's argument, that every Wieferich number is a Crandall number. We shall next show that asymptotically all odd numbers are Wieferich numbers, so that the set of numbers for which Crandall's conjecture is true has relative density 1 in the odd numbers.

In fact, for Wieferich numbers, a stronger version of Crandall's conjecture is true. Call an odd number q a "strong Crandall number" if the set of integers m for which $C_q^j(m) = 1$ for some j has asymptotic density zero. Our proof shows that every Wieferich number is a strong Crandall number, so that almost all odd numbers q are strong Crandall numbers.

Proposition 3. *Every Wieferich number is a Crandall number.*

Proof. If $C_q(m) = 1$, then clearly $m = (2^{k \cdot \ell(q)} - 1)/q$ for some positive integer k. Thus, if $C_q^2(m) = 1$, then

$$qm + 1 = 2^t(2^{k \cdot \ell(q)} - 1)/q$$

for some positive integers t and k. If q is a Wieferich number, then the right side above, being divisible by $(2^{\ell(q)} - 1)/q$, is not coprime to q. But the left side obviously is coprime to q, so that if q is a Wieferich number, then $C_q^2(m) = 1$ has no solutions m. It follows that if q is a Wieferich number, then $C_q^j(m) = 1$ forces $j = 1$ and $m = (2^{k \cdot \ell(q)} - 1)/q$ for some positive integer k. In particular, the set of such integers m has asymptotic density 0, so that q is a (strong) Crandall number.

Theorem 4. *The Wieferich numbers have relative density 1 in the odd numbers.*

Proof. For each number $B > 0$ let

$$\mathscr{W}(B) = \{q \text{ odd} : \text{there is some prime } p_0 \leq B \text{ with } p_0 \| q\}.$$

By $p \| q$ we mean that p divides q, but p^2 does not divide q. It is easy to see that the asymptotic density of $\mathscr{W}(B)$ in the odd numbers is

$$d_B := 1 - \prod_{2 < p \leq B} \left(1 - \frac{p-1}{p^2}\right),$$

where the product is over primes. Since the sum of the reciprocals of the primes diverges, d_B tends to 1 as B tends to infinity.

For p_0 a prime, let $\mathscr{P}(p_0)$ denote the set of primes $p \equiv 1 \bmod p_0$ with 2 mod p not a p_0th power in $(\mathbb{Z}/(p\mathbb{Z}))^*$. The splitting field K of $x^{p_0} - 2$ has degree $p_0(p_0 - 1)$ over $\mathbb{Q}$. Thus, by the prime ideal theorem of Landau [4] (also see Marcus [6, Theorem 43]), the proportion of rational primes p which split completely in the ring of integers of K is $1/(p_0(p_0 - 1))$. These are the rational primes p for which $p \equiv 1 \bmod p_0$ and 2 mod p is a p_0th power in $(\mathbb{Z}/(p\mathbb{Z}))^*$. By the same theorem applied to the splitting field of $x^{p_0} - 1$ (or by the prime number theorem for arithmetic progressions), the proportion of rational primes p for which $p \equiv 1 \bmod p_0$ is $1/(p_0 - 1)$. Thus, the relative density of $\mathscr{P}(p_0)$ in the rational primes is

$$1/(p_0 - 1) - 1/(p_0(p_0 - 1)) = 1/p_0.$$

This result on the relative density of $\mathscr{P}(p_0)$ is true in the strong sense of asymptotic density; that is, we take the proportion among all primes up to x of the members of $\mathscr{P}(p_0)$, and let $x \to \infty$. But we shall only need it in the weaker sense of polar or Dirichlet density, which essentially considers the ratio of the sum of the reciprocals of the members of $\mathscr{P}(p_0)$ up to x and the sum of the reciprocals of all of the primes up to x, and then lets $x \to \infty$. In particular, the sum of the reciprocals of the primes in $\mathscr{P}(p_0)$ diverges. Thus, the set of integers not divisible by any member of $\mathscr{P}(p_0)$ has asymptotic density zero. Indeed, the asymptotic density of the set of such numbers is at most $\prod_{p \in \mathscr{P}(p_0), p \leq T}(1 - 1/p)$ for any T. But as T tends to infinity, this product tends to zero by the result on the sum of reciprocals of the members of $\mathscr{P}(p_0)$. Thus, the asymptotic density in the odd numbers of

$$\mathscr{W}^*(B) := \{q \in \mathscr{W}(B) : \text{ there are primes } p_0 \leq B, \ p \in \mathscr{P}(p_0) \text{ with } p_0\|q, \ p|q\}$$

is also d_B.

Note that $\mathscr{W}^*(B)$ is contained in the set of Wieferich numbers. Indeed, if $q \in \mathscr{W}^*(B)$, then there is some prime $p_0 \leq B$ with $p_0\|q$ and some prime $p \in \mathscr{P}(p_0)$ with $p|q$. We have $\ell(p_0)|\ell(q)$ and $\ell(p)|\ell(q)$. But since $p \in \mathscr{P}(p_0)$, we have $p_0|\ell(p)$. Thus, $p_0\ell(p_0)|\ell(q)$. Note that $\ell(p_0^2)|p_0\ell(p_0)$. Thus, $p_0^2|2^{\ell(q)} - 1$, and since p_0^2 does not divide q, it follows that q is a Wieferich number as claimed.

It thus follows that the relative density of the Wieferich numbers in the odd numbers is at least d_B for any $B > 0$. But, as noted above, d_B tends to 1 as B tends to infinity, which proves the theorem.

Remarks. Our proof shows that for each fixed prime p_0, the set of odd numbers q with p_0 not dividing $\ell(q)$ has asymptotic density zero. By replacing the number field K in the proof with the splitting field of $(x^{p_0} - 2)(x^{p_0^a} - 1)$, one can show the density of the odd numbers q with p_0^a not dividing $\ell(q)$ is zero. As a corollary we have the following result of independent interest:

Theorem 5. *For each fixed positive integer n, the set of odd numbers q with $\ell(q)$ not a multiple of n has asymptotic density zero.*

If $r = C_q(m)$ for some m, then every positive odd number congruent to r mod q is also in the range of C_q. Thus, the range of C_q has an asymptotic density. In fact, this density (in the odd numbers) is $\ell(q)/q$. To see this, note that the odd number r is in the range of C_q if and only if $2^j r \equiv 1 \bmod q$ for some integer j, so that the residue classes in the range are precisely those in the subgroup of $(\mathbb{Z}/(q\mathbb{Z}))^*$ generated by 2 mod q. From [2, Theorem 2], it follows that there is a set $\mathscr{C}$ of odd numbers of relative asymptotic density 1 such that $\ell(q)/q$ tends to zero as $q \in \mathscr{C}$ tends to infinity. That is, for almost all odd numbers q, the density of the range of C_q tends to 0 as q tends to infinity.

The Crandall numbers 5 and 181, found by Crandall himself, are not Wieferich numbers. We do not know if there are infinitely many Crandall numbers that are not Wieferich numbers, or even if there are any more after 5 and 181.

We list below the Wieferich numbers below 1000:

21, 39, 57, 105, 111, 147, 155, 165, 171, 183, 195, 201, 203, 205, 219, 231, 237, 253, 273, 285, 291, 301, 305, 309, 327, 333, 355, 357, 385, 399, 417, 429,

453, 465, 483, 489, 495, 497, 505, 507, 525, 543, 555, 579, 597, 605, 609, 615, 627, 633, 651, 655, 657, 663, 689, 715, 723, 735, 737, 741, 755, 759, 777, 791, 813, 855, 861, 889, 897, 903, 905, 915, 921, 935, 939, 955, 969, 975, 979, 981, 987, 993.

ACKNOWLEDGMENT

We take this opportunity to thank Jeff Lagarias for his helpful comments and encouragement.

BIBLIOGRAPHY

1. R. E. Crandall, *On the $3x + 1$ problem*, Math. Comp. **32** (1978), 1281–1292.

2. P. Erdős, C. Pomerance, and E. Schmutz, *Carmichael's lambda function*, Acta Arith. **58** (1991), 363–385.

3. J. C. Lagarias, *The $3x + 1$ problem and its generalizations*, Amer. Math. Monthly **92** (1985), 3–23.

4. E. Landau, *Einführung in die elementare und analytische Theorie der algebraischen Zahlen und der Ideale*, Chelsea Reprint, 1949.

5. D. H. Lehmer, *On Fermat's quotient, base two*, Math. Comp. **36** (1981), 289–290.

6. D. A. Marcus, *Number fields*, Springer-Verlag, New York, 1977.

DEPARTMENT OF MATHEMATICS, TEXAS A & M UNIVERSITY, KINGSVILLE, TEXAS 78363
E-mail address: zac@taiu.edu

DEPARTMENT OF MATHEMATICS, UNIVERSITY OF GEORGIA, ATHENS, GEORGIA 30602
E-mail address: carl@ada.math.uga.edu

3. A search for Wieferich and Wilson primes

Discussion

A Wieferich prime is an odd prime p enjoying the (evidently quite rare) property

$$2^{p-1} \equiv 1 \pmod{p^2},$$

involving the interesting historical thread that if Fermat's "last theorem" equation

$$x^p + y^p + z^p = 0$$

holds, and p does not divide xyz, then p must be a Wieferich prime.

Heuristically, one might guess—as in the paper—that the number of Wieferich primes between X, Y is roughly

$$\log\left(\frac{\log Y}{\log X}\right),$$

which means we might need to search to 10^{20} or beyond to find the next Wieferich prime after the only known cases $p = 1093, 3511$.

This 1997 paper—now 15 years old—took the search to $4 \cdot 10^{12}$. Modern work, as with colleagues such as M. Rodenkirch and workers on *PrimeNet*, have taken the limit to at least 10^{16}, and still with no new Wieferich primes > 3511 discovered. More important than these advancing numerical efforts, papers such as the one presented here have helped form a model of how to efficiently advance the search limit.

Challenges

The primary computational challenge is to find the next Wieferich prime—if it exists. Of course, the "log-log" heuristic above also implies an infinitude of Wieferichs, albeit occurring very sparsely on the set of integers.

Source

Crandall, R. E., Dilcher, K. and Pomerance, P., "A search for Wieferich and Wilson primes," *Math. Comp.*, **66**, , 433–449 (1997).

MATHEMATICS OF COMPUTATION
Volume 66, Number 217, January 1997, Pages 433–449
S 0025-5718(97)00791-6

A SEARCH FOR WIEFERICH AND WILSON PRIMES

RICHARD CRANDALL, KARL DILCHER, AND CARL POMERANCE

ABSTRACT. An odd prime p is called a *Wieferich prime* if
$$2^{p-1} \equiv 1 \pmod{p^2};$$
alternatively, a *Wilson prime* if
$$(p-1)! \equiv -1 \pmod{p^2}.$$
To date, the only known Wieferich primes are $p = 1093$ and 3511, while the only known Wilson primes are $p = 5, 13$, and 563. We report that there exist no new Wieferich primes $p < 4 \times 10^{12}$, and no new Wilson primes $p < 5 \times 10^8$. It is elementary that both defining congruences above hold merely (mod p), and it is sometimes estimated on heuristic grounds that the "probability" that p is Wieferich (independently: that p is Wilson) is about $1/p$. We provide some statistical data relevant to occurrences of small values of the pertinent Fermat and Wilson quotients (mod p).

0. INTRODUCTION

Wieferich primes figure strongly in classical treatments of the first case of Fermat's Last Theorem ("FLT(I)"). For an odd prime p not dividing xyz, Wieferich [26] showed that $x^p + y^p + z^p = 0$ implies $2^{p-1} \equiv 1 \pmod{p^2}$. Accordingly, we say that an odd prime p is a Wieferich prime if the Fermat quotient

$$q_p(2) = \frac{2^{p-1} - 1}{p}$$

vanishes (mod p). The small Wieferich primes $p = 1093$ and 3511 have long been known. Lehmer [16] established that there exist no other Wieferich primes less than 6×10^9, and David Clark [6] recently extended this upper bound to 6.1×10^{10}. This paper reports extension of the search limit to 4×10^{12}, without a single new Wieferich prime being found. The authors searched to 2×10^{12}. David Bailey used some of our techniques (and some machine-dependent techniques of his own; see §4: Machine considerations) to check our runs to 2×10^{12}, and to extend the search to the stated limit of 4×10^{12}. Richard McIntosh likewise verified results over several long intervals.

Wilson's classical theorem, that if p is prime, then $(p-1)! \equiv -1 \pmod{p}$, and Lagrange's converse, that this congruence characterizes the primes, are certainly

Received by the editor May 19, 1995 and, in revised form, November 27, 1995 and January 26, 1996.

1991 *Mathematics Subject Classification.* Primary 11A07; Secondary 11Y35, 11–04.

Key words and phrases. Wieferich primes, Wilson primes, Fermat quotients, Wilson quotients, factorial evaluation.

The second author was supported in part by a grant from NSERC. The third author was supported in part by an NSF grant.

elegant as a definitive primality test, but are also certainly difficult to render practical. Even in the domain of factoring, if arbitrary $M!$ (mod N) were sufficiently easy to evaluate, one would be able to take the GCD of such factorials with N routinely to produce factors of N. One therefore expects factorial evaluation to be problematic. But their difficulty is not the only interesting aspect of factorial evaluations. Indeed, FLT(I) results can be cast in terms of certain binomial coefficients and the Wilson quotients

$$w_p := \frac{(p-1)! + 1}{p}$$

(E. Lehmer [17]) whose vanishing (mod p) signifies the Wilson primes.

Though our Wieferich search involved successive refinements of one basic algorithm, namely the standard binary powering ladder, we found that many fascinating and disparate options abound for the factorial evaluation. What might be called the "straightforward" approach to a Wilson search is simply to accumulate the relevant product: set $s = 1$, then for $n = 2$ through $p - 1$, accumulate $s := s * n$ (mod p^2). This straightforward approach turns out to admit of dramatic enhancements. By careful selection and testing amongst a wide range of options, we were able to extend W. Keller's search limit of 3×10^6 (see Ribenboim [22]), and a limit by Gonter and Kundert [13] of 1.88×10^7. Our conclusion is: save $p = 5, 13$, and 563 there are no Wilson primes less than 5×10^8. Again, Richard McIntosh verified our results in various disjoint regions below our search limit.

1. WIEFERICH PRIMES

Note the following amusing numerological observation: if the positions of all the ones in the binary representation of p lie in arithmetic progression, then p cannot be a Wieferich prime. (To see this, note that if $p = 1 + 2^k + \cdots + 2^{(t-1)k} = (2^{kt} - 1)/(2^k - 1)$, then the exponent that 2 belongs to modulo p is kt, so kt divides $p - 1$. Raising $2^{kt} = 1 + p(2^k - 1)$ to the $(p-1)/(kt)$ power, we see that $2^{p-1} \equiv 1 + p(2^k - 1)(p-1)/(kt) \not\equiv 1 \pmod{p^2}$.) Thus the set of Wieferich primes cannot contain Fermat primes ($p = 10 \ldots 01$ (binary)) or Mersenne primes ($p = 11 \ldots 11$ (binary)) or primes such as

$$p = 1000000100000010000001000000100000010000001 \text{ (binary)}$$

(see also Ribenboim [21, p.154]). But such special forms are rare indeed, hardly affecting any exhaustive search for Wieferich primes. Generally speaking, there is no known way to resolve 2^{p-1} (mod p^2), other than through explicit powering computations. Let us denote a "straightforward" Wieferich search scheme as one in which a standard binary power ladder is employed, with all arithmetic done in standard high-precision fashion (mod p^2), meaning in particular that standard long division (by p^2) is employed for the mod operations. In this straightforward scheme, intermediate integer results would at times be nearly as large as p^4, always to be reduced, of course, (mod p^2). Beyond this straightforward scheme there exist various enhancements, to which we next turn.

The first enhancement we employed has been used by previous investigators (Lehmer [16], Montgomery [19]). This enhancement is useful if the low-level multiplication in the computing machinery cannot handle products of magnitude p^4. The idea is to invoke base-p representations and thereby "split" the multiplication of two numbers (mod p^2). Let any $x = a + bp$ (mod p^2) be represented by $\{a, b\}$, with both a, b always reduced (mod p). Then the requisite doubling and squaring operations within a standard powering ladder may use the following formulae. We use the operation "%p" to denote modular reduction to a residue between 0 and $p - 1$ inclusive, and [ ] to denote greatest integer part:

(1.1)
$$2x = \{(2a)\%p, (2b + [(2a)/p])\%p\},$$
$$x^2 = \{(a^2)\%p, (2ab + [a^2/p])\%p\}.$$

For primes $p \approx 10^{12}$, our machinery benefitted considerably from this base-p method. On some machines, in fact, the base-p scheme (1.1) is about twice as fast as the straightforward scheme.

A second, and in practice a telling enhancement, is to invoke what we shall call "steady-state" division. In this technique, which exploits the fact that a denominator (for example p or p^2) is fixed, a divide operation can be performed in roughly the same time as a single multiply. The basic idea is described with respect to large-integer arithmetic in [8], and was noticed within a floating-point scenario by Montgomery [19]. To divide by some N repeatedly, one computes and stores a fixed "reciprocal" of N, then uses this systematically to resolve arbitrary values of $[x/N]$. Choose some integer s such that $2^s > N^2$. Then consider, for arbitrary $0 \le x < N^2$, the quantity

$$\left[\frac{\left[\frac{2^s}{N} \right] x}{2^s} \right] = \left[\frac{\left(\frac{2^s}{N} - \theta \right) x}{2^s} \right] = \left[\frac{x}{N} - \frac{x\theta}{2^s} \right],$$

where $0 \le \theta < 1$. Clearly, the far right-hand side is either $[x/N]$ or $[x/N] - 1$. The point is, for given N the "reciprocal" $[2^s/N]$ need be computed only once. And once it is computed, the far left-hand side may be evaluated with a single multiplication and a shift. Let "$>> s$" denote a right-binary-shift by s bits. We express the steady-state division arithmetic in the following way:

Theorem 1. *Let $0 \le x < N^2$, and let $r = [2^s/N]$, where $2^s > N^2$. Then $[x/N]$ is either $(rx) >> s$ or $((rx) >> s) + 1$.*

Armed with the base-p arithmetic of (1.1) and with Theorem 1, we may now exhibit an efficient pseudocode sequence for the squaring of a representation $x = \{a, b\}$:

(1.2)

```
(* Assume r = [2^s/p], 2^s > p^2, has been computed once for given p.  *)
        b := b * a;
        c := (b * r) >> s;
        b := b - p * c;
        if (b >= p) b := b - p;
        a := a * a;
        d := (a * r) >> s;
        a := a - p * d;
        if (a >= p) {
                a := a - p;
                d := d + 1;
        }
        b := c + c + d;
        while (b >= p) b := b - p;
```

436   RICHARD CRANDALL, KARL DILCHER, AND CARL POMERANCE

After this pseudocode sequence, the pair a, b has become in-place its correct, modulo-reduced base-p square. Note that there are no explicit divisions and there are six multiplies, though two of the multiplies involve multiplication of the reciprocal r, which is the size of p, by a number the size of p^2. Again, the magnitude of the speed advantage of (1.2) over a straightforward scheme is machine-dependent.

A third enhancement runs as follows. In Theorem 1, the term $(rx) >> s$ can be obtained using a nonstandard multiplication loop. Observe that, though x will be the size of N^2 (the steady-state denominator) and r the size of N, a count of s bits will be lost after the right-shift. This means that in the "grammar-school parallelogram" generated during long multiplication, many of the entries are meaningless, because they will be shifted into oblivion. By intervening in a detailed manner into the usual long multiplication loop, the value $(rx) >> s$ can be obtained in about half the time it takes to multiply two integers the size of N. We achieved in this way a complete base-p square-and-mod sequence of the form (1.2), that requires about five (size of p) multiply times.

The cumulative advantage of all these various enhancements, compared to the straigthtforward scheme and depending, of course, on the computing machinery used, was typically an order of magnitude in speed increase. Some timing details are given in §4.

The nature of the powering ladder arithmetic is not the only issue, because one wishes to test only actual primes p. The tempting Fermat test, that is whether $2^{p-1} \equiv 1 \pmod{p}$, is generally wasteful for a Wieferich search. For one thing, one may as well always do the $\pmod{p^2}$ arithmetic, which of course contains the Fermat test. We found the most efficient scheme for isolation of testable primes to be an incremental sieve of Eratosthenes. If we are interested in all primes $p < L$, we segment all integers less than L into blocks of common size B, say $B = 10^6$. Then each new block is sieved completely, by all primes $< \sqrt{L}$, so that only primes remain within that block. At any time only one block equivalent of memory is involved, because for the current block we can quickly compute the sieving offsets for each sieving prime, then sieve rapidly. Our overall algorithm for the Wieferich prime search can be described thus:

ALGORITHM FOR WIEFERICH PRIME SEARCH

1) For search limit L, store all primes less than $\sqrt{L}$.
2) For each successive block of length B,
    3) Sieve out, using the stored primes, all composites from the current block,
    4) For each remaining prime p in the block,
       5) Choose $s > 2\log_2 p$, and compute the "reciprocal" $r = [2^s/p]$,
       6) Starting with representation $\{2, 0\}$ in a binary powering ladder, use possible machine-dependent enhancements (such as (1.1), (1.2), Theorem 1) to obtain a final base-p form $2^{(p-1)/2} \pmod{p^2} = \{C, D\}$.
       7) If C is neither 1 nor $p - 1$, exit with fatal error. Otherwise, report any desired D-values. A Wieferich prime must have $\{C, D\} = \{1, 0\}$ or $\{p - 1, p - 1\}$.

In this way we eventually recorded all instances of

$$2^{(p-1)/2} \equiv \pm 1 + Ap \pmod{p^2},$$

A SEARCH FOR WIEFERICH AND WILSON PRIMES   437

TABLE 1. Instances of $2^{(p-1)/2} \equiv \pm 1 + Ap \pmod{p^2}$, with $|A| \leq 100$, for $10^9 < p < 4 \times 10^{12}$. The value $A = 0$ would signify a Wieferich prime

p	$\pm 1 + Ap$	p	$\pm 1 + Ap$
1222336487	$+1 + 60p$	36673326289	$+1 - 45p$
1259662487	$+1 - 71p$	46262476201	$+1 + 5p$
1274153897	$+1 - 86p$	47004625957	$-1 + 1p$
1494408397	$-1 + 52p$	49819566449	$+1 + 27p$
1584392531	$-1 - 24p$	53359191887	$+1 + 50p$
1586651309	$-1 - 24p$	58481216789	$-1 + 5p$
1662410923	$-1 - 70p$	76843523891	$-1 + 1p$
1817972423	$+1 - 56p$	82834772291	$-1 + 82p$
1890830857	$+1 + 69p$	108058158839	$+1 + 58p$
2062661389	$-1 + 55p$	130861186019	$-1 - 97p$
2244893621	$-1 + 47p$	138528575509	$-1 - 23p$
2332252547	$-1 - 33p$	239398882511	$+1 - 72p$
2416644757	$-1 + 67p$	252074060191	$+1 + 61p$
2461090421	$-1 + 47p$	252137567497	$+1 - 31p$
2566816313	$+1 + 52p$	299948374351	$+1 - 19p$
2570948153	$+1 - 41p$	405897532891	$-1 + 61p$
2589186937	$+1 - 85p$	443168739911	$+1 + 64p$
2709711233	$+1 + 50p$	504568016327	$+1 + 55p$
2760945133	$-1 - 77p$	703781283787	$-1 - 49p$
2954547209	$+1 + 32p$	955840782881	$+1 - 84p$
3027263587	$-1 - 95p$	980377925057	$+1 - 90p$
3133652447	$+1 + 87p$	981086885117	$-1 + 85p$
3303616961	$+1 - 20p$	1095406033573	$-1 + 42p$
3520624567	$+1 - 6p$	1104406423781	$-1 + 54p$
3606693551	$+1 + 21p$	1180032105761	$+1 - 6p$
4449676157	$-1 - 15p$	1722721869859	$-1 - 31p$
5045920247	$+1 + 76p$	1730418792409	$+1 - 46p$
5409537149	$-1 + 66p$	1780536689159	$+1 + 84p$
8843450093	$-1 - 20p$	2207775149407	$+1 - 99p$
10048450537	$+1 + 82p$	2424653846701	$-1 - 51p$
10329891503	$+1 + 79p$	2610372685663	$+1 - 28p$
11214704947	$-1 + 56p$	3667691800441	$+1 + 27p$
20051397221	$-1 - 46p$	3713054321579	$-1 - 51p$
20366156849	$+1 - 95p$	3729819224423	$+1 + 77p$

where A is allowed to be bipolar but $|A| \leq 100$, over the primes less than 4×10^{12}. Note that A is essentially the Fermat quotient; or more precisely,

$$q_p(2) \equiv \pm 2A \pmod{p}.$$

Table 1 shows all primes between 10^9 and 4×10^{12} that enjoy the small $|A|$ values. In §3 we discuss some statistical considerations pertinent to these and some analogous data for the Wilson quotients.

The large-integer arithmetic mentioned above is discussed in [9] and [10]. Except for some of our more obscure algorithmic enhancements, relevant code (giants.

[ch]) is stored on the disk supplement of [9]; it can also be obained over the World-Wide Web at ftp://ftp.telospub.com in the directory
   /pub/ScientificComputing/TopicsAdvSciComp/AppendixCode.
The book [10] specifically deals with computational issues related to the Wieferich and Wilson searches.

2. WILSON PRIMES

During factorial calculations all arithmetic $(\bmod \ p^2)$ is subject to possible refinements of the previous section. In particular, base-p arithmetic is appropriate on most machines. But for the current range of interest, namely p on the order of 10^8, most modern machinery allows low-level arithmetic $(\bmod \ p)$ or efficient long-long (double-precision) arithmetic, so the steady-state division of Theorem 1 need not apply (though that theorem becomes useful for much larger p). These arithmetic issues having been settled, we proceeded to analyze various theoretical relations with a view to minimization of computation time.

A useful and interesting identity is due to Granville [14]; for any integer m with $1 < m < p$, we have

$$(2.1) \qquad \prod_{j=1}^{m-1} \binom{p-1}{\left[\frac{jp}{m}\right]} \equiv (-1)^{(p-1)(m-1)/2}(m^p - m + 1) \pmod{p^2}.$$

This congruence for $m = 2$ was known in the nineteenth century and has been rediscovered a few times since; for $m = 3$, 4, 6 the congruence follows from (50), (51), (52) in [17]. For example, (50) in [17] is the congruence

$$(2.2) \qquad \binom{p-1}{[p/3]} \equiv (-1)^{[p/3]}(3^p - 1)/2 \pmod{p^2}$$

for all primes $p > 3$, which when squared gives (2.1) for $m = 3$.

Granville's identity for $m = 2$ shows right off that one never need evaluate the full factorial of $p - 1$, but may evaluate the factorial of $(p-1)/2$ instead. A different kind of identity, when combined with (2.1), yields a considerable algorithm enhancement. Consider the representation of certain primes p by quadratic forms:

$$(2.3) \qquad \begin{aligned} p \equiv 1 \pmod 4: \quad & p = a^2 + b^2; & a \equiv 1 \pmod 4 \\ p \equiv 1 \pmod 3: \quad & 4p = c^2 + 27d^2; & c \equiv 1 \pmod 3 \\ p \equiv 1 \pmod 3: \quad & 4p = u^2 + 3v^2; & u \equiv (-1)^{(p-1)/6} \pmod 3, \\ & & v \equiv 0 \pmod 6 \text{ if possible,} \\ & & v \equiv \pm 1 \pmod 6 \text{ if not .} \end{aligned}$$

It is known (see [3], [5], [11] , and also [7], [28]) that for $p \equiv 1 \pmod 4$,

$$(2.4) \qquad \binom{\frac{p-1}{2}}{\frac{p-1}{4}} \equiv (2^{p-1} + 1)\left(a - \frac{p}{4a}\right) \pmod{p^2},$$

and that for $p \equiv 1 \pmod 3$,

$$(2.5) \qquad \binom{\frac{2p-2}{3}}{\frac{p-1}{3}} \equiv -c + \frac{p}{c} \pmod{p^2}, \qquad \binom{\frac{p-1}{3}}{\frac{p-1}{6}} \equiv \frac{-u}{3}(2^p + 1) + \frac{p}{u} \pmod{p^2}.$$

The results (2.1)-(2.5) can be combined in various ways to yield the following identities, one of which is always applicable to a given odd prime p:

For $p \equiv 1 \pmod 3$, we use (2.2) and (2.5) to get

$$(2.6) \quad (p-1)! \equiv \left(\left(\frac{p-1}{6}\right)!\right)^6 (-u^3(2^p-1)+3pu)\left(-c+\frac{p}{c}\right)\frac{3^p-1}{2} \pmod{p^2}.$$

In (2.6) it is possible to ignore the condition on v in (2.3) since the different values of u produced have $-u^3(2^p-1)+3pu \equiv$ constant $\pmod{p^2}$.

For $p \equiv 5 \pmod{12}$, we use (2.1) with $m=2$ and (2.4) to get

$$(p-1)! \equiv \left(\left(\frac{p-1}{4}\right)!\right)^4 (2^p-1)(2^{p-1}+1)^2\left(a-\frac{p}{4a}\right)^2 \pmod{p^2},$$

so that,

$$(2.7) \qquad (p-1)! \equiv \left(\left(\frac{p-1}{4}\right)!\right)^4 (3\cdot 2^p-4)(2a^2-p) \pmod{p^2}.$$

For $p \equiv 11 \pmod{12}$, we use (2.1) with $m=2$ to get

$$(2.8) \qquad (p-1)! \equiv \left(\left(\frac{p-1}{2}\right)!\right)^2 (1-2^p) \pmod{p^2}.$$

So expensive are the relevant factorial calculations in practice, that one may with impunity ignore fast methods and obtain any of a,b,c,d,u,v by "brute force"; say, by looping through squares a^2, c^2, or u^2. Ignoring the relatively inconsequential operations of powering, inverting, and extraction of $a,b,c,d,u,$ or v, the identities (2.6)-(2.8) involve, for $p \equiv 1 \pmod 3$, $p \equiv 5 \pmod{12}$, $p \equiv 11 \pmod{12}$, respectively, about

$$p/6, \ p/4, \ p/2$$

multiplies $\pmod{p^2}$, amounting to an average reduction, over many primes, of $48/13$ in complexity, as compared to the naive evaluation of $(p-1)!$. As we shall see presently, complexity reduction may be taken further.

An algebraic refinement is to observe that the multiplicands comprising an arbitrary $N!$ generally admit of a certain redundancy. For example, all the even multiplicands can be extracted and written in the form of some power of 2 times a smaller factorial. We start our analysis of what might be called factorial sieving, by defining a generalized factorial:

$$(2.9) \qquad P(q,m) = \prod_{\substack{k=1\\(k,q)=1}}^{m} k = \prod_{\substack{k=1\\(k,q)=1}}^{q-1}\prod_{\substack{j=1\\j\equiv k\pmod q}}^{m} j.$$

For example, $P(1,m)=m!$ and for m odd, $P(2,m)=m!!$, the product of the odd numbers up to m. Now our previous statement concerning the even multiplicands of $N!$, can be written, for N even, as

$$(2.10) \qquad N! = 2^{N/2}P(2,N)\left(\frac{N}{2}\right)!$$

From this identity $N!$ can actually be evaluated in about $3N/4$ multiplications. One multiplies together the even integers $\le N/2$, then the odd integers $\le N/2$, squares the latter; then multiplies everything by all odd integers in $(N/2,N]$. But more can be done along these lines. For example, iterating (2.10) one gets the identity

$$(2.11) \qquad N! = 2^{e_2} \prod_{j \geq 0} P\left(2, \left[\frac{N}{2^j}\right]\right),$$

where $2^{e_2} \| N!$ and the apparent infinite product may be truncated as soon as $2^j > N/3$. The identity (2.11) allows the computation of $N!$ in $(1/2 + o(1))N$ multiplications.

There are some interesting theoretical branches to take at this point. One approach is to generalize (2.10) and (2.11) by sieving the factorial with, say, all primes $p \leq R$. This idea leads to the following general identity:

$$(2.12) \qquad N! = \prod_{p \leq R} p^{e_p} \prod_{\substack{q \text{ is } R-\text{smooth}}}^{N} P\left(\pi, \left[\frac{N}{q}\right]\right),$$

where $p^{e_p} \| N!$, π is the product of all primes not exceeding R, and the q are R-smooth, i.e., not divisible by any prime exceeding R. We have found this formula difficult to use in practice. As just one extra complication, one must find all R-smooth numbers not exceeding N. However, this general factorial-sieve identity may find use in theoretical treatments of factorial complexity. For example, choosing $R = N^{1/\ln \ln N}$, it can be shown that (2.12) allows the computation of $N!$ in $O(N/\ln \ln N)$ multiplications.

A different approach is to partially recurse on some fixed set of sieve identities. By experimentation, the most pragmatic identity which we could find takes $R = 3$ in (2.12), but does not use the full recursion over all 3-smooth numbers. In particular, when $N \equiv 0 \pmod{6912}$ we have

$$(2.13) \qquad N! = 2^{\frac{255N}{256}} 3^{\frac{185N}{432}} \left(\frac{N}{256}\right)! \prod_w P\left(2, \frac{N}{w}\right) \prod_v P\left(6, \frac{N}{v}\right),$$

where w runs through the set $\{16,18,24,27,32,36,64,128\}$ and v runs through the set $\{1,2,3,4,6,8,9,12\}$. Analysis of the P products reveals that, on the basis of (2.13), one may evaluate $N!$ by way of an asymptotic count of $(283/768)N$ multiplies.

Regardless of what factorial reduction formulae are in force, there is a powerful, universal enhancement that removes most of the multiplication in favor of addition. It may be observed that any product of interest in our previous identities can be cast as a product of the products of the terms in disjoint arithmetic progressions. For example, $P(q, m)$ by its definition (2.9) is manifestly such a product. Thus, if we can implement an algorithm for rapid evaluation of the product of terms in an arithmetic progression, we can call such a routine as desired. Happily there exists a suitable such algorithm, which will evaluate any product of n terms in arithmetic progression in $O(n^{\phi+\epsilon})$ multiplies, where $\phi = (\sqrt{5}-1)/2$ is the "golden ratio", and $n + O(n^{3-\sqrt{5}+\epsilon})$ adds. For the Wilson search, it is understood that mods must also be taken. However, since the number of adds will generally far exceed the number of multiplies, most of the mod operations involve only "if" statements and their ensuing subtractions.

An algorithm for evaluating a polynomial along arithmetic progression values is given in [15, p. 469], and uses the fact that if enough successive differences are taken, the difference tableau suffices to determine the required values of the polynomial. We express here, for the purposes of the Wilson search, a particular variant that yields the product of all terms in arithmetic progression:

ALGORITHM TO EVALUATE THE PRODUCT OF TERMS IN ARITHMETIC PROGRESSION

To evaluate

$$f(x) = x(x + d)(x + 2d) \cdots (x + (n - 1)d).$$

at a given point x:

1) Choose $G < n$ (an optimal choice of G is discussed later) and set $K := [\frac{n}{G}] - 1$.
2) Create a_0 through a_G:
   For $j = 0$ to G

$$a_j := \prod_{q=0}^{G-1} (x + (q + Gj)d)$$

3) Create difference tableau:
   For $q = 1$ to G {
         For $j = G$ down to q
              $a_j := a_j - a_{j-1}$
   }

4) Manipulate differences and accumulate:
   $f := a_0$
   For $j = 1$ to K {
         $a_0 := a_0 + a_1$
         $f := f a_0$
         For $q = 1$ to $G - 1$
              $a_q := a_q + a_{q+1}$
   }

5) Finish tail end of product:

$$f := f \prod_{j=G(K+1)}^{n-1} (x + jd).$$

For investigations such as the Wilson search, all sums, differences and products are to be reduced (mod p^2); whence the final output f will be the desired product of arithmetic progression terms, that is $f(x)$ (mod p^2). If base-p arithmetic is used, all of the arithmetic is to be performed in its natural way amongst representations; thus for example each of the initial a_j will be a representation pair.

Analysis shows that the algorithm requires $O(G^2 + n/G)$ multiplies and $n + O(G^2)$ adds. The optimal G is thus in the neighborhood of $n^{1/3}$, yielding $O(n^{2/3})$ multiplies and $n + O(n^{2/3})$ adds. But the algorithm's step (2) itself involves products of arithmetic progression terms. One may thus recurse on step (2), embedding the algorithm within itself to render ultimately the aforementioned operation count of $O(n^{\phi+\epsilon})$ multiplies and $n + O(n^{3-\sqrt{5}+\epsilon})$ adds. However, we found in practice that not even the first recursive level is really necessary. For primes $p \approx 10^8$, the basic algorithm without recursion already has a total multiply count more than two orders of magnitude below the add count.

Armed with these various refinements, we arrived at the following practical algorithm:

ALGORITHM FOR WILSON PRIME SEARCH

1) For search limit L, store all primes less than $\sqrt{L}$.
2) For each successive block of length B,
    3) Sieve out, using the stored primes, all composites from the current block,
    4) For each remaining prime p in the block,
        5) As $p \equiv 1 \pmod 3$, $p \equiv 5 \pmod{12}$, $p \equiv 11 \pmod{12}$ adopt one of the identities (2.6)-(2.8). If p is not 11 $\pmod{12}$ obtain any required quadratic elements a, b, c, d, u, v.
        6) Use a factorial sieve such as (2.13) together with successive arithmetic progression products for the P product terms. In this way obtain a final base-p representation for $(p-1)! = \{C, D\}$.
        7) If C is not $p-1$, exit with fatal error. Otherwise, report any desired D-values. A Wilson prime must have $\{C, D\} = \{p-1, p-1\}$.

In this way we eventually recorded all instances of

$$(p-1)! \equiv (p-1) + p(p-1-B) \pmod{p^2},$$

where $0 \le B \le 100$, over the primes less than 5×10^8. Note that B is essentially the Wilson quotient; or more precisely,

$$w_p \equiv -B \pmod{p}.$$

Table 2 shows all primes between 10^7 and 5×10^8 that admit of such small B values.

Incidental to our search effort, we noted some alternative factorial evaluation schemes. Because of their theoretical attraction we shall briefly mention these alternatives, though for the p-ranges of the reported search we were unable to bring any of these alternatives up to the execution speed of the arithmetic progression algorithm written out above.

We implemented a recursive polynomial remaindering scheme for factorials; this method requiring no worse than $O(p^{1/2}(\ln p)^2)$ arithmetic operations to resolve the Wilson quotient. The idea is to evaluate a polynomial of degree m, namely:

$$f(x) = (x+1)(x+2) \cdots (x+m),$$

at the m points $x = 0, m, 2m, \ldots, (m-1)m$ and multiply together all of these evaluations to yield $(m^2)!$. This polynomial evaluation problem is known to require no more than $O(m(\ln m)^2)$ arithmetic operations [4], [20], [23]. For m a power of two, the recursion proceeds as follows. Define polynomials each of degree $m/2$:

$$g_0(x) = x(x-m)(x-2m) \cdots (x-(m/2-1)m),$$

$$g_1(x) = (x-(m/2)m)(x-(m/2+1)m) \cdots (x-(m-1)m).$$

Then it is immediate that $(m^2)!$ is the product of the evaluations of $f(x) \pmod{g_0(x)}$ at the zeros of g_0, times the product of the evaluations of $f(x) \pmod{g_1(x)}$ at the zeros of g_1. But each of the two reduced polynomials $f(x) \pmod{g_i(x)}$ can be evaluated in the same way, modulo appropriate members of a set of four degree-$m/4$ polynomials, and so on recursively. The recursion hits bottom at a chosen level

TABLE 2. Instances of $(p-1)! \equiv -1-Bp \pmod{p^2}$, with $0 \leq B \leq 100$, for $10^7 < p < 5 \times 10^8$. The value $B = 0$ would signify a Wilson prime

p	$-1 - Bp$
10746881	$-1 - 7p$
11465149	$-1 - 62p$
11512541	$-1 - 26p$
11892977	$-1 - 7p$
12632117	$-1 - 27p$
12893203	$-1 - 53p$
19344553	$-1 - 93p$
21561013	$-1 - 90p$
27783521	$-1 - 51p$
39198017	$-1 - 7p$
45920923	$-1 - 63p$
53188379	$-1 - 54p$
56151923	$-1 - p$
57526411	$-1 - 66p$
72818227	$-1 - 27p$
87467099	$-1 - 2p$
91926437	$-1 - 32p$
93445061	$-1 - 30p$
93559087	$-1 - 3p$
94510219	$-1 - 69p$
101710369	$-1 - 70p$
117385529	$-1 - 43p$
212911781	$-1 - 92p$
216331463	$-1 - 36p$
327357841	$-1 - 62p$
411237857	$-1 - 84p$
479163953	$-1 - 50p$

for which direct evaluation, say by Horner's rule, of the current reduced polynomials proceeds efficiently.

In our implementation, we used polynomials of degree about $(p/k)^{1/2}$, where k is an appropriate integer gleaned from the reduction formulae (2.6)-(2.8). Each coefficient of each polynomial was a base-p representation. The polynomial remaindering used a Newton method for polynomial inversion, with large polynomial multiplication performed by way of a Nussbaumer convolution scheme in which signal array elements are base-p representations. The resulting experiments showed that, indeed, the arithmetic scheme is bested by this remaindering scheme for sufficiently large p. Our implementation of the remaindering scheme begins to be superior in speed for p in the neighborhood of 10^{11}. Though remaindering did not apply over our stated Wilson search region, we were able to use that scheme to resolve isolated, huge factorials. For example, for $p = 1099511628401 \ (= 2^{40} + 5^4)$, we calculated

$$(p - 1)! \equiv -1 - 533091778023p \pmod{p^2}.$$

444   RICHARD CRANDALL, KARL DILCHER, AND CARL POMERANCE

To our knowledge this is the largest prime for which the Wilson quotient is known $(\mathrm{mod}\ p)$; and also the largest integer to have been proven prime, as we have, on the basis of Lagrange's converse of Wilson's classical theorem. We believe that, given the computing machinery of today and the favorable asymptotic complexity of the remaindering algorithm, such calculations for p as high as 10^{20} are not out of the question.

Another scheme for Wilson testing was proposed to us by J. P. Buhler, and is one of two Wilson tests we noted that involve identities merely $(\mathrm{mod}\ p)$. For g a primitive root of an odd prime p, denote by $\{a_k, b_k\}$ the base-p representation of $g^k \pmod{p^2}$. Then

$$(p-1)! = \prod_{k=1}^{p-1} a_k \equiv \prod_{k=1}^{p-1} (g^k - pb_k) \pmod{p^2}$$

$$\equiv g^{\frac{p(p-1)}{2}} \left(1 - p\sum_{k=1}^{p-1} b_k a_{p-1-k}\right) \equiv -1 + p\sum_{k=1}^{p-1} b_k a_{p-1-k} \pmod{p^2},$$

which establishes

Theorem 2. *Let p be an odd prime possessed of a primitive root g. Then the Wilson quotient satisfies*

$$w_p \equiv \sum_{k=1}^{p-1} \left[\frac{g^k}{p}\right] g^{p-1-k} \pmod{p}.$$

It is interesting that the Fermat quotient $q_p(g)$ associated with the Wieferich problem appears in the theorem, in the guise of the final summand. At first glance the convolution in Theorem 2 appears to require at least $O(p)$ multiplies. This is not so in general; in fact we found means by which Theorem 2 may be applied with no multiplies, and $O(p \ln g)$ adds. The main idea is to think of the sum in the theorem as a polynomial in g, and invoke Horner's evaluation rule. We give here the surprisingly simple pseudocode loop in the case that $g = 2$ happens to be a primitive root:

```
(* Assume 2 is a primitive root of the prime p.*)
        a = 1;
        b = s = 0;
        loop{
            a := a + a;
            b := b + b;
            if(a >= p) {
                a := a - p;
                b := b + 1;
            }
            if(b >= p) b := b - p;
            s := s + s + b;
            while(s >= p) s := s - p;
            if(a == 1) break;
        }
(* p is a Wilson prime if and only if s = 0.  *)
```

Note that this pseudocode contains no multiplies and uses negligible memory. When 2 is not a primitive root, one may replace the various doubling steps in the loop with binary addition ladders to effect multiplication by $g > 2$. In spite of its extreme simplicity, this scheme still could not be made competitive with the arithmetic progression scheme. The basic reason for this failure is that the factorial schemes starting from (2.6)-(2.8) involve immediate reduction of the factorial term count, whereas we do not yet know any means by which the number of summands in Theorem 2 can be similarly reduced.

During attempts to reorder factorial products we found a second mere (mod p) algorithm, which we believe to be new, and which is expressed as follows.

Theorem 3. *For an odd prime p and any integer q with $1 < q < p$, denote by $(1/p)_q$ the unique inverse of p (mod q) lying in $[1, q-1]$. Then p is a Wilson prime if and only if*

$$\sum_{q=2}^{p-1} \left(\frac{1}{p}\right)_q \equiv -1 \quad (\text{mod } p).$$

This theorem follows immediately upon the observation that the product of terms $(p(1/p)_q - 1)/q$, over all q in the stated range, happens to be $(p-2)!$.

Theorem 3 has so far been difficult to render practical. However, if sufficient memory is available, one might consider storing tables of inverses (mod q) for many small primes q, and attempting to reconstruct the required terms $(1/p)_q$ rapidly. In addition, there is the possibility of parallelism. One could perhaps evaluate quickly sets of inverse terms for many primes p at once.

Finally, we mention the beautiful relation between Bernoulli numbers and the Wilson quotients due to N. G. W. H. Beeger [2]:

$$pB_{p-1} \equiv p - 1 + pw_p \quad (\text{mod } p^2).$$

It is interesting that Beeger actually used his congruence and a table of Bernoulli numbers to show that 5 and 13 are the only Wilson primes up to 113.

We have observed that the left-hand side of the Beeger congruence can be evaluated via polynomial algebra (mod p^2). The somewhat intricate polynomial manipulations can be summarized briefly, as follows. First, note that upon formal series expansion of

$$f(x) = \frac{2x \sinh x - x^2}{2(\cosh x - 1)}$$

in even powers of x, the coefficient of x^n is $(n+1)B_n/n!$. Define a coefficient T_{p-1} implicitly by

$$\frac{1 + \sum_{k=1}^{(p-3)/2} 2x^{2k}/(2k+1)!}{1 + \sum_{k=1}^{(p-3)/2} 2x^{2k}/(2k+2)!}$$

$$= 1 + \cdots + T_{p-1}x^{p-1} + O(x^{p+1}),$$

Note that the k-dependent factorials can all be inverted (mod p^2), so that T_{p-1} can be calculated unambiguously via polynomial division (mod p^2). Now, to obtain the desired coefficient $pB_{p-1}/(p-1)!$, which is not quite T_{p-1}, the latter must be corrected on the basis of the missing ($k = (p-1)/2$) terms, the result being:

$$pB_{p-1} \equiv T_{p-1}(p-1)! + 2 - 2p \quad (\text{mod } p^2).$$

446   RICHARD CRANDALL, KARL DILCHER, AND CARL POMERANCE

In this way Wilson quotients can be obtained via power series manipulation (mod p^2). We are aware that the multiplier $(p-1)!$ in this last relation gives the appearance of recourse, as do the factorials that enter into the evaluation of the hyperbolic polynomials. But we cannot yet rule out the possibility of a polynomial algorithm that does not involve recourse to factorials (for example, there are other, nonhyperbolic expansions involving Bernoulli numbers [10]). Such an algorithm might yield entirely new means for rapid evaluation of Wilson quotients.

In a converse spirit, we note that the Beeger congruence, together with Theorem 2, gives a clear, simple, multiply-free algorithm for evaluation of $pB_{p-1} \pmod{p^2}$. This algorithm involves $O(p \ln g)$ adds and, as we have seen, is especially efficient when $g = 2$ is the primitive root.

3. STATISTICAL CONSIDERATIONS

Are there infinitely many Wieferich primes, or *any* larger than 3511? Is 563 the last Wilson prime? Since $A = A(p)$ is an integer in the interval $(-p/2, p/2)$, one might view the "probability" of A assuming any particular value, say the value 0, to be $1/p$. And one might view experiments for different primes p as "independent" events. Thus, heuristically we might argue that the number of Wieferich primes in an interval $[x, y]$ is expected to be

$$\sum_{x \le p \le y} 1/p \approx \ln(\ln y / \ln x).$$

If this is the case, we would only expect to find one Wieferich prime above Lehmer's search bound of 6×10^9 and below 3.8×10^{26}. It is thus not too shocking that no new instances occur below 4×10^{12}. A similar heuristic suggests that there should be just one Wilson prime above the search limit 1.88×10^7 of [13] and below 5.9×10^{19}. Again it is no surprise that we did not find any new Wilson primes. (The "expected" number of Wieferich primes in our Wieferich search interval was about 0.25, and the "expected" number of Wilson primes in our Wilson search interval was about 0.18, so it was not *a priori* completely hopeless to embark upon such a search.)

However, we did find some "near" Wieferich and Wilson primes, where, respectively, the A-values and B-values are small. Tables 1 and 2 show all instances of small-magnitude even Fermat quotients, and small-negative Wilson quotients; more precisely, all occurrences of $0 \le |A|, B \le 100$ over the stated search regions. One might test the heuristic model above with the actual evidence. The "expected" number of primes p in $[10^9, 4 \times 10^{12}]$ with $|A| \le 100$ is $201 \ln(\ln(4 \times 10^{12}) / \ln(10^9)) \approx 67.7$. In fact we found exactly 68 such near Wieferich primes p, a very close fit indeed. The "expected" number of primes p in $[10^7, 5 \times 10^8]$ with B-value satisfying $0 \le B \le 100$ is $101 \ln(\ln(5 \times 10^8) / \ln(10^7)) \approx 21.9$. In fact we found exactly 27 near Wilson primes in the interval.

4. MACHINE CONSIDERATIONS

For our Wieferich and Wilson searches, we used various 32-bit processors, such as 68040, 486/586, HP7000, SPARC, and i860. (We inspected the software for the infamous, sometimes faulty "fdiv" instruction on 586 Pentium; and finding one occurrence of that instruction, we double checked all of the relevant primes.) Typically, every individual machine would handle all the primes in some particular region. We allowed each machine to set up its own, equivalent incremental sieve over that machine's region of responsibility, since this was not too wasteful. Under the algorithm of §1, such 32-bit processors sustained Wieferich test rates of approximately 100 to 1000 primes per second per machine, in the $p \approx 10^{12}$ region. We have noted already that the straightforward scheme (size p^4 products and long divide) to obtain Fermat quotients runs about ten times slower. With the aid of Richard McIntosh we tested the base-p arithmetic on a 64-bit (DEC) machine that allowed low-level divide/mod, finding that such a machine can sustain close to 10,000 primes per second in the stated region. Bailey's computations, as mentioned in §1, were performed on an IBM SP-2 parallel computer at NASA Ames Research Center, utilizing otherwise idle machine cycles. Each of his one hundred individual machines worked with its own set of primes, performing arithmetic in our base-p fashion. For the base-p arithmetic, Bailey used floating-point double precision (64-bit) arithmetic, including clever use of "floating multiply-add" operations, which on the IBM RS6000/590 processors maintain internal 106-bit accuracy. The resulting code achieved an impressive sustained testing rate of over one million primes per second.

The Wilson search proceeded, after all the refinements noted in the arithmetic progression based algorithm, at processor-dependent rates between 1/20 and 1/2 prime per second, in the $p \approx 10^8$ region. That the fastest 32-bit machines would require several seconds to resolve the Wilson quotient for a single p near 100 million is not unimpressive when one considers that the straightforward multiply-accumulate-(mod p^2) method, with multi-precision division also assumed, would require the equivalent of perhaps 10^9 size p multiplies. In fact the straightforward method is roughly 100 times slower than the rate we achieved. The polynomial remaindering scheme, which dominates for $p > 10^{11}$, is roughly ten times faster than our arithmetic progression based scheme for p in the region of 10^{12}. Nevertheless, using polynomial remaindering, it still took a SPARC processor about one day to resolve the Wilson quotient of $p = 1099511628401$ (see §3).

5. RELATED SEARCHES

In the absence of future theoretical results, one is moved to attach the probability $1/p$ also to certain other properties (mod p^2). What might be called Wall-Sun-Sun primes (see [25] and [24]) are those $p > 5$ satisfying

$$(4.1) \qquad F_{p-\left(\frac{p}{5}\right)} \equiv 0 \pmod{p^2},$$

where F_n is the nth Fibonacci number. This congruence is, again, always satisfied merely (mod p). Williams [27] found no Wall-Sun-Sun primes whatsoever, below 10^9, and Montgomery [19] extended this to 2^{32}. On the basis of a search being conducted by R. McIntosh [18], we have learned that there exist no Wall-Sun-Sun primes less than 2×10^{12}. Beyond the $1/p$ statistics, it is of interest that, as Sun and Sun [24] proved, if p is a failing exponent in FLT(I), then p satisfies (4.1).

There are Wieferich composites, namely composite integers N such that

$$2^{\varphi(N)} \equiv 1 \pmod{N^2},$$

where φ is Euler's function, see [1]. (For N odd, the left side is always 1 (mod N).)

448   RICHARD CRANDALL, KARL DILCHER, AND CARL POMERANCE

For example, 3279 ($= 3 \times 1093$) is such a number N. This congruence is one way to generalize the Wieferich prime definition. Another is the weaker condition that $(N, (2^l - 1)/N) > 1$, where l is the least positive integer with N dividing $2^l - 1$. For this latter generalization it has been shown in [12], perhaps surprisingly, that asymptotically all odd numbers N are Wieferich.

Finally, there are the Wilson composites, being composite N such that

$$P(N, N) = \prod_{\substack{j=1 \\ (j,N)=1}}^{N} j \equiv \pm 1 \pmod{N^2}.$$

This congruence is one way to generalize the Wilson prime definition. Analysis and search is underway for such composites [1]. We were able to offer aid to that project, by using the factorial sieve and arithmetic progression based algorithms of §2 to calculate $P(N, N)$ rapidly, finding the new instances $N = 558771$, 1964215, 8121909 and 12326713. For example, the product of all positive integers less than and coprime to 12326713 is 1 (mod 12326713^2).

ACKNOWLEDGMENTS

We wish to thank D. Bailey, J. P. Buhler, T. E. Crandall, R. Evans, A. Granville, and H. W. Lenstra, Jr., for valuable discussions pertaining to this work. Special thanks to Thomas Trappenberg (Dalhousie University and Optimax Software) whose generous contributions of personal time and computing power are responsible for a great deal of the entire Wieferich search; and to Richard McIntosh for his aid at many levels to the search project.

REFERENCES

1. T. Agoh, K. Dilcher and L. Skula, *Fermat and Wilson quotients for composite moduli*, Preprint (1995).
2. N. G. W. H. Beeger, *Quelques remarques sur les congruences $r^{p-1} \equiv 1 \pmod{p^2}$ et $(p-1)! \equiv -1 \pmod{p^2}$*, The Messenger of Mathematics **43** (1913–1914), 72–84.
3. B. Berndt, R. Evans and K. Williams, *Gauss and Jacobi sums*, Wiley-Interscience, to appear.
4. J. M. Borwein and P. B. Borwein, *Pi and the AGM*, John Wiley & Sons, Inc., 1987. MR **89a**:11134
5. S. Chowla, B. Dwork, and R. Evans, *On the mod p^2 determination of $\binom{(p-1)/2}{(p-1)/4}$*, J. Number Theory **24** (1986), 188–196. MR **88a**:11130
6. D. Clark, *Private communication*.
7. M. Coster, *Generalisation of a congruence of Gauss*, J. Number Theory **29** (1988), 300–310. MR **89i**:11005
8. R. Crandall and B. Fagin, *Discrete weighted transforms and large-integer arithmetic*, Math. Comp. **62** (1994), 305–324. MR **94c**:11123
9. R. Crandall, *Projects in Scientific Computation*, TELOS/Springer-Verlag, Santa Clara, CA, 1994. MR **95d**:65001
10. ———, *Topics in Advanced Scientific Computation*, TELOS/Springer-Verlag, Santa Clara, CA, 1995.
11. R. Evans, *Congruences for binomial coefficients*, Unpublished manuscript (1985).
12. Z. Franco and C. Pomerance, *On a conjecture of Crandall concerning the $qx + 1$ problem*, Math. Comp. **64** (1995), 1333–1336. MR **95j**:11019
13. R. H. Gonter and E. G. Kundert, *All prime numbers up to 18,876,041 have been tested without finding a new Wilson prime*, Preprint (1994).
14. A. Granville, *Binomial coefficients modulo prime powers*, Preprint.
15. D. E. Knuth, *The art of computer programming*, Vol.2, Addison-Wesley, Reading, Massachusetts, 2nd ed. 1973. MR **44**:3531

A SEARCH FOR WIEFERICH AND WILSON PRIMES   449

16. D. H. Lehmer, *On Fermat's quotient, base two*, Math. Comp. **36** (1981), 289–290. MR **82e**:10004

17. E. Lehmer, *On congruences involving Bernoulli numbers and the quotients of Fermat and Wilson*, Ann. of Math. **39** (1938), 350–360.

18. R. McIntosh, *Private communication.*

19. P. Montgomery, *New solutions of $a^{p-1} \equiv 1 \pmod{p^2}$*, Math. Comp. **61** (1991), 361–363. MR **94d**:11003

20. J. Pollard, *Theorems on factorization and primality testing*, Proc. Cambridge Phil. Soc. **76** (1974), 521–528. MR **50**:6992

21. P. Ribenboim, *13 lectures on Fermat's last theorem*, Springer-Verlag, New York, 1979. MR **81f**:10023

22. ———, *The book of prime number records*, Springer-Verlag, New York, 1988. MR **89e**:11052

23. V. Strassen, *Einige Resultate über Berechnungskomplexität*, Jahresber. Deutsch. Math.-Verein. **78** (1976/77), 1–8. MR **55**:11713

24. Zhi-Hong Sun and Zhi-Wei Sun, *Fibonacci numbers and Fermat's last theorem*, Acta Arith. **60** (1992), 371–388. MR **93e**:11025

25. D. D. Wall, *Fibonacci series modulo m*, Amer. Math. Monthly **67** (1960), 525–532. MR **22**:10945

26. A. Wieferich, *Zum letzten Fermat'schen Theorem*, J. Reine Angew. Math. **136** (1909), 293–302.

27. H. C. Williams, *The influence of computers in the development of number theory*, Comput. Math. Appl. **8** (1982), 75–93. MR **83c**:10002

28. K. M. Yeung, *On congruences for binomial coefficients*, J. Number Theory **33** (1989), 1–17. MR **90i**:11143

CENTER FOR ADVANCED COMPUTATION, REED COLLEGE, PORTLAND, OREGON 97202
E-mail address: `crandall@reed.edu`

DEPARTMENT OF MATHEMATICS, STATISTICS AND COMPUTING SCIENCE, DALHOUSIE UNIVERSITY, HALIFAX, NOVA SCOTIA, B3H 3J5, CANADA
E-mail address: `dilcher@cs.dal.ca`

DEPARTMENT OF MATHEMATICS, UNIVERSITY OF GEORGIA, ATHENS, GEORGIA 30602
E-mail address: `carl@ada.math.uga.edu`

4. Parallelization of Pollard-rho factorization

Discussion

This Pollard-rho paper is a preprint, unpublished. It is controversial in the sense that some scholars have expressed skepticism as to the parallelism heuristics (see Challenges). The eminent J. M. Pollard himself had some doubts back in 1999, and such are well taken. (Pollard did, though, offer many insights that have been installed in this paper.)

Challenges

Someone should do an elaborate, incisive experiment to see if the parallel heuristic—namely factoring time $\sim \frac{\log^2 m}{m}$ for m machines—works out in real life. The small experiments discussed in this paper were encouraging, but certainly not enough to firmly establish even the heuristic estimates.

Source

R. Crandall, "Parallelization of Pollard-rho factorization," (preprint 1999).

Parallelization of Pollard-rho factorization

Richard E. Crandall

Center for Advanced Computation, Reed College, Portland OR

Abstract. It is an established heuristic that if each of m machines carries out an independent (in a very specific statistical sense) "Pollard-rho"—sometimes called "Monte Carlo"—factorization sequence, then the expected time to discover a prime factor p of N is on the order of

$$\tau_N \frac{\sqrt{p}}{\sqrt{m}},$$

where τ_N is the time for a multiplication (mod N). This gain of $\sqrt{m}$ from such parallelization of m machines is of course disappointing. Herein we show that, by invoking appropriate fast algorithms to resolve certain correlation products, one may obtain (if the heuristics be valid) virtually ideal parallelism, in the sense that the time to discover p is now estimated as:

$$\tau_N \frac{\sqrt{p} \log^2 m}{m},$$

with somewhat lower complexities obtainable in certain cases via known modifications of the Pollard iteration itself. On this heuristic for parallelism, various factoring labors are assessed in regard to their practicality.

DRAFT 23 Jun 1999

1. Motivation for parallel Pollard-rho

The "Pollard-rho" factorization method is based on an idea of [Pollard 1975] to employ random, or "Monte Carlo" sequences to discover, by virtue of a deterministic periodicity inherent to certain such sequences, hidden factors of a given number N. The method has enjoyed a certain vogue, peaking in a sense with the impressive factorization of $F_8 = 2^{2^8} + 1$ about two decades ago [Brent and Pollard 1981]. In general, the Pollard-rho method or any of its modern variants and enhancements [Montgomery 1987][Brent 1980] discovers a hidden prime factor p of N in time

$$O(\tau_N \sqrt{p}), \tag{1.1}$$

where τ_N is the time for a ring multiplication in Z_N. We remind ourselves right off that this $p^{1/2}$ behavior is heuristic, yet seems to be quite the case in practice. All of these established enhancements alter only the implied big-O constant, so we can think of any Pollard-rho variant as requiring $O(p^{1/2})$ operations in some statistical sense, with worst-case factoring of composite N involving therefore $O(N^{1/4})$ operations.

Though the elliptic curve method (ECM) of H. Lenstra has become a dominant scheme for extracting relatively small factors of large numbers, there remains a genuine niche for Pollard-rho and its variants. The Pollard-rho amounts in a certain sense to a "statistical sieve" (of course not technically so, we mean that the rho may be used to ferret out, if you will, "most" factors up to some limit). Indeed, with the rho method one can typically attain sieving bounds as high as 10^{12} or more on possible factors p, at least with reasonable probability that all relevant p have thus been checked. One infers from the above heuristic big-O estimate that only a few million Pollard iterations are required to saturate with high chance such a limit of 10^{12}. As pointed out by [Brent 1990], the use of m machines in an elementary attempt to parallelize the Pollard-rho, say by running each machine independently (meaning, with independent iteration polynomials and independent random starting values, as we later clarify) stopping the entire ensemble of machines when some machine finds a factor p, should result in expected discovery time

$$O(\tau_N \frac{\sqrt{p}}{\sqrt{m}}). \tag{1.2}$$

There is an easy heuristic argument that explains the disappointing acceleration factor of only $\sqrt{m}$, as we shall see.

Because it is a kind of simple sieve, the Pollard-rho method still has applications in computational number theory. For example, pre-factoring of residues from other problems, quick factorization of elliptic curve orders in elliptic-curve primality proving (ECPP), and so on. What is more, there are other domains in which a parellelized rho method might apply. For example, as known to Brent and Pollard, if the hidden prime factor of N is known to be of the form $p \equiv 1 \pmod{2K}$ then the time to discover p is reduced further by about $(\log_2 K)/\sqrt{2K-1}$, the logarithm appearing due to extra operations in a modified, K-dependent rho sequence. So for such as Fermat numbers $F_n = 2^{2^n} + 1$, with $n > 1$, any prime factor of which being $\equiv 1 \pmod{2^{n+2}}$, the rho method enjoys a substantial boost; witness the success on F_8 as mentioned above, for which Brent and Pollard used $K = 512$.

For example, the recent (April 1999) discovery by R. McIntosh and C. Tardif, namely the factor

$$81274690703860512587777 \quad | \; F_{18}$$

though found via a particular ECM variant [Crandall and Pomerance 1999][Brent et. al. 1999], could in principle have been found on the same type of machinery, as we argue later, via the new parallel realization of the rho method.

Recently, [van Oorschot and Wiener 1999] reported a clever way of applying a rho method to the discrete logarithm (DL) problem in parallel fashion. The DL problem has two primary manifestations: first, given g, y, p, to solve $g^x \equiv y \pmod{p}$ for x; or second, to solve $xg = y$ on an elliptic curve. Note that whereas the former problem has various known attacks, the rho method is the most powerful known general-purpose scheme for the elliptic case. In those authors' use of "distinguished points" the utilization of m machines is ideal, in that the time to solve DL is reduced by a factor of m, so that a time bound of the type (1.2) is obtained, but with m in the denominator instead of $\sqrt{m}$. However, as pointed out by [Wiener 1999], there is so far no known extension of their parallel-DL scheme to factorization *per se*.

The purpose of the present paper is to show how Pollard-rho factoring can, with m machines operating in parallel fashion, and under heuristic assumptions about the statistics of the relevant iterations, indeed be achieved in time

$$O(\tau_N \, \frac{\sqrt{p} \log^2 m}{m}), \tag{1.3}$$

using a method which appears on the face of it to be unrelated to the DL parallelism of van Oorschot and Wiener.

2. Statistical and algebraic pictures

The original—and in many ways still the most elegant—manifestation of the idea of Pollard is, for given N to be factored, to start with random x_0, a and iterate a quadratic map:

$$x_{n+1} := (x_n^2 + a) \bmod N, \tag{2.1}$$

where here and elsewhere the final operation on the right is reduction to least nonnegative residue $\pmod{N}$. Normally one uses an a parameter $\neq 0, -2$, to ensure a suitable pseudorandom sequence $\{x_n\}$. If p be a hidden prime factor of N, an occurrence of $x_i = x_j \pmod{p}$ for $i \neq j$ is sought, in the sense that p (sometimes along with yet other prime factors of N) might be the result of an operation $\gcd(x_i - x_j, N)$. Thus the Pollard sequence (2.1) is likely to succeed if we can detect such an i, j-collision via the gcd. The celebrated Floyd expedient is to observe that the fact of a collision also means there must be an instance of $x_{2k} = x_k$, some k; indeed the least such successful k does not exceed the first i for which there is a collision of x_i and $x_{j<i}$. This means one need not check pairs $x_i - x_j$ directly, testing instead various $\gcd(x_{2i} - x_i, N)$. Now one sees how very simple the rho method is; one only need initialize random $x = w$, and iterate:

$$x := (x^2 + a) \bmod N, \tag{2.2}$$

3

$$w := (w^2 + a) \bmod N,$$

$$w := (w^2 + a) \bmod N,$$

where the w iteration happens intentionally twice, so that w values amount to the x_{2i}. Thus the memory usage is about as low as can be for any factorization method. One need not even perform a gcd operation on every iteration. Instead, simply accumulating the product of many differences $(w - x)$, seldom taking one gcd of the accumulation with N, is the common practice. Various authors [Brent 1980][Montgomery 1987] have reduced the necessary arithmetic, for example by noting other ways to detect collision and stating simplifications of the iteration algebra, but it remains that the (heuristic) expected time behaves as $\sqrt{p}$, in the manner of relation (1.1).

There are two interesting ways to look at the underlying mechanism of the Pollard-rho method. One we shall call the "statistical picture" and the other the "algebraic picture." In the statistical picture we think of the sequence $\{x_i : i = 1, ..., n\}$ as somehow random modulo a hidden factor p of N. How far should this sequence be taken before it replicates a previous value? Let us estimate crudely the probability that a random sequence traversing a set of P elements has its first collision on the n-th iterate. Starting with random x_0, the probability that x_1 misses x_0 is $(P - 1)/P$, the probability x_2 collides with neither x_0 nor x_1 is $(P - 2)/P$ and so on, so that the first collision happens for x_n with probability:

$$(1 - \frac{1}{P})(1 - \frac{2}{P})...(1 - \frac{n-1}{P})\frac{n}{P}$$

which we further approximate as a particular, Poisson density function

$$f(n) = \frac{n}{P}e^{-n^2/(2P)},$$

whose integral over $n \in (0, \infty)$ is in fact unity. Crude as this line of argument may be, it explains well the basic statistical measures encountered in actual Pollard-rho applications. For example, we find the expected number of iterates required for a collision to be:

$$<n> \sim \int_0^\infty nf(n)dn = \sqrt{\frac{\pi P}{2}},$$

Now as to the comparison of the effective set size P with the prime p, we have, at least roughly speaking, $P \sim p$. There is a delicate issue here—a paradox, if you will—that runs as follows. Because every element in our set (save perhaps for x_0) is a number of the form (quadratic residue + a) modulo p, it would appear that $P = p/2$ is a good choice for the effective cardinality of the set. But as explained by [Brent and Pollard 1981], this notion is faulty, in that the empirical expectation seems to be $<n> \sqrt{\pi p/2}$ for the collision index (i of the first $x_i = x_{j<i}$). To paraphrase [Pollard 1999], the iterative map behaves as if it acts on a set of the full p elements. More precise heuristics are also presented in [Cohen 1993], who gives for example the expected length of the actual *period* of the $\{x_i\}$ sequence to be $<n> /2 \sim \sqrt{\pi p/8}$; yet in that work the paradox is not addressed directly. Note that the period is not the same as the expected collision time; for one thing, the collision happens after an initial "pretail" and then a full period—hence the origins of the algorithm label "rho" whose very symbol shape reminds one of a typical collision scenario. Regardless of which analysis is used and what is the resolution of the aforementioned paradox, the established Pollard-rho heuristic is that the expectation of necessary iterations to discover p is: $< n > \sim c\sqrt{p}$ for constant c.

Now to the algebraic picture, which has been analyzed before in such treatments as [Riesel 1994]. Observe the chain of factors that accrues from *every* index k of the x_{2k} iterates:

$$x_{2k} - x_k = x_{2k-1}^2 + a - x_{k-1}^2 - a = x_{2k-1}^2 - x_{k-1}^2$$

$$= (x_{2k-1} + x_{k-1})(x_{2k-1} - x_{k-1})$$

$$= (x_{2k-1} + x_{k-1})(x_{2k-2} + x_{k-2})(x_{2k-3} + x_{k-3})... \qquad (2.3)$$

and we conclude that the difference $x_{2k} - x_k$ is possessed of about k algebraic factors. This means that if we run the index k from 1 through some n, we accumulate about $n^2/2$ algebraic factors. Though the rest of this line of argument is fraught with complications— such as $\log\log$ factors in the theory of the number of prime factors of a number, not to mention the interference between algebraic factors—it is reasonable that $O(n^2)$ "random" factors have a fair chance of involving p, when n is $O(\sqrt{p})$. Though less precise even than the already heuristic statistical picture, this algebraic picture lends itself easily to considerations of parallelization.

The important special instance when one has foreknowledge of $p \equiv 1(\text{mod } 2K)$ has interesting interpretations in the statistical and algebraic pictures. Statistically, the iteration:

$$x := (x^{2K} + a) \text{ mod } p$$

runs over a restricted set, since for the class of p in question the set of $2K$-th powers has cardinality $P = (p-1)/(2K)$. Thus an operation reduction by $1/\sqrt{K}$ might be expected to appear in the relevant expectation formulae. Note however, that this thinking leads to the aforementioned paradox in the case of $K = 1$, and we recall the discovery of [Brent and Pollard 1981] that the better reduction factor is $1/\sqrt{2K-1}$. On the other heuristic hand, that is in the algebraic picture, we have

$$x_{2i} - x_i = x_{2i-1}^{2K} - x_{i-1}^{2K}$$

and the right-hand side can be expected to have amplified probability of containing a factor p, because in the field $F_p[X, Y]$ the binomial $X^{2K} - Y^{2K}$ has factors $X - g^\gamma Y$, where g is a primitive $2K$-th root of unity. For either picture, the heuristic reduction in total time is (avoiding the more correct $2K - 1$ factor for the moment) $O(\log_2 K/\sqrt{K})$, the logarithmic factor for the extra powering inherent to the $2K$-power iteration. Incidentally, even if one uses iteration $x := x^{2K} + a$ without any known caveat on the hidden p, it is an established heuristic [Montgomery 1987][Brent and Pollard 1981], and it can be inferred roughly from either of our heuristic pictures, that the reduction factor in expected number of ring operations is:

$$1/\sqrt{\gcd(p-1, 2K) - 1},$$

where here we have given what is believed to be the correct radicand as a function of K.

3. Parallelization

Let us use the statistical picture to infer the observation of [Brent 1990], that the use of m machines in parallel—if running independent (in both iteration polynomial and initial seeds) Pollard-rho processes—reduces the expected factoring time only by $\sqrt{m}$. Denote by $x_i^{(k)}$ the i-th iterate on machine k of m machines. For the moment, we assume that the k-th machine uses an iteration $x := x^2 + a_k$, with the a_k chosen independently (except $\neq 0, -2$ say); and also each machine is independently seeded with an $x_0^{(k)}$. Along the lines of the statistical picture of the previous section, the probability that none of the m machines has, upon its i-th iterate, a collision $x_i^{(k)} = x_j^{(k)}$ for any $j < i$ would be

$$(1 - \frac{i}{P})^m,$$

and we arrive at an approximate probability density

$$f(n) = \frac{2mn}{P} e^{-mn^2/(2P)} \tag{3.1}$$

for the first collision on some machine to happen at iterate n. This leads immediately to the expectation

$$<n> \sim \sqrt{\frac{\pi p}{2m}},$$

showing the disappointing reduction by only $1/\sqrt{m}$.

From the algebraic picture we can deduce the same heuristic for this weak parallelism. Since we have algebraic factors as in (2.3) for each of the machines, a total of $O(mn^2/2)$ factors are generated. Thus the heuristic is that $n \sim \sqrt{p/m}$ steps will be required to discover p.

This having been said about noncoupled machines, let us consider a more elaborate construct from the algebraic picture, and hope that stronger machine coupling will allow the construct to be efficiently evaluated. At this juncture it is important to note that we shall have to use, in the following parallel construction, the *same a-parameter across machines*. Later in Section 5 we bring up the difficult issue of whether said parameter can be thus frozen.

The new parallel model on which we shall base the new heuristic has, at its core, the correlation product

$$Q = \prod_{i=1}^{n} \prod_{k=0}^{m-1} \prod_{j=0}^{m-1} (x_{2i}^{(k)} - x_i^{(j)}). \tag{3.2}$$

Now assume that each machine has the *same* iteration parameter a, yet the initial seedings $\{x_0^{(k)} : k = 0...m-1\}$ are independently random. We should then have, in the manner of the formal expansion (2.3), $O(m^2n^2)$ algebraic factors comprising Q, and thus the n required to discover p should be $O(\frac{\sqrt{p}}{m})$. From the statistical picture, the same heuristic obtains, as the probability density for the first collision *amongst* machines is just (3.1) but with $m \to m^2$.

The reduction by $1/m$ of the expected n is fine and good, but what about the time to evaluate the elaborate product (3.2)? An answer is, the evaluation of product Q takes, in an appropriate asymptotic sense, just a little longer than the calculation of the usual, basic rho sequence through index $2n$. In particuler, let us say for convenience m divides n, whence machine μ of m machines can compute the partial correlation product

$$Q_\mu = \prod_{i=\mu n/m}^{(\mu+1)n/m-1} \prod_{k=0}^{m-1} \prod_{j=0}^{m-1} (x_{2i}^{(k)} - x_i^{(j)}) \tag{3.3}$$

in just n/m applications of a fast polynomial evaluation algorithm. In fact, machine μ can evaluate the polynomial

$$q_i(x) = \prod_{k=0}^{m-1} (x_{2i}^{(k)} - x)$$

at the points $x \in \{x_i^{(j)} : j = 0, ..., m-1\}$ in $O(m \log^2 m)$ ring operations [Montgomery 1992][Crandall and Pomerance 1999]. Thus the total number of ring operations, on the μ-th machine, is

$$O(\frac{n}{m} m \log^2 m) = O(n \log^2 m).$$

This basic idea, together with the observation that all other calculations, such as accumulation of the actual product in (3.3), are $O(n)$ ring operations, leads to the heuristic estimate (1.3) of total *time* to discover p, with the effective reduction factor being $O(\log^2 m/m)$.

Subject as usual to the validity of heurism, the advantages of using iteration $x := (x^{2K} + a) \bmod N$ when it be given that a prime divisor p of N is $1 \pmod{2K}$, or at least has a fairly large $\gcd(p-1, 2K)$, likewise apply in this new parallel mode. That is to say, each machine would use the degree-$(2K)$ iteration, with a absolutely fixed across machines, but $x_0^{(k)}$ random on the k-th machine as before. With all of these considerations in place, the total time to discover a factor with our parallel algorithm should follow the heuristic estimate:

$$O\left(\tau_N \ \frac{\sqrt{p}}{m} \ \frac{1 + \log K + \log^2 m}{\sqrt{\gcd(p-1, 2K) - 1}}\right). \tag{3.4}$$

We are now positioned to summarize a parallelization algorithm, as follows:

Algorithm for parallelization of Pollard-rho

0) We are given N to be factored, and m machines with which to do this. We assume a number K which will determine the degree (as $2K$) of the Pollard iteration, knowing that a statistical gain from large $\gcd(p-1, 2K)$ is possible. For example, if it be known that a hidden odd prime factor p of N must have $p \equiv 1 \pmod{2K'}$, then K should be a multiple of K'.

1) Choose a trial n, being a multiple of m, noting that primes up to a bound $p \sim n^2 m^2 (\gcd(p-1, 2K) - 1)$ have a good chance of being discovered via this algorithm, and

under the constraint that acyclic convolution of length-m sequences modulo N be possible on each machine.

2) For each machine k of m, establish an initial independent random seed $x_0^{(k)}$ and set $w_0^{(k)}$ equal to that same respective seed. Then choose a parameter a which is the *same* across all machines.

3) For $i \in [1, ..., n]$, iterate on each machine k of the m machines simultaneously:

$$x_i^{(k)} = ((x_{i-1}^{(k)})^{2K} + a) \bmod N,$$

$$w = ((w_{i-1}^{(k)})^{2K} + a) \bmod N,$$

$$w_i^{(k)} = (w^{2K} + a) \bmod N,$$

so that $w_i^{(k)} = x_{2i}^{(k)}$ always.

4) On each machine μ of the m machines, evaluate modulo N the product

$$Q_\mu = \prod_{i=\mu n/m}^{(\mu+1)n/m-1} \prod_{k=0}^{m-1} \prod_{j=0}^{m-1} (w_i^{(k)} - x_i^{(j)})$$

via n/m polynomial evaluations of degree m each, so that this step requires $O(n \log^2 m)$ ring operations on each machine.

5) On each machine, take the $\gcd(Q_\mu, N)$ and report any nontrivial factor arising from any machine.

6) On failure of step (5), go to (1) to increase n, or continue iterations in (3), or effect some other means of effectively modifying n and/or parameters such as a, K.

4. Open questions, Fermat numbers, quantum computation

First we remind ourselves that little if anything has been proven so far, and mention some open issues in regard to the kind of heuristics we have so brazenly invoked. The "final" heuristic relation (3.4) for the parallel scheme might be viewed with suspicion, along the following lines. As J. M. Pollard himself has pointed out to the present author, the assumption of fixed a-parameter leads to certain difficulties in the heuristic analysis. In the scenario of weak parallelism as enunciated in the very opening of Section 3 (i.e., independent seeds and independent a-parameters), different machines can be expected to have widely varying iterative cycles in force. But if a is fixed across machines, there is a very small expected number (actually $O(\log p)$) of cycles at work [Pollard 1999]. Pollard cites the small example $p = 257$, $a = -1$, for which there are only three periods 12, 7, 2 (and we note that the number of iterates to discovery is the least multiple of said period that is not less than the tail). For example, the initial value $x_0 = 0$ leads to the 2-cycle $\{0, -1\}$, while the initial value $x_0 = 6$ gives rise to the sequence

$$\{6, 35, 196, 122, 234, 14, 195, 245, 143, 145, 207, 186, 157, 233, 61, 122\}$$

whose period—after a tail of 3 iterates—is one of the possible ones cited, namely 12. The component counts for the possible cycles decompose respectively as $142 + 98 + 17 = 257$. For example, the probability that an initial seed x_0 will eventually exhibit the period 12 is $142/257$. Already, in this small example with three possible periods, it is not clear until one does some detailed combinatorics how the *expected* minimum period across m machines decays; although the expected period will of course be monotonic nondecreasing in m.

What happens in practice is that when p and a fixed a are chosen, with larger and larger machine counts m emulated (say m is much less than say $\sqrt{p}$ itself) the tendency for some machine to pick up one of the smaller possible periods naturally increases. However in such experiments with fixed a one does witness the aforementioned quantization of cycle periods. In order to get the theoretically correct expectation of minimum period for fixed a across machines therefore, one would have to know something about the relatively small set of $O(\log p)$ periods, and furthermore know how random initial seeds deposit themselves into the respective cycles. We repeat here the reason that it is not so hard to infer the $1/\sqrt{m}$ acceleration in the Brent scenario—in which each machine has a differant a_k–namely, widely varying periods will be available to different machines; in effect the space of possible cycle periods will be widely sampled. Instead of $O(\log p)$ possible cycles, there would be perhaps as many as $O(m \log p)$ cycles available on the parallel network.

Faced with such intricate statistical issues, one might be suspicious of the full parallel scheme. The present author carried out the following experiment on the Q product (3.2) for which a is fixed across machines. Choosing:

$p = 2^{31} - 1$,

$n = 250$ iterations,

$m = 250$ machines,

and one random a parameter for every full parallel run, the empirical probability that Q contains the factor p was a healthy 0.61. This was based on one hundred random choices of a and a complete evaluation of Q for each choice of a. Now for these choices we have $n^2m^2 \quad p$, so this brief experiment supports—at least in order-of-magnitude fashion—the parallel heuristic based on the Q product. So even in the face of the the complicated quantization of cycle periods when a is fixed across machines, this experiment is encouraging.

Next, regardless of the open issues we turn to speculations stemming from the heuristic (3.4). Let us attempt some estimates of actual run time for hypothetical factorizations. We shall use formula (3.4) with the big-O cavalierly stripped away, using $\log_2$ for any logarithms (actually such is a realistic substitution), with the effective coefficients of the $\log K$, $\log^2 m$ terms both unity (actually a reasonable guess in actual practice), and a time for one multiplication modulo $N = F_{18}$ say, to be 0.1 seconds (today this is a realistic estimate on such a large-integer operation, in fact multiply times an order of magnitude lower than this have been reported for certain machine architectures [Mayer 1999]). Then to discover the 23-digit McIntosh-Tardif factor mentioned in Section 1 we need an (operation $\to$ time) equivalent count of

$$\sqrt{p} \quad \to \quad 1000 \text{ years}$$

using basic, unadorned Pollard-rho. Using the fact that any factor of F_{18} is of the form $p \equiv 1 (\text{mod } 2^{20})$ we set $2K = 2^{20}$ and note that the better iteration $x := x^{2K} + a$, for which the estimate of (operations $\to$ time) to discover the factor is:

$$\sqrt{p} \, \frac{\log_2 2K}{\sqrt{2K - 1}} \quad \to \quad 17 \text{ years.}$$

On the other hand, we assess the impact of parallelism as follows. In this age of massive volunteer network computations, let us be generous and take $m = 10000$ machines. Now in the "weak" parallel mode, where there is essentially no intermachine communication, the (operation $\to$ time) estimate comes out as:

$$\sqrt{p} \, \frac{1}{\sqrt{m}} \frac{\log_2 2K}{\sqrt{2K - 1}} \quad \to \quad 2 \text{ months.}$$

The new parallel scheme though has estimate:

$$\frac{\sqrt{p}}{m\sqrt{2K - 1}} (\log_2 2K + \log_2^2 m) \quad \to \quad 4 \text{ days.}$$

As with this example, it is generally true that hundreds or thousands of machines would be required to bring the new parallel method beyond the efficiency of the weak one, due to the $\log^2 m$ factor. But let us at least establish that the machine parameters are not too absurd in such regions of advantage. From the algorithm of the previous section, we see that for n parallel iterates, per-machine, of Pollard-rho sequences the approximate relation

$$p \quad n^2m^2K$$

is expected. This means that each of the $m = 10000$ machines would only do about $n \quad 40000$ iterations, or on the order 10^6 multiplies modulo F_{18}. So the number of iterates in step (3) of the algorithm is entirely realistic, and along with the time for product evaluation in step (4), is on the order of days. It may be possible to bring overall times down yet further by allowing the parallel machines to cooperate even more in the evaluation of the products. (For example it is unknown to the present author whether a degree-m polynomial can be evaluated at m points via m coupled machines in $O(m^\epsilon)$ ring operations per machine, yet this may well be possible.)

As it stands the new algorithm requires extensive sharing of stored iterates. Let us see if such storage is even possible for our example of F_{18}. Each machine must be capable of fast polynomial evaluation for degree m, so via convolution techniques (such as Nussbaumer convolution with memory optimizations [Crandall 1995]) that would involve a small multiple of the memory required for 10^4 copies of F_{18} (a 32-kbyte entity). This comes out to about 700 megabytes per machine, which although stultifying is not out of the question even for personal computers of today. It is also the case that some kind of memory pool must exist for all the stored iterates from step (3) of the algorithm. The total memory required is about $2n$ copies of size-F_{18} residues on each machine μ, where we might assume that the sequence $\{x_i^{(\mu)}\}$ is actually stored. In our example that is about 7 gigabytes per machine, again unfortunate but not out of the question. An interesting line of future research is to determine whether alterations of the precise of order of doing things between algorithm steps (3) and (4) will save memory. For example, when machine $\mu = 0$ does its n/m polynomial evaluations each of degree m, at m points, perhaps that effort should be shared amongst all machines, and an accumulation (or gcd) operation performed, so that the indices $i \in [0, n/m - 1]$ once processed may theretofore be ignored. If such memory-reduction schema can be made to work, then the dominant memory is that of the polynomial evaluation itself, or as we have said, $O(m)$ copies of size-N residues, per machine.

Another interesting line of possible optimization is the following. Observe that the choice of K can be arbitrary, on the idea that a time reduction of about $1/(\gcd(p-1, 2K) - 1)^{1/2}$ results. There is nothing preventing one from trying various K values at the start of the algorithm. It is not yet known whether using different K values *per machine* in this new parallel scenario affords any advantage. It would be interesting to analyze this question from both statistical and algebraic pictures. But since the McIntosh-Tardif factor is

$$81274690703860512587777 = 1 + 2^{23} \cdot 29 \cdot 293 \cdot 1259 \cdot 905678539,$$

any machine that happens to have an extra factor 2^3 or even $2^3 \cdot 29$ (beyond the known efficiency factor $2K' = 2^{20}$) contained in its $2K$ value will likely contribute somewhat more powerfully than otherwise. Indeed, the appearance of the $(\log K + \log^2 m)$ factor in the overall time estimate (3.4) shows that for large numbers m of machines, one will usually obtain a definite acceleration from a lucky K value, and in any case one may use various small K values with relative impunity.

Aside from the issue of whether it makes sense to "blindly" adopt various K accelerators, and whether to do this on a per-machine basis, there is the question of the Pollard iteration itself, and how its effective traversal set can be restricted. Though linear iterations $x := ax + b$ are known to be unsatisfactory (e.g. they might have periods of length $O(p)$), it would be good to know the effective periods for $x := a \bullet x + b$ where $\bullet$ is an elliptic multiplication on an elliptic curve, with x now being a point. Such an elliptic iteration was once suggested by V. Miller, in regard to pseudorandom number generation. In fact, [Miller 1999] points out that the period of such an elliptic generator can again be p (good for cryptography, bad for rho-factoring), depending on the group structure at hand; i.e. whether the elliptic group is cyclic. Thus the elliptic generator—with each step involving expensive elliptic arithmetic—may be of no use in rho-based factoring; yet the elliptic generator should still be studied completely in this regard. Then, too, it would be of interest to apply various of the ideas herein to the Pollard-$(p-1)$ method. For the moment, we note that [Montgomery and Silverman 1990][Montgomery 1992] have shown the importance of polynomial-evaluation acceleration to Pollard-$(p-1)$ and ECM, respectively. It should be remarked that the aforementioned McIntosh-Tardif factor, as well as other ECM-based discoveries such as the factor

$$3603109844542291969 = 1 + 2^{19} \cdot 3 \cdot 4363 \cdot 525050549$$

of F_{13}, could be found rather quickly with Pollard-$(p-1)$ with suitable second stage, perhaps in the "FFT" style of Montgomery and Silverman (to pick out the factor 525050549). Certainly the corresponding time estimates [Mayer 1999] show that such an effort would clearly be more lucrative than single-machine Pollard-rho. But this comparison is not at all so one-sided in a parallel-rho scenario for which some lucky machine is performing an iteration: $x := x^{2^{19} \cdot 3 \cdot 4363} + a$, say. For that machine should only require about $n \sim 10^5$ iterates all by itself, to discover the given 19-digit factor of F_{13}, and this would only take a matter of hours. Again this raises the open question of whether there be genuine gain in having machines carry out separate Pollard iterations. One may look longingly at the older results of [Brent 1985] on ECM-rho comparisons—which results imply a kind of crossover at about 11 digit factors—and wonder what higher crossover might accrue from both parallelism and use of hidden K values.

Now, if the present author may speculate in unrestrained fashion: whether or not further reduction in the complexity estimate (3.4) can be effected in future, there is the issue of quantum computation. It is now known that factorization via quantum Turing machine (QTM) can proceed, at least in principle, at unprecedented speed, as in the Shor factoring algorithm [Ekert and Jozsa 1996]. In such an approach, quantum cells using natural quantum interference would be used to perform gargantuan fast Fourier transforms (FFTs), and so look for a certain kind of periodicity of powers modulo an N to be factored. By the same token, we expect the new parallel scheme for Pollard-rho to have a reasonably direct analogue in the QTM world. After all, the notions of parallelism, inter-processor communication, and large, fast transforms are all present. In fact, transforms tend to figure into the best polynomial evaluation schemes. If there be a valid analogue of the basic Pollard-rho on QTMs, we can say now that there should also be a *parallel* analogue. If yet more speculation can be extended from this already speculative posture, let us be truly generous and think of $m = 10^{24}$ (one "mole") of quantum cells. Not to stretch the notion too far, but a nanotechnological society might one day view such a cell count as, well, commonplace. Then for this m,

$$\frac{\log_2^2 m}{m} \sim 10^{-20},$$

so that formula (3.4) allows 70-digit factors p of N to be obtained in about 10^{15} ring operations in Z_N—certainly a feasible operation count by comparison with many of the deeper computational successes of the present era of conventional TMs. Needless to say, 70-digit extractions are right about at the current state of the factoring art (via the number field sieve (NFS), say). Of course one could respond that, by the time a quantum-nano technology is upon us, other dominant factorization methods will have been correspondingly accelerated to unprecedented performance levels. But to this we can say, Pollard-rho is so very simple, it may be one of the first implementations in any really new technology. After all, it has already been, once for a time the best available method for certain classes of N. True, there is polynomial evaluation of the Q products indicated for our stated algorithm, but for all this author knows, a future machine would be better off doing brute-force $O(m^2)$ product evaluations, and *that* is a manifestly simple procedure. Because of allowed special K values, something like the elusive F_{14} or one of its relatives might finally be demolished in this way (it is an interesting exercise to estimate how many new Fermat factors would accrue at a 70-digit search limit). It would therefore seem that, even though more dominant factorization schemes have migrated to the forefront over the last two decades, the last chapter in the story of the Pollard-rho method is not yet written.

5. Acknowledgments

The author is indebted to M. Wiener for commentary and insights relevant to the theoretical issues herein, to R. Brent for remarks pertaining to the manuscript, and to C. Pomerance for his particular insight into Pollard-rho and exponential factoring algorithms in general. J. M. Pollard himself corrected several misconceptions on the present author's part, thereby rendering this a (relatively) more professional paper.

6. References

Brent R 1980, "An improved Monte Carlo facorization algorithm," *BIT*, 20, 176-184.

Brent R 1985, "Some integer factorization algorithms using elliptic curves," manuscript, ftp://pub/Documents/techpapers/Richard.Brent/rpb102.dvi.gz

Brent R 1990, "Parallel algorithms for integer facorization," in Loxton J H Ed., *Number Theory and Cryptography, London Math. Soc. Lect. Note Series*, Cambridge University Press, Ca.b., 154, 26-37

Brent R and Pollard J 1981, "Factorization of the eight Fermat number," *Math. Comp.*, 36, 627-630.

Brent R, Crandall R, Dilcher K, and van Halewyn C 1999, "New Factors of Fermat Numbers," *Math. Comp.*, to appear

Cohen H 1993, *A course in computational algebraic number theory*, Graduate Texts in Math. 138, Springer-Verlag, Berlin Heidelberg New York

Crandall R 1995, *Topics in Advanced Scientific Computation*, Springer-Verlag, New York.

Crandall R and Pomerance C 1999, "Prime numbers: a computational perspective," Springer-Verlag (mansuscript t.b.p.)

Ekert A and Jozsa R 1996, "Quantum computation and Shor's factoring algorithm," *Rev. Mod. Phys.* 68, 3, 733-753

Mayer E 1999, private communication

Miller V 1999, private communication

Montgomery P L 1987, "Speeding the Pollard and Elliptic Curve Methods of Factorization," *Math. Comp.*, 48, 177, 243-264

Montgomery P L 1992, " An FFT Extension of the Elliptic Curve Method of Factorization," Ph. D. Dissertation, University of California, Los Angeles

Montgomery P and Silverman R 1990, Math. Comp., 54, 839-854.

Pollard J M 1975, "A Monte Carlo method for factorization," *BIT*, 15, 331-334.

Pollard J M 1999, private communication.

Riesel H 1994, *Prime Numbers and Computer Methods for Factorization*, Birkhauser, Boston.

van Oorschot P C and Wiener M J 1999, "Parallel Collision Search with Cryptanalytic Applications," *J. Cryptology*, 12, 1-28.

Wiener M 1999 1998, private communication

5. Three new factors of Fermat numbers

Discussion

As with the Wieferich–Wilson-primes paper previous, the numerical results herein for new Fermat factors are not as important as is the method. This manner of applying the elliptic-curve method (ECM) is largely due to R. Brent, and involves various clever algorithm optimizations.[1]

Challenges

The natural challenge is to continue the discovery of Fermat factors. T. Rajala found (2010), using software from G. Woltman et al., the remarkable factor

$$1169280858730743698290359938345963713403867034233733313 \text{ of } F_{14} = 2^{2^{14}} + 1.$$

This factor could be called "lucky," because ECM is after all a statistical method with random seeds and so on. Indeed, the van Halewyn 33-digit factor of F_{15} in the paper here is likewise lucky.

But such remarks should not take away anything from the marvelous achievements of these discoverers: As the great chess champion José Capablanca once said, when confronted with the critique that his play often ended up in lucky end-game scenarios, "a good [chess] player is always lucky." An immediate challenge here is to attack the remaining cofactor of F_{14}, which cofactor is known to be composite. Beyond this, probably many factors remain undiscovered amongst the proven composites such as F_{20}, F_{24} (D. Bessell and G. Woltman found in 2010 a factor of F_{22}, namely 6465870599459185100905774868504577).

Source

Brent, R. P., Crandall, R. E., Dilcher, K. and van Halewyn, C., "Three new factors of Fermat numbers," *Math. Comp.*, **69**, 1297–1304 (2006).

---

[1]See also Chapter 7, R. Crandall and C. Pomerance, *Prime numbers: A computational perspective*, Springer New York (2002).

THREE NEW FACTORS OF FERMAT NUMBERS

R. P. BRENT, R. E. CRANDALL, K. DILCHER, AND C. VAN HALEWYN

ABSTRACT. We report the discovery of a new factor for each of the Fermat numbers F_{13}, F_{15}, F_{16}. These new factors have 27, 33 and 27 decimal digits respectively. Each factor was found by the elliptic curve method. After division by the new factors and previously known factors, the remaining cofactors are seen to be composite numbers with 2391, 9808 and 19694 decimal digits respectively.

1. INTRODUCTION

For a nonnegative integer n, the n-th *Fermat number* is $F_n = 2^{2^n} + 1$. It is known [12] that F_n is prime for $0 \le n \le 4$, and composite for $5 \le n \le 23$. For a brief history of attempts to factor Fermat numbers, we refer to [3, §1] and [5].

In recent years several factors of Fermat numbers have been found by the elliptic curve method (ECM). Brent [2, 3, 4] completed the factorization of F_{10} (by finding a 40-digit factor) and F_{11}. He also "rediscovered" the 49-digit factor of F_9 and the five known prime factors of F_{12}. Crandall [10] discovered two 19-digit factors of F_{13}.

This paper reports the discovery of 27-digit factors of F_{13} and F_{16} (the factor of F_{13} was announced in [3, §8]) and of a 33-digit factor of F_{15}. All three factors were found by ECM, although the implementations and hardware differed between F_{13} and F_{15}, F_{16}. In fact, we used Dubner Crunchers (see §3) on F_n for $12 \le n \le 14$, and Sun workstations with DWT multiplication (see §4.2) for $16 \le n \le 21$, as well as a Pentium Pro for $n = 15$, again with DWT multiplication. Details of the computations on F_{13}, F_{16} and F_{15} are given in §5, §6 and §7 respectively.

F_{16} is probably the largest number for which a nontrivial factor has been found by ECM. Factors of larger numbers are customarily found by trial division [16, 18].

2. THE ELLIPTIC CURVE METHOD

ECM was invented by H. W. Lenstra, Jr. [23]. Various practical refinements were suggested by Brent [1], Montgomery [24, 25], and Suyama [32]. We refer to [3, 14, 22, 26, 31] for a description of ECM and some of its implementations.

In the following, we assume that ECM is used to find a prime factor $p > 3$ of a composite number N, not a prime power [21, §2.5]. The first-phase limit for ECM is denoted by B_1.

Although p is unknown, it is convenient to describe ECM in terms of operations in the finite field $K = GF(p) = \mathbf{Z}/p\mathbf{Z}$. In practice we work modulo N (or sometimes

1991 *Mathematics Subject Classification.* 11Y05, 11B83, 11Y55; Secondary 11-04, 11A51, 11Y11, 11Y16, 14H52, 65Y10, 68Q25.

Key words and phrases. discrete weighted transform, DWT, ECM, elliptic curve method, factorization, Fermat number, F_{13}, F_{15}, F_{16}, integer factorization.

2   R. P. BRENT, R. E. CRANDALL, K. DILCHER, AND C. VAN HALEWYN

modulo a multiple of N, if the multiple has a convenient binary representation), and occasionally perform GCD computations which will detect a nontrivial factor of N.

The computations reported here used two parameterizations of elliptic curves. These are the symmetrical Cauchy form [9, §4.2]:

$$x^3 + y^3 + z^3 = \kappa xyz ,\tag{1}$$

and the homogeneous form recommended by Montgomery [24]:

$$by^2 z = x^3 + ax^2 z + xz^2 .\tag{2}$$

Here κ, a and b are constants satisfying certain technical conditions. For details we refer to [3, §2.1].

The points (x, y, z) satisfying (1) or (2) are thought of as representatives of elements of $P^2(K)$, the projective plane over K, i.e. the points (x, y, z) and (cx, cy, cz) are regarded as equivalent if $c \not\equiv 0 \bmod p$. We write $(x : y : z)$ for the equivalence class containing (x, y, z). When using (2) it turns out that the y-coordinate is not required, and we can save work by not computing it. In this case we write $(x :: z)$.

2.1. **The starting point.** An advantage of using (2) over (1) is that the group order is always a multiple of four (Suyama [32]; see [24, p. 262]). It is possible to ensure that the group order is divisible by $8, 12$ or 16. For example, if $\sigma \notin \{0, 1, 5\}$,

$$
\begin{aligned}
u &= \sigma^2 - 5, \quad v = 4\sigma, \\
x_1 &= u^3, \quad z_1 = v^3, \\
a &= \frac{(v-u)^3(3u+v)}{4u^3 v} - 2,
\end{aligned}
\tag{3}
$$

then the curve (2) has group order divisible by 12. As starting point we can take $(x_1 :: z_1)$. It is not necessary to specify b or y_1. When using (2) we assume that the starting point is chosen as in (3), with σ a pseudo-random integer.

3. THE DUBNER CRUNCHER

The Dubner Cruncher [8, 15] is a board which plugs into an IBM-compatible PC. The board has a digital signal processing chip (LSI Logic L64240 MFIR) which, when used for multiple-precision integer arithmetic, can multiply two 512-bit numbers in 3.2 μsec. A software library has been written by Harvey and Robert Dubner [15]. This library allows a C programmer to use the Cruncher for multiple-precision integer arithmetic. Some limitations are:

1. Communication between the Cruncher and the PC (via the PC's ISA bus) is relatively slow, so performance is much less than the theoretical peak for numbers of less than say 1000 bits.
2. Because of the slow communication it is desirable to keep operands in the on-board memory, of which only 256 KByte is accessible to the C programmer.

The combination of a cheap PC and a Cruncher board ($US2,500) is currently very cost-effective for factoring large integers by ECM. The effectiveness of the Cruncher increases as the integers to be factored increase in size. However, due to memory limitations, we have not attempted to factor Fermat numbers larger than F_{15} on a Cruncher. A number the size of F_{15} requires 4 KByte of storage.

4. Arithmetic

4.1. Multiplication and division.

Most of the cost of ECM is in performing multiplications mod N. Our Cruncher programs all use the classical $O(w^2)$ algorithm to multiply w-bit numbers. Karatsuba's algorithm [19, §4.3.3] or other "fast" algorithms [11, 13] are preferable for large w on a workstation. The crossover point depends on details of the implementation. Morain [27, Ch. 5] states that Karatsuba's method is worthwhile for $w \geq 800$ on a 32-bit workstation. On a Cruncher the crossover is much larger because the multiplication time is essentially linear in w for $w < 10000$ (see [3, Table 4]).

Our programs avoid division where possible. If the number N to be factored is a composite divisor of F_n, then the elliptic curve operations are performed mod F_n rather than mod N. At the end of each phase we compute a GCD with N. The advantage of this approach is that we can perform the reductions mod F_n using binary shift and add/subtract operations, which are much faster than multiply or divide operations. Thus, our Cruncher programs run about twice as fast on Fermat (or Mersenne) numbers as on "general" numbers.

4.2. Use of the discrete weighted transform.

For F_n with $n > 14$ we found it more efficient overall to employ standard workstations with an asymptotically fast multiplication algorithm rather than special hardware. For these larger F_n we employed the "discrete weighted transform" (DWT) of Crandall and Fagin [10, 13]. In this scheme, one exploits the fact that multiplication modulo F_n is essentially a negacyclic convolution [11] which can be effected via three DWTs. For two integers x, y to be multiplied modulo F_n, one splits each of x, y into D digits in some base W, with $D \log_2 W = 2^n$. We actually used $W = 2^{16}$, and employed the "balanced digit" scheme which is known to reduce floating-point convolution errors [10]. The three length-$D/2$ DWTs were then performed using a split-radix complex-FFT algorithm. The operation complexity is $O(D \log D)$, and 64-bit IEEE floating point arithmetic is sufficiently precise to attack Fermat numbers at least as large as F_{21} in this way. Our DWT approach becomes more efficient than "grammar-school" $O(D^2)$ methods in the region $n \sim 12$. However, the Cruncher hardware is so fast that a Cruncher performs faster than a 200 Mhz Pentium Pro workstation for $n \leq 14$.

The advantage of DWT methods is not restricted to multiplication. The elliptic curve algebra using the Montgomery parameterization (2) can be sped up in a fundamental way via transforms. The details are given in [10]; for present purposes we give one example of this speedup. For the point-adding operation

$$\frac{x_{m+n}}{z_{m+n}} = \frac{z_{|m-n|}(x_m x_n - z_m z_n)^2}{x_{|m-n|}(x_m z_n - x_n z_m)^2}$$

it is evident that one can compute the transforms of x_m, x_n, z_m, z_n, then compute the relevant cross-products in spectral space, then use the (stored) transforms of $x_{|m-n|}, z_{|m-n|}$ to obtain x_{m+n}, z_{m+n} in a total of 14 DWTs, which is equivalent to $14/3 \simeq 4.67$ multiplies. Similar enhancements are possible for point-doubling.

Memory capacity is a pressing concern for the largest Fermat numbers under consideration (F_{16} through F_{21}). Another enhancement for the ECM/DWT implementation is to perform the second stage of ECM in an efficient manner. We note that a difference of x-coordinates can be calculated from:

$$x_m z_n - x_n z_m = (x_m - x_n)(z_m + z_n) - x_m z_m + x_n z_n,$$

and a small table of $x_m z_m$ can be stored. Thus the coordinate difference requires only one multiply, plus one multiply for accumulation of all such differences. Again, if DWTs are used, one stores the transforms of: $x_m, z_m, x_m z_m$, whence the difference calculation comes down to 2/3 of a multiply, plus the accumulation multiply. The accumulation of differences can likewise be given a transform enhancement, with the result that each coordinate difference in stage two consumes only 4/3 of a multiply. In practice, this second stage efficiency allows the choices of stage two limit B_2 at least as large as $50B_1$.

4.3. **GCD computation.** It is nontrivial to compute GCDs for numbers in the F_{21} region. We used a recursive GCD implementation by J. P. Buhler [6], based on the Schönhage algorithm [7, 28]. The basic idea is to recursively compute a 2×2 matrix M such that if $v = (a, b)^T$ is the column vector containing the two numbers whose GCD we desire, then $Mv = (0, \gcd(a, b))^T$. The matrix M is a product of 2×2 matrices and is computed by finding the "first half" of the product recursively. The first-half function calls itself twice recursively (for details see [7]). In practice it is important to revert to a classical algorithm (such as Euclid's) for small enough integers. We found that GCDs taken during factorization attempts on numbers as small as F_{13} could be speeded up by using the recursive algorithm. In the region of F_{21} the recursive approach gives a speedup by a factor of more than 100 over the classical GCD.

The Cruncher programs use the classical (non-recursive) GCD but only perform two GCDs per curve (one at the end of each phase). This is possible, at a small cost in additional multiplications, because the programs use the homogeneous forms (1) and (2) and never divide by the z-coordinate.

5. A NEW FACTOR OF F_{13}

Our first Cruncher ECM program [3, Program F] was implemented and debugged early in December 1994. It used the Cauchy form (1) with a "birthday paradox" second phase. In the period January – June 1995 we used a Cruncher in an 80386/40 PC to attempt to factor F_{13} (and some other numbers). We mainly used phase 1 limit $B_1 = 100000$. On F_{13} each curve took 137 minutes (91 minutes for phase 1 and 46 minutes for phase 2). At the time three prime factors of F_{13} were known:

$$F_{13} = 2710954639361 \cdot 2663848877152141313 \cdot 3603109844542291969 \cdot c_{2417} \, .$$

The first factor was found by Hallyburton and Brillhart [17]. The second and third factors were found by Crandall [10] on Zilla net (a network of about 100 workstations) in January and May 1991, using ECM.

On June 16, 1995 our Cruncher program found a fourth factor

$$p_{27} = 319546020820551643220672513 = 2^{19} \cdot 51309697 \cdot 11878566851267 + 1$$

after a total of 493 curves with $B_1 = 100000$. The overall machine time was about 47 days. We note that $p_{27} + 1 = 2 \cdot 3 \cdot 7^3 \cdot 59 \cdot p_{22}$. The factorizations of $p_{27} \pm 1$ explain why Pollard's $p \pm 1$ methods could not find the factor p_{27} in a reasonable time.

The successful curve was of the form (1), with initial point $(x_1 : y_1 : z_1) = (1504005881887334009298847531 : 2771949085106764625878880207 : 1) \bmod p_{27}$ and group order

$$g = 3^2 \cdot 7^2 \cdot 13 \cdot 31 \cdot 3803 \cdot 6037 \cdot 9887 \cdot 28859 \cdot 274471 \, .$$

NEW FACTORS OF FERMAT NUMBERS   5

Using Fermat's little theorem [5, p. lviii], we found the 2391-digit quotient c_{2417}/p_{27} to be composite. Thus, we now know that

$$F_{13} = 2710954639361 \cdot 2663848877152141313 \cdot$$
$$3603109844542291969 \cdot 3195460208205516432206 72513 \cdot c_{2391} \, .$$

At about the time that the p_{27} factor of F_{13} was found, our Cruncher ECM program was modified to use the Montgomery form (2) with the "improved standard continuation" second phase [3, §3.2]. Testing the new program with $B_1 = 500000$ and second-phase limit $B_2 = 35B_1$, we found p_{27} seven times, with a total of 579 curves. The expected number of curves, predicted as in [3, § 4.4], is $7 \times 137 = 959$. The successful curves are defined by (3), with σ and the group order g given in Table 1.

TABLE 1. Some curves finding the p_{27} factor of F_{13}

σ	g
1915429	$2^4 \cdot 3 \cdot 29 \cdot 857 \cdot 12841 \cdot 42451 \cdot 48299 \cdot 10173923$
2051632	$2^3 \cdot 3 \cdot 17 \cdot 19 \cdot 1031 \cdot 23819 \cdot 65449 \cdot 86857 \cdot 295277$
2801740	$2^2 \cdot 3 \cdot 7 \cdot 79 \cdot 157 \cdot 19813 \cdot 89237 \cdot 122819 \cdot 1412429$
4444239	$2^2 \cdot 3 \cdot 23 \cdot 173 \cdot 191 \cdot 907 \cdot 1493 \cdot 3613 \cdot 4013 \cdot 1784599$
6502519	$2^2 \cdot 3 \cdot 23 \cdot 131 \cdot 14011 \cdot 305873 \cdot 433271 \cdot 4759739$
8020345	$2^3 \cdot 3 \cdot 17 \cdot 23 \cdot 41 \cdot 113 \cdot 271 \cdot 3037 \cdot 10687 \cdot 12251 \cdot 68209$
8188713	$2^2 \cdot 3^2 \cdot 17 \cdot 41 \cdot 47 \cdot 139 \cdot 181 \cdot 34213 \cdot 265757 \cdot 1184489$

The fact that our programs found the same 27-digit factor many times suggests (but does not prove) that the unknown factors of F_{13} are larger than p_{27}.

When testing our program with $B_1 = 500000$, we also "rediscovered" both of Crandall's 19-digit factors using the *same* elliptic curve (mod F_{13}). In fact, our program returned the 39-digit product of Crandall's factors. Taking $\sigma = 6505208$ in (3), the group corresponding to the factor 2663848877152141313 has order $2^2 \cdot 3^2 \cdot 1879 \cdot 2179 \cdot 3677 \cdot 4915067$, and the group corresponding to the factor 3603109844542291969 has order $2^4 \cdot 3 \cdot 7^2 \cdot 22003 \cdot 79601 \cdot 874661$, so both factors will be found if $B_1 \geq 79601$ and $B_2 \geq 4915067$.

6. A NEW FACTOR OF F_{16}

The DWT/ECM program was run on a small network of SPARCstations at Dalhousie University from June 1996, in an attempt to find factors of $F_{16}, \ldots, F_{20}$. It was known that

$$F_{16} = 825753601 \cdot c_{19720}$$

where the 9-digit factor was found by Selfridge [29].

Over the period September to December 1996 an average of 6 SPARCstations ran the DWT/ECM program exclusively on F_{16} and in December found a new factor

$$p_{27} = 188981757975021318420037633$$

of F_{16}. Since

$$p_{27} - 1 = 2^{20} \cdot 3^2 \cdot 31 \cdot 37 \cdot 13669 \cdot 1277254085461$$

6   R. P. BRENT, R. E. CRANDALL, K. DILCHER, AND C. VAN HALEWYN

and

$$p_{27} + 1 = 2 \cdot 240517 \cdot 1389171559 \cdot 282805744939,$$

it would have been very difficult to find p_{27} by Pollard's $p \pm 1$ methods.

We remark that p_{27} was found twice – the first time with $B_1 = 400000$, $B_2 = 50B_1$, $\sigma = 1944934539$, and group order

$$g = 2^2 \cdot 3 \cdot 5^3 \cdot 7 \cdot 13 \cdot 19 \cdot 83 \cdot 113 \cdot 2027 \cdot 386677 \cdot 9912313 \,,$$

and the second time with $B_1 = 200000$, $B_2 = 50B_1$, $\sigma = 125546653$, and group order

$$g = 2^2 \cdot 3^2 \cdot 7^2 \cdot 109 \cdot 761 \cdot 2053 \cdot 20297 \cdot 101483 \cdot 305419 \,.$$

The ECM limits were set so that each curve required roughly four days of CPU. Altogether, we ran 130 curves with various $B_1 \in [50000, 400000]$.

The quotient $q = c_{19720}/p_{27}$ was a 19694-digit number. We computed $x = 3^q \bmod q$ and found $x \neq 3$. Thus, q is composite, and we now know that

$$F_{16} = 825753601 \cdot 188981757975021318420037633 \cdot c_{19694}$$

As a check, the computation of q and x was performed independently by Brent and Crandall (using different programs on different machines in different continents). In both cases the computations found $x \bmod 2^{16} = 12756$.

7. A NEW FACTOR OF F_{15}

Using the same DWT/ECM program, run on a 200 MHz Pentium Pro, a search for a new factor of F_{15} was attempted during the Spring and early Summer of 1997. It was known that

$$F_{15} = 1214251009 \cdot 2327042503868417 \cdot c_{9840},$$

where the 13- and 16-digit prime factors were found by Kraitchik (1925; see [20]) and Gostin (1987; see [16]) respectively.

On July 3, 1997 we found the new factor

$$p_{33} = 168768817029516972383024127016961$$

after running only three curves with $B_1 = 10^7$ and $B_2 = 50B_1$. Each curve took approximately 920 hours of CPU time (a Cruncher would have taken about 1250 hours per curve). The successful curve had $\sigma = 253301772$ and group order

$$g = 2^5 \cdot 3 \cdot 4889 \cdot 5701 \cdot 9883 \cdot 11777 \cdot 5909317 \cdot 91704181.$$

As before, we remark that

$$p_{33} - 1 = 2^{17} \cdot 5 \cdot 7 \cdot 53 \cdot 97 \cdot 181 \cdot 199 \cdot 1331471 \cdot 149211834097$$

and

$$p_{33} + 1 = 2 \cdot 3 \cdot 61 \cdot 5147 \cdot 9835373 \cdot 9108903846900395897.$$

To determine whether the 9808-digit cofactor $q' = c_{9840}/p_{33}$ is composite, we computed $x' = 3^{q'} \bmod q'$ and found $x' \neq 3$; in fact, the least positive residue $x' \bmod 2^{16}$ is 557. As before, q' and x' were computed independently by Brent and Crandall.

NEW FACTORS OF FERMAT NUMBERS   7

8. ACKNOWLEDGEMENTS

Harvey Dubner provided a Dubner Cruncher and encouraged one of us to implement ECM on it. Dennis Andriolo and Robert Dubner provided assistance with aspects of the Cruncher hardware and software. We are indebted to J. P. Buhler for his inestimable aid in the optimization of various large-integer algorithms. The factor of F_{15} was found on one of several Pentium Pro machines which were generously donated by Intel to the Oregon Graduate Institute.

Note added in proof. On April 16, 1999, Richard McIntosh and Claude Tardif of the University of Regina found the new 23-digit factor

$$81274690703860512587777 = 2^{23} \cdot 29 \cdot 293 \cdot 1259 \cdot 905678539 + 1$$

of F_{18}, using the same method and software as described in Section 4. McIntosh and Tardif report that they were successful after having run about a dozen curves with $B_1 = 100\,000$, $B_2 = 40B_1$ on a Sparc Ultra 1; the successful σ was $\sigma = 731185968$.

REFERENCES

[1] R. P. Brent, *Some integer factorization algorithms using elliptic curves*, Australian Computer Science Communications **8** (1986), 149–163. Also Report CMA-R32-85, Centre for Mathematical Analysis, Australian National University, Canberra, Sept. 1985, 20 pp.

[2] R. P. Brent, *Factorization of the eleventh Fermat number (preliminary report)*, AMS Abstracts **10** (1989), 89T-11-73.

[3] R. P. Brent, *Factorization of the tenth and eleventh Fermat numbers*, Report TR-CS-96-02, Computer Sciences Laboratory, Australian National Univ., Canberra, Feb. 1997. `ftp://nimbus.anu.edu.au/pub/Brent/rpb161tr.dvi.gz` .

[4] R. P. Brent, *Factorization of the tenth Fermat number*, Math. Comp., to appear.

[5] J. Brillhart, D. H. Lehmer, J. L. Selfridge, B. Tuckerman, and S. S. Wagstaff, Jr., *Factorizations of $b^n \pm 1$, $b = 2, 3, 5, 6, 7, 10, 11, 12$ up to high powers*, 2nd ed., Amer. Math. Soc., Providence, RI, 1988.

[6] J. P. Buhler, personal communication to Crandall, 1993.

[7] P. Bürgisser, M. Clausen and M. A. Shokrollahi, *Algebraic Complexity Theory*, Springer-Verlag, 1997.

[8] C. Caldwell, *The Dubner PC Cruncher – a microcomputer coprocessor card for doing integer arithmetic*, review in J. Rec. Math. **25**(1), 1993.

[9] D. V. and G. V. Chudnovsky, *Sequences of numbers generated by addition in formal groups and new primality and factorization tests*, Adv. in Appl. Math. **7** (1986), 385–434.

[10] R. E. Crandall, *Projects in scientific computation*, Springer-Verlag, New York, 1994.

[11] R. E. Crandall, *Topics in advanced scientific computation*, Springer-Verlag, New York, 1996.

[12] R. Crandall, J. Doenias, C. Norrie, and J. Young, *The twenty-second Fermat number is composite*, Math. Comp. **64** (1995), 863–868.

[13] R. Crandall and B. Fagin, *Discrete weighted transforms and large-integer arithmetic*, Math. Comp. **62** (1994), 305–324.

[14] B. Dixon and A. K. Lenstra, *Massively parallel elliptic curve factoring*, Proc. Eurocrypt '92, Lecture Notes in Computer Science **658**, Springer-Verlag, Berlin, 1993, 183–193.

[15] H. Dubner and R. Dubner, *The Dubner PC Cruncher: Programmers Guide and Function Reference*, February 15, 1993.

[16] G. B. Gostin, *New factors of Fermat numbers*, Math. Comp. **64** (1995), 393–395.

[17] J. C. Hallyburton and H. Brillhart, *Two new factors of Fermat numbers*, Math. Comp. **29** (1975), 109–112. Corrigendum, *ibid* **30** (1976), 198.

[18] W. Keller, *Factors of Fermat numbers and large primes of the form $k \cdot 2^n + 1$*, Math. Comp. **41** (1983), 661–673. Also part II, preprint, Universität Hamburg, Sept. 27, 1992 (available from the author).

[19] D. E. Knuth, *The art of computer programming, Volume 2: Seminumerical algorithms* (2nd ed.), Addison-Wesley, Menlo Park, CA, 1981.

[20] M. Kraitchik, *On the factorization of $2^n \pm 1$*, Scripta Math. **18** (1952), 39–52.

8   R. P. BRENT, R. E. CRANDALL, K. DILCHER, AND C. VAN HALEWYN

[21] A. K. Lenstra, H. W. Lenstra, Jr., M. S. Manasse, and J. M. Pollard, *The factorization of the ninth Fermat number*, Math. Comp. **61** (1993), 319–349.

[22] A. K. Lenstra and M. S. Manasse, *Factoring by electronic mail*, Proc. Eurocrypt '89, Lecture Notes in Computer Science **434**, Springer-Verlag, Berlin, 1990, 355–371.

[23] H. W. Lenstra, Jr., *Factoring integers with elliptic curves*, Ann. of Math. (2) **126** (1987), 649–673.

[24] P. L. Montgomery, *Speeding the Pollard and elliptic curve methods of factorization*, Math. Comp. **48** (1987), 243–264.

[25] P. L. Montgomery, *An FFT extension of the elliptic curve method of factorization*, Ph. D. dissertation, Mathematics, University of California at Los Angeles, 1992. `ftp://ftp.cwi.nl/pub/pmontgom/ucladissertation.psl.Z` .

[26] P. L. Montgomery, *A survey of modern integer factorization algorithms*, CWI Quarterly **7** (1994), 337–366.

[27] F. Morain, *Courbes elliptiques et tests de primalité*, Ph. D. thesis, Univ. Claude Bernard – Lyon I, France, 1990. `ftp://ftp.inria.fr/INRIA/publication/Theses/TU-0144.tar.Z` .

[28] A. Schönhage, *Schnelle Berechnung von Kettenbruchentwicklungen*, Acta Inf. **1** (1971), 139–144.

[29] J. L. Selfridge, *Factors of Fermat numbers*, MTAC **7** (1953), 274–275.

[30] J. H. Silverman, *The arithmetic of elliptic curves*, Graduate Texts in Mathematics **106**, Springer-Verlag, New York, 1986.

[31] R. D. Silverman and S. S. Wagstaff, Jr., *A practical analysis of the elliptic curve factoring algorithm*, Math. Comp. **61** (1993), 445–462.

[32] H. Suyama, *Informal preliminary report* (8), personal communication to Brent, October 1985.

COMPUTER SCIENCES LABORATORY, AUSTRALIAN NATIONAL UNIVERSITY, CANBERRA, ACT 0200, AUSTRALIA
E-mail address: `Richard.Brent@anu.edu.au`

CENTER FOR ADVANCED COMPUTATION, REED COLLEGE, PORTLAND, OR 97202, USA
E-mail address: `crandall@reed.edu`

DEPARTMENT OF MATHEMATICS AND STATISTICS, DALHOUSIE UNIVERSITY, HALIFAX, NOVA SCO-TIA B3H 3J5, CANADA
E-mail address: `dilcher@cs.dal.ca`

DEPARTMENT OF COMPUTER SCIENCE AND ENGINEERING, OREGON GRADUATE INSTITUTE
Current address: Deutsche Bank AG, London, England
E-mail address: `Christopher.van-halewyn@db.com`

6. Random generators and normal numbers

Discussion

This paper explores the fascinating relationship between the digits of a real number and certain defining sums for said number. Hypothesis A—from an earlier paper by Bailey and Crandall[1]—in the present paper Hypothesis 3.1—states that for any rational function $r_n := p(n)/q(n)$ satisfying certain constraints, the real sequence (for integer $b \geq 2$)

$$x_n := (bx_{n-1} + r_n) \mod 1$$

is either equidistributed or has a finite attractor. On this unproven hypothesis, it follows that $\pi, \log 2, \zeta(3)$ are all 2-normal, meaning the binary digits of each are "random" in a quantifiable sense.

Thus the paper explores in a rigorous fashion the notion that the stream of digits from many fundamental constants is the output of a PRNG (pseudo-random number generator). A side result of some modern interest is that the Erdős–Borwein constant

$$E := \sum_{n \geq 1} \frac{1}{2^n - 1}$$

is 2-normal if the PRNG

$$x_d := \sum_{k=1}^{d} \frac{2^d - 1}{2^k - 1}$$

is equidistributed mod 1.[2]

Challenges

To this day, not a single fundamental constant having non-artificial construction has been proven b-normal for any b. This would seem most likely assailable for an algebraic constant such as $\sqrt{2}$, and yet, the paper here seems to favor transcendentals such as $\log 2$. But "favor" is a weak word in this context of global failure of mathematics to prove a genuinely interesting normality case.

Source

D. H. Bailey and R. Crandall, "Random generators and normal numbers," *Exp. Math.*, **11**, 527–547 (2002).

---

[1] D. H. Bailey and R. Crandall, "On the Random Character of Fundamental Constant Expansions," *Experimental Mathematics* 10, 175-190 (2001).

[2] This author recently found the googol-th binary bit of E using some number-theoretical results together with the kinds of ideas found in such papers as the enclosed. See R. Crandall, "The googol-th bit of the Erdős–Borwein constant," included in the present volume.

Random Generators and Normal Numbers

David H. Bailey and Richard E. Crandall

CONTENTS

2000 AMS Subject Classification: Primary 11K16, 11K06;
Secondary 11J81, 11K45

Keywords: Normal numbers, transcendental numbers,
pseudo-random number generators

Pursuant to the authors' previous chaotic-dynamical model for random digits of fundamental constants [Bailey and Crandall 01], we investigate a complementary, statistical picture in which pseudorandom number generators (PRNGs) are central. Some rigorous results are achieved: We establish b-normality for constants of the form $\sum_i 1/(b^{m_i} c^{n_i})$ for certain sequences $(m_i), (n_i)$ of integers. This work unifies and extends previously known classes of explicit normals. We prove that for coprime $b, c > 1$ the constant $\alpha_{b,c} = \sum_{n=c,c^2,c^3,\dots} 1/(nb^n)$ is b-normal, thus generalizing the Stoneham class of normals [Stoneham 73a]. Our approach also reproves b-normality for the Korobov class [Korobov 90] $\beta_{b,c,d}$, for which the summation index n above runs instead over powers $c^d, c^{d^2}, c^{d^3}, \dots$ with $d > 1$. Eventually we describe an uncountable class of explicit normals that succumb to the PRNG approach. Numbers of the α, β classes share with fundamental constants such as π, $\log 2$ the property that isolated digits can be directly calculated, but for these new classes such computation tends to be surprisingly rapid. For example, we find that the googol-th (i.e., 10^{100}-th) binary bit of $\alpha_{2,3}$ is 0. We also present a collection of other results—such as digit-density results and irrationality proofs based on PRNG ideas—for various special numbers.

1. INTRODUCTION

We call a real number b-normal if, qualitatively speaking, its base-b digits are "truly random." For example, in the decimal expansion of a number that is 10-normal, the digit 7 must appear 1/10 of the time, the string 783 must appear 1/1000 of the time, and so on. It is remarkable that in spite of the elegance of the classical notion of normality, and the sobering fact that almost all real numbers are absolutely normal (meaning b-normal for *every* $b = 2, 3, \dots$), proofs of normality for fundamental constants such as $\log 2$, π, $\zeta(3)$ and $\sqrt{2}$ remain elusive. In [Bailey and Crandall 01] we proposed a general "Hypothesis A" that connects normality theory with a certain aspect of chaotic dynamics. In a subsequent work, J. Lagarias [Lagarias 01] provided some additional interesting viewpoints and analyses using the dynamical approach.

528 Experimental Mathematics, Vol. 11 (2002), No. 4

There is a fascinating historical thread in normality theory, from "artificial" or "unnatural" normals to "natural" normals. By the adjective "artificial" or "unnatural," we mean that a number's construction is relatively nonalgebraic and nonanalytic, in contrast to a "natural" normal number, which is given by some reasonably analytic formulation, such as a conveniently defined series. Such talk is of course qualitative and heuristic; yet, in the last few decades we have seen numbers, provably normal to some base, and via elegant series descriptions looking more like fundamental constants.

Since the 1930s we have known of artificial constructions, such as the 10-normal, binary Champernowne constant [Champernowme 33]:

$$C_{10} \;=\; 0.(1)(2)(3)(4)(5)(6)(7)(8)(9)(10)(11)(12)\cdots,$$

(or the 2-normal $C_2 = 0.(1)(10)(11)(100)(101)(110)$ $(111)\cdots_2$), with the $(\cdot)$ notation meaning the expansion is constructed via mere *concatenation* of registers.

The migration toward more natural constructions was intensified by the work of Korobov and Stoneham from the 1950s into the 1990s, with Levin [Levin 99] and others working even more recently on statistical properties of normals. Those investigations were rooted in the recurring-decimal constructions of Good [Good 46] (see, e.g., the interesting historical discussions in [Stoneham 76]). Let us now highlight what we shall call Korobov and Stoneham classes of normals, referring further details to the works of those authors, [Stoneham 83], [Stoneham 70], [Stoneham 73b], [Stoneham 74], [Stoneham 76], [Korobov 92], [Korobov 90], and [Korobov 72]. Although Korobov achieved normal number construction in the 1950s by parlaying Good's ideas [Stoneham 76], and Stoneham gave some explicit—yet rather recondite—series construction of normals in 1970 [Stoneham 70], an elegant and easily described representative class of "natural" normals was exhibited in 1973 by Stoneham [Stoneham 73a]. We shall denote these Stoneham numbers by $\{\alpha_{b,c}\}$, with $b, c > 1$ coprime:

$$\alpha_{b,c} \;=\; \sum_{n=c^k>1} \frac{1}{b^n n} \;=\; \sum_{k=1}^{\infty} \frac{1}{b^{c^k} c^k}.$$

Stoneham proved that $\alpha_{b,c}$ is b-normal whenever c is an odd prime and b is a primitive root of c^2. We shall in the present paper generalize this class of normals by removing Stoneham's restrictions, demanding only coprime $b, c > 1$. Another class of normals we shall be able to cover with the present techniques is the Korobov class whose members we denote (for $b, c > 1$ again coprime

and $d > 1$):

$$\beta_{b,c,d} \;=\; \sum_{n=c,c^d,c^{d^2},c^{d^3},\ldots} \frac{1}{nb^n},$$

Korobov showed in 1990, via a clever combinatorial argument, that $\beta_{b,c,d}$ is b-normal [Korobov 90]. We shall reprove this result, and do so with a general method that encompasses also the Stoneham class and generalizations. It should be remarked that these pioneers were not merely concerned with the aforementioned thread from artificial to natural constructions. For example, Stoneham used the representations

$$\sqrt{2} \;=\; 2 \prod_{d \text{ odd}} \left(1 - \frac{1}{4d^2}\right),$$

$$\pi \;=\; 4 \prod_{d \text{ odd} >1} \left(1 - \frac{1}{d^2}\right),$$

to creatively demonstrate (for either constant) that for a *fixed* number of multiplicands, certain digit strings must appear in the resulting rational period [Stoneham 76]. Unfortunately this does not on the face of it lead to rigorous results about the exact constants π and $\sqrt{2}$. (It is also puzzling, in that, whereas π falls squarely under the rubric of the present authors' Hypothesis A [Bailey and Crandall 01], the constant $\sqrt{2}$ does not, and one can only wonder whether the two constants should ultimately be treated in the same fashion as regards normality.) As for Korobov, his work actually included explicit continued fractions for the $\beta_{b,c,d}$ and related normals. For example,

$$\beta_{2,3,2} \;=\; \sum_{i \geq 0} \frac{1}{3^{2^i} 2^{3^{2^i}}} \;=\; \frac{1}{23 + \frac{1}{1 + \frac{1}{7 + \cdots}}},$$

with the precise algorithm for the fraction's ensuing elements given in the reference [Korobov 90] (and note that the fraction elements soon grow extremely rapidly, the 11-th element being $2^{6399} - 1$).

In the present paper, the way we generalize and extend such normality classes is to adopt a complementary viewpoint to Hypothesis A, focusing upon pseudo-random number generators (PRNGs), with relevant analyses of these PRNGs carried out via exponential-sum and other number-theoretical techniques. One example of success along this pathway is the establishment of large (indeed, uncountable) classes of "natural" normal numbers. Looking longingly at the fundamental constants

$$\log \frac{b}{b-1} \;=\; \sum_{n>0} \frac{1}{b^n n}$$

whose normality—for any $b \geq 2$ and to any base, b or not—remains to this day unresolved, we use PRNG concepts to prove b-normality for sums involving sparse filtering of the logarithms' summation indices. Of specific interest to us are sums

$$\alpha_{b,c,m,n} = \sum_{i \geq 1} \frac{1}{b^{m_i} c^{n_i}}$$

for certain integer pairs b, c and sequences $m = (m_i), n = (n_i)$ that enjoy certain growth properties. Note that our definition of Stoneham numbers $\alpha_{b,c}$ is the case $n_i = i$, $m_i = c^{n_i}$, while the Korobov numbers $\beta_{b,c,d}$ arise from sequence definitions $n_i = d^i$, $m_i = c^{n_i}$.

It is tantalizing that Stoneham and Korobov numbers both involve restrictions on the summation indices in the aforementioned logarithmic expansion, in the sense that numbers of either class enjoy the general form $\sum_{n \in S} 1/(nb^n)$ for some subset $S \subset Z^+$. Our generalizations include the sums

$$\sum_{n=c^{f(1)}, c^{f(2)}, \ldots} \frac{1}{nb^n},$$

for suitable integer-valued functions f; so again we have a restriction of a logarithmic sum to a sparse set of indices.

In addition to the normality theorems applicable to the restricted sums mentioned above, we present a collection of additional results on irrationality and b-density (see ensuing definitions), these side results having arisen during our research into the PRNG connection.

2. NOMENCLATURE AND FUNDAMENTALS

We first give some necessary nomenclature relevant to base-b expansions. For a real number $\alpha \in [0, 1)$ we shall assume uniqueness of base-b digits, b an integer ≥ 2; i.e., $\alpha = 0.b_1 b_2 \cdots$ with each $b_j \in [0, b-1]$, with a certain termination rule to avoid infinite tails of digit values $b - 1$. One way to state the rule is simply to define $b_j = \lfloor b^j \alpha \rfloor$; another way is to convert a trailing tail of consecutive digits of value $b - 1$, as in $0.4999 \cdots \to 0.5000 \cdots$ for base $b = 10$. Next, denote by $\{\alpha\}$, or $\alpha \bmod 1$, the fractional part of α, and denote by $\|\alpha\|$ the closer of the absolute distances of $\alpha \bmod 1$ to the interval endpoints $0, 1$, i.e., $\|\alpha\| = \min(\{\alpha\}, 1 - \{\alpha\})$. Denote by (α_n) the ordered sequence of elements $\alpha_0, \alpha_1, \ldots$. Of interest will be sequences (α_n) such that $(\{\alpha_n\})$ is equidistributed in $[0, 1)$, meaning that any subinterval $[u, v) \subseteq [0, 1)$ is visited by $\{\alpha_n\}$ for a (properly defined) limiting fraction $(v - u)$ of the n indices; i.e., the members of the sequence fall in a "fair" manner. We sometimes consider a weaker

condition that $(\{a_n\})$ be merely dense in $[0, 1)$, noting that equidistributed implies dense.

Armed with the above nomenclature, we paraphrase from [Bailey and Crandall 01] and references [Kuipers and Niederreiter 74], [Hardy and Wright 79], [Niven 56], and [Korobov 92] in the form of a collective definition.

Definition 2.1. (Collection.) The following pertain to real numbers α and sequences of real numbers ($\alpha_n \in [0, 1) : n = 0, 1, 2, \ldots$). For any base $b = 2, 3, 4 \ldots$ we assume, as enunciated above, a unique base-b expansion of whatever real number is in question.

(1) α is said to be b-dense iff in the base-b expansion of α every possible finite string of consecutive digits appears.

(2) α is said to be b-normal iff in the base-b expansion of α every string of k base-b digits appears with (well-defined) limiting frequency $1/b^k$. A number that is b-normal for every $b = 2, 3, 4, \ldots$ is said to be absolutely normal. (This definition of normality differs from, but is provably equivalent to, other historical definitions [Hardy and Wright 79], [Niven 56].)

(3) The discrepancy of (α_n), essentially a measure of unevenness of the distribution in $[0, 1)$ of the first N sequence elements, is defined (when the sequence has at least N elements) as

$$D_N = \sup_{0 \leq a < b < 1} \left| \frac{\#(n < N : \alpha_n \in [a, b))}{N} - (b - a) \right|.$$

One may also speak of a number α's b-discrepancy, as the discrepancy of the sequence $(b^n \alpha)$, which sequence being relevant to the study of b-normality.

(4) The gap-maximum of (α_n), the largest gap "around the mod-1 circle" of the first N sequence elements, is defined (when the sequence has at least N elements) as

$$G_N = \max_{k=0, \ldots, N-1} \|\beta_{(k+1) \bmod N} - \beta_{k \bmod N}\|,$$

where (β_n) is a sorted (either in decreasing or increasing order) version of the first N elements of $(\alpha_n \bmod 1)$.

On the basis of such definition we next give a collection of known results in regard to b-dense and b-normal numbers:

Theorem 2.2. (Collection.) *In the following we consider real numbers and sequences as in Definition 2.1. For any*

530   Experimental Mathematics, Vol. 11 (2002), No. 4

base $b = 2, 3, 4 \ldots$ we assume, as enunciated above, a unique base-b expansion of whatever number in question.

(1) If α is b-normal then α is b-dense.

Proof: If every finite string appears with well-defined, fair frequency, then it appears perforce. □

(2) If, for some b, α is b-dense then α is irrational.

Proof: The base-b expansion of any rational is ultimately periodic, which means some finite digit strings never appear. □

(3) Almost all real numbers in $[0, 1)$ are absolutely normal (the set of non-absolutely-normal numbers is null).

Proof: See [Kuipers and Niederreiter 74, page 71, Corollary 8.2], [Hardy and Wright 79]. □

(4) α is b-dense iff the sequence $(\{b^n \alpha\})$ is dense.

Proof: See [Bailey and Crandall 01]. □

(5) α is b-normal iff the sequence $(\{b^n \alpha\})$ is equidistributed.

Proof: See [Kuipers and Niederreiter 74, page 70, Theorem 8.1] □

(6) Let $m \neq k$. Then α is b^k-normal iff α is b^m-normal.

Proof: See [Kuipers and Niederreiter 74, page 72, Theorem 8.2] □

(7) Let q, r be rational, $q \neq 0$. If α is b-normal then so is $q\alpha + r$, while if $c = b^q$ is an integer then α is also c-normal.

Proof: The b-normality of $q\alpha$ is a consequence of the Birkoff ergodic theorem—see [Bailey and Rudolph 02]; see also [Kuipers and Niederreiter 74, page 77, Exercise 8.9]. For the additive ($+r$) part, see end of the present section. For the c-normality see [Kuipers and Niederreiter 74, page 77, Exercise 8.5]. □

(8) (Weyl criterion.) A sequence $(\{\alpha_n\})$ is equidistributed iff for every integer $h \neq 0$

$$\sum_{n=0}^{N-1} e^{2\pi i h \alpha_n} = o(N).$$

Proof: See [Kuipers and Niederreiter 74, page 7, Theorem 2.1]. □

(9) (Erdős–Turan discrepancy bound.) There exists an absolute constant C such that for any positive integer m the discrepancy of any sequence $(\{\alpha_n\})$ satisfies (again, it is assumed that the sequence has at least N elements):

$$D_N \;<\; C\left(\frac{1}{m} + \sum_{h=1}^{m} \frac{1}{h} \left| \frac{1}{N} \sum_{n=0}^{N-1} e^{2\pi i h \alpha_n} \right| \right).$$

Proof: See [Kuipers and Niederreiter 74, pages 112–113], where an even stronger Theorem 2.5 is given. □

(10) Assume (x_n) is equidistributed (dense). If $y_n \to c$, where c is constant, then $(\{x_n + y_n\})$ is likewise equidistributed (dense). Also, for any nonzero integer d, $(\{d x_n\})$ is equidistributed (dense).

Proof: For normality (density) of $(\{x_n + y_n\})$ see [Bailey and Crandall 01], [Kuipers and Niederreiter 74, Exercise 2.11] (one may start with the observation that $(x_n + y_n) = (x_n + c) + (y_n - c)$ and $(\{x_n + c\})$ is equidistributed iff (x_n) is). The equidistribution of $(\{d x_n\})$ follows immediately from parts (5) and (8) above. As for density of $(\{d x_n\})$, one has $\{d x_n\} = \{d\{x_n\}\}$ for any integer d, and the density property is invariant under any dilation of the mod-1 circle, by any real number of magnitude ≥ 1. □

(11) Given a number α, define the sequence $(\alpha_n) = (\{b^n \alpha\})$. Then α is b-dense iff $\lim_{N \to \infty} G_N = 0$.

Proof: The only-if is immediate. Assume, then, the vanishing limit, in which case for any $\epsilon > 0$ and any point in $[0, 1)$ some sequence member can be found to lie within $\epsilon/2$ of said point, hence we have density. □

(12) Consider α and the corresponding sequence (α_n) of the previous item. Then α is b-normal iff $\lim_{N \to \infty} D_N = 0$.

Proof: See [Kuipers and Niederreiter 74, page 89, Theorem 1.1]. □

Some of the results in the above collection are simple, some are difficult; the aforementioned references reveal the difficulty spectrum. This collective Theorem 2.2 is a starting point for many interdisciplinary directions. Of special interest in the present treatment is the interplay between normality and equidistribution.

We focus first on the celebrated Weyl result, Theorem 2.2(8). Observe the little-o notation, essentially saying that the relevant complex vectors will on average exhibit significant cancellation. An immediate textbook application of the Weyl theorem is to show that for any irrational α, the sequence $(\{n\alpha\})$ is equidistributed. Such elementary forays are of little help in normality studies, because we need to contemplate not multiples $n\alpha$ but the rapidly diverging constructs $b^n\alpha$.

We shall be able to put the Weyl theorem to some use in the present treatment. For the moment, it is instructive to look at one nontrivial implication of Theorem 2.2(8). We selected the following example application of the Weyl sum to foreshadow several important elements of our eventual analyses. With Theorem 2.2((5),(6), and (8)) we can prove part of Theorem 2.2(7), namely: If α is b-normal and r is rational then $\alpha + r$ is b-normal. Let $r = p/q$ in lowest terms. The sequence of integers $(b^m \bmod q)$ is eventually periodic, say with period T. Thus for some fixed integer c and any integer n we have $b^{nT} \bmod q = c$. Next, we develop an exponential sum, assuming nonzero h:

$$S = \sum_{n=0}^{N-1} e^{2\pi i h b^{nT}(\alpha+p/q)} = e^{2\pi i h c p/q} \sum_{n=0}^{N-1} e^{2\pi i h b^{nT}\alpha}.$$

Now a chain of logic finishes the argument: α is b-normal so it is also b^T-normal by Theorem 2.2(6). But this implies $S = e^{2\pi i h c p/q} o(N) = o(N)$ so that $\alpha + p/q$ is b^T-normal, and so by Theorem 2.2(6) is thus b-normal.

3. PSEUDO-RANDOM NUMBER GENERATORS

We consider pseudo-random number generators (PRNGs) under the iteration

$$x_n = (bx_{n-1} + r_n) \bmod 1,$$

which is a familiar congruential form, except that the perturbation sequence r_n is not yet specified (in a conventional linear-congruential PRNG this perturbation is constant). Much of the present work is motivated by the following hypothesis from [Bailey and Crandall 01].

Hypothesis 3.1. (Bailey–Crandall "Hypothesis A.") If the perturbation $r_n = p(n)/q(n)$, a nonsingular rational-polynomial function with $\deg q > \deg p \geq 0$, then (x_n) is either equidistributed or has a finite attractor.

It is unknown whether this hypothesis be true, however a motivation is this: The normality of many fundamental constants *believed* to be normal would follow from Hypothesis 3.1. Let us now posit an *unconditional* theorem that leads to both conditional and unconditional normality results:

Theorem 3.2. (Unconditional.) *Associate a real number*

$$\beta = \sum_{n=1}^{\infty} \frac{r_n}{b^n}$$

where $\lim_{n\to\infty} r_n = c$, *a constant, with a PRNG sequence* (x_n) *starting* $x_0 = 0$ *and iterating*

$$x_n = (bx_{n-1} + r_n) \bmod 1.$$

Then (x_n) *is equidistributed (dense) iff* β *is* b-*normal (b-dense).*

Proof: Write

$$
\begin{aligned}
b^d\beta - x_d &= \sum_{n=1}^{\infty} b^{d-n} r_n - (b^{d-1}r_1 + b^{d-2}r_2 + \cdots + r_d) \\
&= \frac{r_{d+1}}{b} + \frac{r_{d+2}}{b^2} + \cdots \to c',
\end{aligned}
$$

with c' a constant. Therefore by Theorem 2.2(10), (x_n) equidistributed (dense) implies β is b-normal (b-dense). Now assume b-normality (b-density). Then (x_d) is the sequence $(\{b^d\beta\})$ plus a sequence that approaches constant, and again by Theorem 2.2(10), (x_d) is equidistributed (dense). $\square$

In our previous work [Bailey and Crandall 01] this kind of unconditional theorem led to the following (conditional) result:

Theorem 3.3. (Conditional.) *On Hypothesis 3.1, each of the constants*

$$\pi, \ \log 2, \ \zeta(3)$$

is 2-*normal. Also, in Hypothesis 3.1, if* $\zeta(5)$ *is irrational, then it likewise is* 2-*normal.*

Theorem 3.3 works, of course, because the indicated fundamental constants admit polylogarithm-like expansions of the form $\sum r_n b^{-n}$ where r_n is rational-polynomial. The canonical example is

$$\log 2 = \sum_{n=1}^{\infty} \frac{1}{n 2^n}$$

532   Experimental Mathematics, Vol. 11 (2002), No. 4

and 2-normality of $\log 2$ comes down to the question of whether (for $x_0 = 0$)

$$x_n = \left(2x_{n-1} + \frac{1}{n}\right) \bmod 1$$

gives rise to an equidistributed (x_n). The main results of the present paper will be to establish equidistribution for generators reminiscent of, but not quite the same as, this one for $\log 2$.

With a view to ultimate achievement of normality results, let us take a brief tour of some other (not rational-polynomial) perturbation functions. The iteration

$$x_n = \left(2x_{n-1} + \frac{n}{2^{n^2-n}}\right) \bmod 1$$

is associated with the constant

$$\beta = \sum_{n \geq 1} \frac{n}{2^{n^2}},$$

which is 2-dense but *not* 2-normal, as we establish later. Another rather peculiar perturbation, for base $b = 4$, is

$$r_n = \frac{1}{(2n)!} \frac{4n+1}{4n+2}.$$

If the associated PRNG is equidistributed, then $1/\sqrt{e}$ is 2-normal. Likewise, and again for base $b = 4$, a result of equidistribution for a perturbation

$$r_n = \frac{(2n-3)!!}{n!} = \frac{(2n-3)(2n-5)\cdots 3 \cdot 1}{n(n-1)(n-2)\cdots 2 \cdot 1}$$

would prove that $\sqrt{2}$ is 4-normal, hence 2-normal. It might have seemed on the face of it that the decay rate of the perturbation r_n has something to do with normality. But the conditional results in Hypothesis 3.1 involve only polynomial-decay perturbations, while the r_n assignments immediately above involve rapid, factorial decay. On the other hand, there are very slowly decaying perturbation functions for which one still embraces the likelihood of normality. For example, the mysterious Euler constant γ can be associated with the base $b = 2$ and perturbation function r_n that decays like $n^{-1/2}$ (see Section 5 and [Bailey and Crandall 01]).

In a spirit of statistical investigation let us revisit once again the canonical case of the number $\beta = \log 2$ and base $b = 2$. For the purpose of discussion, we write out for $d = 1, 2, 3, \ldots$ an iterate as assembled from d explicit terms:

$$x_d = \left(\frac{2^{d-1} \bmod 1}{1} + \frac{2^{d-2} \bmod 2}{2} + \frac{2^{d-3} \bmod 3}{3} + \right.$$
$$\left. \cdots + \frac{2}{d-1} + \frac{1}{d}\right) \bmod 1,$$

and remind ourselves that

$$2^d \log 2 = x_d + t_d,$$

where t_d is a "tail" term that vanishes in the limit, but is also a kind of source for subsequent generator iterates. (Note that the first term always vanishes modulo 1; we include that term for clarity.) One can think of such a PRNG as a "cascaded" random number generator, in which distinct generators $(2^{d-m} \bmod m)/m$ are added together, with the number of moduli m steadily diverging.

There are difficult aspects of the PRNG analysis for $\log 2$. First, the theory of cascaded PRNGs appears difficult; even the class of generators with *fixed* numbers of summands are not completely understood. Second, even if we succeeded in some form of equidistribution theorem for cascaded generators, we still have the problem that the tail t_d is to be added into the final segment of the generator that has just been started with its power-of-two numerators.

These difficulties may be insurmountable. Nonetheless, there are two separate approaches to altering the $\log 2$ PRNG such that density and normality results accrue. These separate modifications are:

(1) Arrange for some kind of synchronization, in which iterates change number-theoretic character on the basis of a "kicking" pertubation that emerges only at certain iterates.

(2) Arrange somehow for the tail t_d to be so very small that meaningful statistical properties of the first $d + d'$ generator terms are realized before t_d is significantly magnified via d' multiplies by b.

We shall be able to apply both of these qualitative alterations. For the first case (kicking/synchronization) we shall finally achieve normality proofs. For the second kind of alteration (small tail) we shall be able to effect some proofs on density and irrationality.

4. PRNGs ADMITTING OF NORMALITY PROOFS

Herein we exhibit a class of generators—we shall call them (b, c, m, n)-PRNGs, for which normality proofs can be achieved due to the special synchronization such generators enjoy. We begin with some necessary nomenclature (we are indebted to C. Pomerance for his expertise, ideas and helpful communications on nontrivial arithmetic modulo prime powers).

Bailey and Crandall: Random Generators and Normal Numbers   **533**

Definition 4.1. We define a (b, c, m, n)-PRNG sequence $x = (x_0, x_1, x_2, \ldots)$ for coprime integers $b, c > 1$ and strictly increasing sequences $m = (0, m_1, m_2, \ldots), n = (0, n_1, n_2, \ldots)$ as follows: Set $x_0 = 0$ and for $k > 0$ iterate

$$x_k = (bx_{k-1} + r_k) \bmod 1,$$

with the perturbation given by

$$r_{m_i} = \frac{1}{c^{n_i}},$$

with all other r_k vanishing.

Knowing the parameters (b, c, m, n) determines the PRNG sequence x. The perturbation is of the "kicking" variety, not happening unless the index j on r_j is one of the exponents m_i. So the PRNG runs as follows (we denote simply by $\equiv$ an equality on the mod-1 circle):

$$x_0 \equiv 0,$$
$$\cdots,$$
$$x_{m_1} \equiv b^{m_1} \cdot 0 + r_{m_1} \equiv \frac{1}{c^{n_1}},$$
$$x_{m_1+1} \equiv \frac{b}{c^{n_1}},$$
$$\cdots,$$
$$x_{m_2-1} \equiv \frac{b^{m_2-m_1-1}}{c^{n_1}},$$
$$x_{m_2} \equiv \frac{b^{m_2-m_1}}{c^{n_1}} + r_{m_2} \equiv \frac{b^{m_2-m_1}c^{n_2-n_1}+1}{c^{n_2}},$$
$$\cdots,$$

and generally speaking,

$$x_{m_k} \equiv \frac{a_k}{c^{n_k}},$$

where a_k is always coprime to c. It is evident that upon the $1/c^{n_k}$ perturbation, the x_k commence an orbit of length $\mu_{k+1} = m_{k+1} - m_k$ before the next perturbation. Therefore, the first N terms of the sequence x can be envisioned as follows: Observe $m_0 = 0$ and write $N = \mu_1 + \cdots + \mu_K + J$ where $J \in [1, \mu_{K+1}]$ is the (possibly partial) length of the last orbit. Then the first N terms of the (b, c, m, n)-PRNG sequence start with $\mu_1 = m_1$ zeros and appear:

$$(x_n)|_{k=0}^{N-1} \equiv (0, 0, \ldots, 0,$$
$$\frac{a_1}{c^{n_1}}, \frac{ba_1}{c^{n_1}}, \frac{b^2 a_1}{c^{n_1}}, \ldots, \frac{b^{\mu_2-1}a_1}{c^{n_1}},$$
$$\cdots,$$
$$\frac{a_k}{c^{n_k}}, \frac{ba_k}{c^{n_k}}, \frac{b^2 a_k}{c^{n_k}}, \ldots, \frac{b^{\mu_{k+1}-1}a_k}{c^{n_k}},$$
$$\cdots,$$
$$\frac{a_K}{c^{n_K}}, \frac{ba_K}{c^{n_K}}, \frac{b^2 a_K}{c^{n_K}}, \ldots, \frac{b^{J-1}a_K}{c^{n_K}}).$$

We intend to argue, for certain parameter sets (b, c, m, n), that $x = (x_k)$ is equidistributed. For this we shall require some results from the number theory of power moduli, and an important lemma on exponential sums. But first we give a general lemma useful for estimating the discrepancy of an ordered union of finite subsequences. We use notation reminiscent of that above for our (b, c, m, n)-PRNG sequences:

Lemma 4.2. *For an infinite sequence (y_n) built as an ordered union,*

$$((y_0, \ldots, y_{N_1-1}), (y_{N_1}, \ldots, y_{N_1+N_2-1}), \cdots))$$

of subsequences of respective lengths N_i, we have, for $K > 0$, $N = N_1 + N_2 + \cdots + N_K + J$ with $J \in [1, N_{K+1}]$, a discrepancy bound

$$D_N \leq \sum_{i=1}^{K} \frac{N_i}{N} D_{N_i} + \frac{J}{N} D_J,$$

where D_{N_i} are the respective discrepancies of the finite subsequences and D_J is the discrepancy of the partial sequence $(y_{N_1+\cdots+N_K+j} : j = 0, 1, 2, \ldots, J-1)$.

Proof: This is proved simply, in [Kuipers and Niederreiter 74, page 115, Theorem 2.6]. □

Now we can focus upon number-theoretical ideas, in order to bound subsequence discrepancies. We shall make use of the following lemma which gives relations for the order of numbers modulo a given modulus c. We denote the multiplicative order of y modulo c by $\mathrm{ord}(y, c)$ in what follows.

Lemma 4.3. *Let $b, c > 1$ be coprime with c having prime decomposition $c = p_1^{t_1} \cdots p_s^{t_s}$. Let $\tau_1(c) = \mathrm{ord}(b, p_1 \cdots p_s)$ and define β_i by:*

$$p_i^{\beta_i} \parallel b^{(\mu+1)\tau_1} - 1$$

where $\mu = 1$ if c is even and τ_1 is odd and $b \equiv 3 \pmod 4$; otherwise $\mu = 0$. Define further

$$c_1(c) = \prod_{k=1}^{s} p_k^{\min(t_k, \beta_k)}.$$

Then

$$\mathrm{ord}(b, c) = \frac{c}{c_1}\tau',$$

where $\tau' = 2\tau_1$ if $\mu = 1$ and $c \equiv 0 \pmod 4$; otherwise $\tau' = \tau_1$.

534   Experimental Mathematics, Vol. 11 (2002), No. 4

Proof: This lemma is proved in [Korobov 72] and references therein. □

These above order relations lead easily to a key lemma for our present treatment.

Lemma 4.4. *Let $b, c > 1$ be coprime. Then there exist constants A_1, A_2 such that for sufficiently large n both of these conditions hold:*

$$\operatorname{ord}(b, c^n) = A_1 c^n,$$
$$\frac{\operatorname{ord}(b, c^n)}{c_1(c^n)} = A_2 c^n$$

Proof: The simple replacement $c \to c^n$ in Lemma 4.3 leaves the values of the β_i and τ_1 invariant. Thus, for sufficiently large n, we have $c_1(c^n) = \prod p_i^{\beta_i}$ which is fixed, and both large-n results follow. □

Next we state a lemma on exponential sums.

Lemma 4.5. (Korobov, Niederreiter.)   *For $b, c > 1$ coprime, with $c_1(c)$ defined as in Lemma 4.3, and an integer h such that $d = \gcd(h, c) < c/c_1$, and an integer $J \in [1, \operatorname{ord}(b, c)]$, we have*

$$\left| \sum_{j=0}^{J-1} e^{2\pi i h b^j / c} \right| < \sqrt{\frac{c}{d}} \left(1 + \log \frac{c}{d} \right).$$

Proof: The lemma is a direct corollary of results found in [Korobov 92, e.g., page 167, Lemma 32], for odd c, but (earlier) results of Korobov [Korobov 72] are sufficiently general to cover all composite c. See also [Niederreiter 78, pages 1004–1008]. A highly readable proof of a similar result and an elementary description of Niederreiter's seminal work on the topic can be found in [Knuth 81, pages 107–110]. There are also enhancements on the theory of fractional parts for the exponential function, as in [Levin 99] and references therein. □

Lemma 4.5 speaks to the distribution of powers of b modulo general c coprime to b. For our purposes we want to bound the magnitudes of exponential sums when the modulus is a pure power, say c^n. (Incidentally, we shall not be needing the dependence of the lemma's bound on d.) To this end we establish a theorem. Note that in this theorem and thereafter, when we say constants exist we mean always positive constants depending only on b, c, therefore independent of any running indices or growing powers. The idea of the following theorem is not only to make the transition $c \to c^n$ for the exponential sum, but also to unrestrict J.

Theorem 4.6. *For $b, c > 1$ coprime, there exist constants A, B, D such that for any positive integer J and sufficiently large n, the condition $\gcd(H, c^n) < D c^n$ implies*

$$\left| \sum_{j=0}^{J-1} e^{2\pi i H b^j / c^n} \right| < B \left(A c^{n/2} + J c^{-n/2} \right) \log c^n.$$

Proof: Substituting $c \to c^n$ in Lemma 4.5, and using Lemma 4.4, we establish that for sufficiently large n, the indicated exponential sum of the theorem, for any H as indicated, is less in magnitude than a bound $E c^{n/2} \log c^{n/2}$, where E is constant, as long as J does not exceed $\operatorname{ord}(b, c^n)$. But for larger J we have at most $\lceil J/\operatorname{ord}(b, c^n) \rceil$ copies of the exponential sum, and this ceiling is bounded by $1 + J/(A_1 c^n)$, so the result follows. □

We are aware that one could start from Theorem 4.6 and apply the Weyl criterion (Theorem 2.2 (8)) to establish equidistribution for certain (b, c, m, n)-PRNG sequences. We shall prove a little more, by virtue of discrepancy formulae. Again consider the first N terms of the sequence $x = (x_n)$, where $N = N_0 + \mu_{k_1} + \cdots + \mu_K + J$, where $J \in [1, \mu_{K+1}]$ as before, but k_1 is chosen so that the powers c^n for any $n > n_{k_1 - 1}$ are sufficiently large, as in Lemma 4.4 and Theorem 4.6, and so N_0 is constant. Then the discrepancy of the first J' elements of an orbit, namely of the subsequence

$$\left(\frac{a_k}{c^{n_k}}, \frac{b a_k}{c^{n_k}}, \frac{b^2 a_k}{c^{n_k}}, \ldots, \frac{b^{J'-1} a_k}{c^{n_k}} \right)$$

for $k \geq k_1$ is bounded according to the Erdős–Turán Theorem 2.2(9):

$$D_{\mu_{k+1}} < C_1 \left(\frac{1}{M} + \sum_{h=1}^{M} \frac{1}{h} \left| \frac{1}{J'} \sum_{j=0}^{J'-1} e^{2\pi i h a_k b^j / c^{n_k}} \right| \right),$$

where we are at liberty to chose $M = \lfloor D c^{n_k/2} \rfloor$ with the constant D from Theorem 4.6, so that the exponential sum appearing in the discrepancy bound is covered by the theorem—recall a_k and c are coprime so that $\gcd(h a_k, c^{n_k}) = \gcd(h, c^{n_k}) \leq h \leq M < D c^{n_k}$. We then get, for an orbit's discrepancy for J' terms of that orbit,

$$D_{J'} < B' \left(A' \frac{c^{n_k/2}}{J'} + c^{-n_k/2} \right) \log^2 c^{n_k},$$

where A', B' are constants, and we shall take $J' = \mu_{k+1}$ for each complete orbit and observe $J' = J$ in our last orbit (the orbit in which lies the last element x_{N-1}). Now using Lemma 4.2, we can obtain an overall discrepancy formula for the N sequence terms, N sufficiently large:

$$D_N \;<\; \frac{N_0}{N} + \frac{B'}{N}\sum_{k=k_1}^{K}\left(A'c^{n_{k-1}/2} + \frac{\mu_k}{c^{n_{k-1}/2}}\right)\log^2 c^{n_{k-1}}$$
$$+ \frac{B'}{N}\left(A'c^{n_K/2} + \frac{J}{c^{n_K/2}}\right)\log^2 c^{n_K}.$$

We can weaken this bound slightly, in favor of economy of notation, by observing that $N\mu_K$, $J < N$, and the powers c^{n_i} are monotonic in i, so that the following result is thereby established (note that we allow ourselves to rename constants with previously used names when such nomenclature is not ambiguous):

Lemma 4.7. *For the (b,c,m,n)-PRNG sequence $x = (x_n)$, the discrepancy is bounded for sufficiently large N by*

$$D_N(x) \;<\; \left(N_0 + Ac^{n_{K-1}/2} + B\sum_{k=1}^{K}\frac{\mu_k}{c^{n_{k-1}/2}}\right)\frac{\log^2 c^{n_{K-1}}}{\mu_K}$$
$$+ \left(A\frac{c^{n_K/2}}{\mu_K} + B\frac{1}{c^{n_K/2}}\right)\log^2 c^{n_K},$$

where $\mu_k = m_k - m_{k-1}$ and N is decomposed as $N = \mu_1 + \cdots + \mu_K + J$ with $J \in [1, \mu_{K+1}]$, with N_0, A, B constant.

It is now feasible to posit growth conditions on the m, n sequences of our PRNGs such that discrepancy vanishes as $N \to \infty$. One possible result is

Theorem 4.8. *For the (b,c,m,n)-PRNG sequence $x = (x_k)$ of Definition 4.1, assume that the difference sequences $\mu_k = m_k - m_{k-1}$, $\nu_k = n_k - n_{k-1}$ satisfy the following growth conditions:*

(i) (ν_k) is nondecreasing,

(ii) There exists a constant $\gamma > 1/2$ such that for sufficiently large k

$$\frac{\mu_k}{c^{\gamma n_k}} \;\geq\; \frac{\mu_{k-1}}{c^{\gamma n_{k-1}}}.$$

Then x is equidistributed and the number

$$\alpha_{b,c,m,n} \;=\; \sum_{k=1}^{\infty}\frac{1}{b^{m_k}c^{n_k}}$$

is b-normal.

Proof: We bound contributions to the discrepancy bound of Lemma 4.7, using growth condition (ii) as follows:

$$\frac{1}{\mu_K}\sum_{k=1}^{K}\frac{\mu_k}{c^{n_{k-1}/2}} \;\leq\; \frac{1}{c^{n_{K-1}/2}}\sum_{L=0}^{K-1}\frac{c^{\gamma n_{K-L}}}{c^{\gamma n_K}}\frac{c^{n_{K-1}/2}}{c^{n_{K-L-1}/2}}.$$

Now

$$\gamma(n_K - n_{K-L}) + \frac{1}{2}(n_K - n_{K-1} - (n_{K-L} - n_{K-L-1}))$$
$$\geq \left(\gamma - \frac{1}{2}\right)L,$$

this last inequality arising from growth condition (i) and the fact that the n_k are monotone increasing with $n_1 \geq 1$. Thus

$$\frac{1}{\mu_K}\sum_{k=1}^{K}\frac{\mu_k}{c^{n_{k-1}/2}} \;\leq\; \frac{1}{c^{n_{K-1}/2}}\left(\frac{1}{1 - c^{1/2-\gamma}}\right).$$

Actually this component is the only problematic part of the discrepancy bound in Lemma 4.7, and we quickly write, now with more conveniently labeled constants C_i, and for sufficiently large N,

$$D_N(x) \;<\; C_1\frac{1}{c^{n_K\gamma}} + C_2\frac{\log^2 c^{n_{K-1}}}{c^{n_{K-1}(\gamma-1/2)}} + C_3\frac{\log^2 c^{n_{K-1}}}{c^{n_{K-1}/2}}$$
$$+ C_4\frac{\log^2 c^{n_K}}{c^{n_K(\gamma-1/2)}} + C_5\frac{\log^2 c^{n_K}}{c^{n_K/2}}.$$

Since $(\log^2 z)/z^\epsilon$ is, for any $\epsilon > 0$, eventually monotone decreasing as $z \to \infty$, it follows that for sufficiently large N

$$D_N(x) \;<\; C\frac{\log^2 c^{n_{K-1}}}{c^{n_{K-1}\min(\gamma-1/2,1/2)}}.$$

But this means $D_N \to 0$ as $N \to \infty$, so x is equidistributed by Theorem 2.2(12). Theorem 3.1 then establishes the number $\alpha_{b,c,m,n}$ as b-normal.   □

As we have said, equidistribution of sequences x as in Theorem 4.8 would also follow from simpler arguments involving the Weyl criterion. However, the discrepancy bound at the end of the above proof is interesting in its own right. For certain choices of the m, n sequences, such as $m_i = c^i, n_i = i$, we can easily relate the index N to the $c^{n_{K-1}}$ to obtain a discrepancy bound

$$D_N(x) \;<\; C\frac{\log^2 N}{\sqrt{N}}.$$

Now such bounds (logarithmic numerator and square-root denominator) on classical PRNGs have been obtained in brilliant fashion by Niederreiter (see [Niederreiter 78, page 1009], and [Neiderreiter 92, pages 169–170],

536   Experimental Mathematics, Vol. 11 (2002), No. 4

who also gives arguments as to the best possible nature of such bounds in the classical PRNG contexts. It is thus no surprise that our exponent factor $\min(\gamma - 1/2, 1/2)$ at the end of the above proof cannot exceed $1/2$ (however, we see later in this section that other constructions of normals involved number discrepancies of somewhat lower magnitude).

The discrepancy approach, in yielding Theorem 4.8, leads us to specific classes of normals:

Corollary 4.9. *In what follows we assume an underlying* (b, c, m, n)*-PRNG and so* $b, c > 1$ *are coprime.*

(i) The generalized Stoneham numbers

$$\alpha_{b,c} = \sum_{i \geq 1} \frac{1}{c^i b^{c^i}}$$

are each b-normal.

(ii) The generalized Korobov numbers $(d > 1)$*:*

$$\beta_{b,c,d} = \sum_{i \geq 1} \frac{1}{c^{d^i} b^{c^{d^i}}}$$

are each b-normal.

(iii) For positive integers $s < 2r$*, each of the numbers*

$$\sum_{i \geq 1} \frac{1}{c^{ir} b^{c^{is}}}$$

is b-normal. So for example,

$$\sum_{i \geq 1} \frac{1}{8^i 3^{32^i}}$$

is 3-normal.

(iv) If integer $d > \sqrt{c}$*, then each of*

$$\sum_{i \geq 1} \frac{1}{c^i b^{d^i}}$$

is b-normal. So for example,

$$\sum_{i \geq 1} \frac{1}{3^i 2^{2^i}}$$

is 2-normal.

(v) If the integer-valued function $f(i)$ *is strictly increasing, and* $f(i) - f(i-1)$ *is nondecreasing for sufficiently large* n*, then*

$$\sum_{i \geq 1} \frac{1}{c^{f(i)} b^{c^{f(i)}}}$$

is b-normal. So for example,

$$\sum_{i \geq 1} \frac{1}{c^{i^2} b^{c^{i^2}}}$$

and

$$\sum_{i \geq 1} \frac{1}{c^{i!} b^{c^{i!}}}$$

are both b-normal.

Proof: Each of the results (i)–(v) follows easily from Theorem 4.8. For example, in case (iv), we have $d > \sqrt{c}$ so define δ by $c = c^{\delta + 1/2}$ so that for a constant C_1

$$\frac{\mu_K}{c^{\gamma n_K}} = \frac{C_1 c^{(\delta + 1/2)K}}{c^{\gamma K}} \tag{4-1}$$

which, for γ chosen between $1/2$ and $1/2 + \delta$, is monotone increasing. The rest of the results follow in similar fashion. $\qquad\square$

It is natural to ask whether a number $\alpha_{b,c,m,n}$ associated with a (b, c, m, n)-PRNGs is transcendental, which question we answer at least for some such numbers:

Theorem 4.10. *Denote* $\alpha_{b,c,m,n} = \sum_{i \geq 1} 1/(c^{n_i} b^{m_i})$ *where* (b, c, m, n) *are as in Definition 4.1 but without the restriction of coprimality. Let* $\lambda = \log b / \log c$ *and assume that for some fixed* $\delta > 2$ *and sufficiently large* K

$$\frac{n_{K+1} + \lambda m_{K+1}}{n_K + \lambda m_K} > \delta.$$

Then $\alpha_{b,c,m,n}$ *is transcendental.*

Remark 4.11. Some of the $\alpha_{b,c,m,n}$ appearing here are not in the class of b-normal constants relevant to Theorem 4.8, because of the coprimality requirement for our defined PRNGs. Conversely, some of the b-normals in question are not covered by Theorem 4.10; for example, the assignments $b = 3, c = 2, n_k = k, m_k = 2^k$ yield a b-normal, but the inequality above involving δ fails (yet, the b-normal may well be transcendental; we are saying the following application of the Roth theorem is not sufficient).

Proof: For simplicity, denote $\alpha = \alpha_{b,c,m,n}$. The celebrated Roth theorem states [Roth 55] [Cassels 57] that if $|P/Q - \alpha| < 1/Q^{2+\epsilon}$ admits of infinitely many distinct rational solutions P/Q (i.e., if α is approximable to degree $2 + \epsilon$ for some $\epsilon > 0$), then α is transcendental. Write

$$\alpha = P/Q + \sum_{i > k} \frac{1}{c^{n_i} b^{m_i}},$$

where $\gcd(P, Q) = 1$ with $Q = c^{n_k} b^{m_k}$. The sum over i gives

$$|\alpha - P/Q| \;<\; \frac{2}{c^{n_{k+1}} b^{m_{k+1}}}.$$

But the right-hand side is, by virtue of the δ-inequality, less than $2/Q^\delta$ and transcendency follows. $\qquad\square$

Theorem 4.10 immediately repeats transcendency results on Stoneham and Korobov numbers, such results having been known to both of those authors, except for the former class when $c = 2$, as explained in Remark 4.11.

For a different research foray, consider the problem of specifying an uncountable collection of normals. One way to do this is surprisingly easy. In what follows, we define the bits of a real number according to the no-trailer rule: Any infinite trailer of 1s is to be removed via carry; e.g., $0.01111\cdots \to 0.1$ in binary; and then when we ask for the k-th bit of a real number in $[0, 1)$, we shall mean the k-bit to the right of the decimal point.

Theorem 4.12. *Let $b, c > 1$ be coprime and for each real $t \in [0, 1)$ denote by t_k the k-th binary bit of t. Then the collection of numbers*

$$\alpha(t) \;=\; \sum_{i \geq 0} \frac{1}{c^i b^{c^i + t_i}},$$

is uncountable, and each is b-normal.

Remark 4.13. We are creating here what could be called "perturbed Stoneham numbers," yet we could just as easily perturb in this way other kinds of normals.

Proof: That $\alpha(t)$ is always b-normal follows from Theorem 4.8—take $\gamma = 2/3$, say, so that the perturbation of adding t_i to m_i does not harm the required growth condition (ii). It remains to show that the $\alpha(t)$ are all pairwise distinct. Indeed, let $s > t$, with the first occurrence of unequal corresponding bits between s and t being $s_k = 1, t_k = 0$. Then

$$\alpha(t) - \alpha(s) = \frac{1}{c^K} \frac{1}{b^{c^K}} \left(1 - \frac{1}{b}\right)$$
$$+ \sum_{i > k} \frac{1}{c^i} \left(\frac{1}{c^{c^i + t_i}} - \frac{1}{b^{c^i + s_i}}\right)$$
$$> \frac{1}{c^K} \frac{1}{b^{c^K}} \left(1 - \frac{1}{b}\right) \left(1 - \sum_{i > k} \frac{1}{c^{i-k}} \frac{1}{b^{c^i - c^k}}\right)$$
$$> 0. \qquad\qquad\qquad\qquad\qquad\square$$

Inasmuch as this construction is fairly straightforward, one wonders whether there are other simple approaches. As a possible example, for the Champernowne C_{10}, let (u_k) be the ordered set of *positions* of 0s, and likewise let (v_k) be the set of positions of 1s. Clearly, by 10-normality of C_{10}, these two sets are infinite. Now, based on some real number t as was used in Theorem 4.12, either swap (when $t_k = 1$) or do not swap the (0,1) digit pair from respective positions (u_k, v_k) depending on the k-th bit of t. So for $t = 0 = 0.000\ldots$, the Champernowne is left unchanged; while for any other real $t < 1$, the Champernowne is altered and may be 10-normal. We have not finished this argument; we mention it only to note that there may be other means of constructing an uncountable class of normal numbers.

Next we move to a computational issue: Can one efficiently obtain isolated digits of $\alpha_{b,c,m,n}$? It turns out that at least the Stoneham number $\alpha_{b,c}$ admits an individual digit-calculation algorithm, as was established for π, $\log 2$ and some others in the original Bailey–Borwein–Plouffe (BBP) paper [Bailey et al. 97]—the same approach works for the new, b-normal and transcendental constants. Indeed, for $\alpha_{b,c}$ the BBP algorithm is extraordinarily rapid: The overall bit-complexity to resolve the n-th base-b digit of $\alpha_{b,c}$ is

$$O(\log^2 n \log\log n \log\log\log n),$$

which can conveniently be thought of as $O(n^\epsilon)$. By comparison, the complexity for the BBP scheme applied to fundamental constants such as π and $\log 2$ (in general, the constants falling under the umbrella of Hypothesis 3.1) is $O(n^{1+\epsilon})$. As a specific example, in only 2.8 seconds run time on a modern workstation the authors were able to calculate binary bits of $\alpha_{2,3}$, beginning at position one googol (i.e., 10^{100}). The googol-th binary digit is 0; the first ten hexadecimal digits starting at this position are **2205896E7B**. In contrast, C. Percival's recent resolution of the quadrillionth (10^{15}-th) binary bit of π is claimed to be the deepest computation in history for a 1-bit result [Percival 00], finding said bit to be 0 but at the cost of over 10^{18} CPU clocks.

At this point, one might look longingly at the b-normality of $\alpha_{b,p}$ and wonder how difficult it is to relax the constraint on summation indices in $\sum_{n \in S} 1/(nb^n)$ in order finally to resolve logarithmic sums. Some relaxations of the set $S \subset Z^+$ may be easier than others. We conjecture that

$$\alpha \;=\; \sum \frac{1}{p2^p},$$

538   Experimental Mathematics, Vol. 11 (2002), No. 4

where p runs through the set of Artin primes (of which 2 is a primitive root), is 2-normal. It is a celebrated fact that under the extended Riemann hypothesis (ERH) the Artin-prime set is infinite, and in fact—this may be important—has positive density amongst the primes. We make this conjecture not so much because of statistical evidence, but because we hope the fact of 2 being a primitive root for every index p might streamline any analysis. Moreover, any connection whatever between the ERH and the present theory is automatically interesting.

With these results in hand, let us sketch some alternative approaches to normality. We have mentioned in our introduction some of the directions taken by Good, Korobov, Stoneham et al. over the decades. Also of interest is the form appearing in [Korobov 92, Theorem 30, page 162], where it is proven that

$$\alpha = \sum_{n \geq 1} \frac{\lfloor b\{f(n)\}\rfloor}{b^n}$$

is b-normal for any "completely uniformly distributed" function f, meaning that for every integer $s \geq 1$ the vectors $(f(n), f(n+1), \ldots f(n+s))$ are, as $n = 1, 2, 3, \ldots$, equidistributed in the unit s-cube. (Korobov also cites a converse, that *any* b-normal number has such an expansion with function f.) Moreover, Korobov gives explicit functions such as

$$f(x) = \sum_{k=0}^{\infty} e^{-k^5} x^k,$$

for which the number α above is therefore b-normal. It is possible to think of some normals as being "more normal" than others, in the sense of discrepancy measures. We have seen that the normals of our Theorem 4.8 enjoy discrepancy no better than $D_N(x) = O\left(\log^2 N/\sqrt{N}\right)$, while on the other hand we know [Levin 99] that for almost all real x,

$$D_N(\{(b^n x)\}) = O\left(\left(\frac{\log \log N}{N}\right)^{1/2}\right).$$

Yet, researchers have done better than this. Levin gives [Levin 99] constructions of normals based on certain well-behaved sequences—such as quasi-Monte Carlo, low-discrepancy sequences or Pascal matrices—and derives discrepancy bounds as good as

$$D_N(\{(b^n \alpha)\}) = O\left(\frac{\log^k N}{N}\right)$$

for $k = 2, 3$.

For another research direction, there is another exponential-sum result of Korobov [Korobov 92, Theorem 33, page 171] that addresses the distribution of the powers $(b, b^2, b^3, \ldots b^m)$ modulo a prime power p^i, but where m is significantly less than $\sqrt{\operatorname{ord}(b, p^i)}$. It may be possible to use such a result to establish normality of numbers such as $\sum 1/(p^i b^{m_i})$ where the m_i have different growth conditions than we have so far posited via Theorem 4.8.

One also looks longingly at some modern treatments of nonstandard exponential sums, such as the series of papers [Friedlander et al. 01], [Friedlander and Shparlinski 01], and [Friedlander and Shparlinski 02], wherein results are obtained for power generators, which generators having become of vogue in cryptography. The manner in which Friedlander et al. treat exponential sums—for their purposes the summands being such as $\exp(2\pi i g^{g^j}/c)$—is of interest not because of any direct connection to normality, but because of the bounding techniques used.

5. PRNGs LEADING TO DENSITY AND IRRATIONALITY PROOFS

Independent of number theory and special primes, one could ask what is the statistical behavior of truly random points chosen modulo 1; for example, what are the expected gaps that work against uniform point density?

In view of Definition 2.1(4) and Theorem 2.2(11), it behooves us to ponder the expected gap-maximum for *random* points: If N random (with uniform distribution) points are placed in $[0, 1)$, then the probability that the gap-maximum G_N exceeds x is known to be [Jacobsen 78]

$$\operatorname{Prob}(G_N \geq x) = \sum_{j=1}^{\lfloor 1/x \rfloor} \binom{N}{j} (-1)^{j+1} (1 - jx)^{N-1}.$$

The expectation E of the gap-maximum can be obtained by direct integration of this distribution formula, to yield:

$$E(G_N) = \frac{1}{N}(\psi(N+1) + \gamma)$$

where ψ is the standard polygamma function Γ'/Γ. Thus for large N we have

$$E(G_N) = \frac{\log N + \gamma - 1/2}{N} + O(\frac{1}{N^2})$$
$$\sim \frac{\log N}{N}.$$

This shows that whereas the mean gap is $1/N$, the mean *maximum* gap is essentially $(\log N)/N$. In this sense, which remains heuristic with an uncertain implication for our problem, we expect a high-order cascaded PRNG to have gaps no larger than "about" $(\log P)/P$ where P is the overall period of the PRNG.

It turns out that for very specialized PRNGs, we can effect rigorous results on the gap-maximum G_N. One such result is as follows:

Theorem 5.1. *Let $1 = e_1 < e_2 < e_3 < \ldots < e_k$ be a set of pairwise coprime integers. Consider the PRNG with any starting seed $(s_1, \ldots, s_k)$:*

$$x_d = \left(2^d \left(\frac{2^{s_1}}{2^{e_1} - 1} + \frac{2^{s_2}}{2^{e_2} - 1} \cdots \frac{2^{s_k}}{2^{e_k} - 1} \right) \right) \bmod 1.$$

Then the generated sequence (x_d) has period $e_1 e_2 \cdots e_k$ and for sufficiently large N we have

$$G_N < 3/2^{\lfloor k/2 \rfloor}.$$

Proof: Each numerator 2^{d+s_i} clearly has period e_i modulo the respective denominator $2^{e_i} - 1$, so the period is the given product. The given bound on gaps can be established by noting first that the behavior of the PRNG defined by

$$y_{(f_i)} = \frac{2^{f_1} - 1}{2^{e_1} - 1} + \frac{2^{f_2} - 1}{2^{e_2} - 1} + \cdots + \frac{2^{f_k} - 1}{2^{e_k} - 1},$$

as each f_i runs over its respective period interval $[0, e_i - 1]$, is very similar to the original generator. In fact, the only difference is that this latter form has constant offset $\sum 1/(2^{e_i} - 1)$ so that the maximum gap around the mod 1 circle is unchanged. Now consider a point $z \in [0, 1)$ and attempt construction of a set (f_i) such that $y_{(f_i)} \approx z$, as follows. Write a binary expansion of z in the (nonstandard) form:

$$z = \sum_{n=1}^{\infty} \frac{1}{2^{b_n}},$$

i.e., the b_n denote the positions of the 1 bits of z. Now set $f_i = e_i - b_i$ for i from k down to $k - K + 1$ inclusive. Using the following inequality chain for any real $0 < a < b > 1$:

$$\frac{a}{b} - \frac{1}{b} < \frac{a-1}{b-1} < \frac{a}{b},$$

it follows that we can find a PRNG value such that

$$\|y_{(f_i)} - z\| < \left| -\frac{2}{2^{e_{k-K+1}}} + \sum_{j=1}^{K} \frac{1}{2^{b_k}} \right|.$$

Attention to the fact that the e_i are strictly increasing leads directly to the upper bound $3/2^{\lfloor k/2 \rfloor}$ on the maximum gap for the y, and hence the x generator. □

Of course, the maximum-gap theorem just exhibited is weaker than the statistical expectation of the maximum gap, roughly $(\log E)/E$ where $E = e_1 \cdots e_k$, but at least we finally have a rigorously vanishing gap and therefore, as we shall see, some digit-density, hence irrationality results.

Though the previous section reveals difficulties with the PRNG approach, there are ways to apply these basic ideas to obtain irrationality proofs for certain numbers of the form

$$x = \sum_i \frac{1}{m_i 2^{n_i}}.$$

for integers m_i and n_i. A first result is based on our rigorous PRNG gap bound, from Theorem 5.1, as:

Theorem 5.2. *Let $1 = e_1 < e_2 < \ldots$ be a strictly increasing set of integers that are pairwise coprime. Let (d_i) be a sequence of integers with the growth property:*

$$d_{k+1} > \prod_{i=1}^{k} d_i + \prod_{i=1}^{k} e_i.$$

Then the number:

$$x = \sum_{m=1}^{\infty} \frac{1}{2^{d_m}(2^{e_m} - 1)}$$
$$= \frac{1}{2^{d_1}(2^{e_1} - 1)} + \frac{1}{2^{d_2}(2^{e_2} - 1)} + \cdots$$

is 2-dense and hence irrational.

Proof: Fix a k, define $D = \prod d_i$, $E = \prod e_i$, and for $0 \leq g < E$ consider the fractional part of a certain multiple of x:

$$\{2^{g+D} x\} = \sum_{i=1}^{k} \frac{2^{f_i} - 1}{2^{e_i} - 1} + \sum_{i=1}^{k} \frac{1}{2^{e_k} - 1} + T,$$

where $f_i = 2^{g+D-d_i}$ and error term $|T| < 1/2^{e_k}$. By the Chinese remainder theorem, we can find, in the stated range for g, a g such that the PRNG values of Theorem 5.1 are attained. Thus the maximum gap between successive values of the sequence $\{2^n x\}$ vanishes as $k \to \infty$, so the sequence is dense by Theorem 2.2(11) and desired results follow. □

540   Experimental Mathematics, Vol. 11 (2002), No. 4

Of course, there should be an alternative—even easy—means to establish such an irrationality result. In fact, there are precedents arising from disparate lines of analysis. Consider what we call the Erdős–Borwein number: The sum of the reciprocals of all Mersenne numbers, namely:

$$E = \sum_{n=1}^{\infty} \frac{1}{2^n - 1}.$$

This still-mysterious number is known to be irrational, as shown by Erdős [Erdős 48] with a clever number-theoretical argument. More recently, P. Borwein [Borwein 92] established the irrationality of more general numbers $\sum 1/(q^n - r)$ when $r \neq 0$, using Pade approximant techniques. Erdős also once showed that the sum of terms $1/(b_n 2^{2^n})$ is *always* irrational for any positive integer sequence (b_n). Such binary series with reciprocal terms have indeed been studied for decades.

The Erdős approach for the E number can be sketched as follows. It is an attractive combinatorial exercise to show that

$$E = \sum_{a=1}^{\infty} \sum_{b=1}^{\infty} \frac{1}{2^{ab}} = \sum_{n=1}^{\infty} \frac{d(n)}{2^n},$$

where $d(n)$ is the number of divisors of n (including 1 and n). To paraphrase the Erdős method for our present context, consider a relevant fractional part:

$$\{2^m E\} =$$
$$\left(\frac{d(m+1)}{2} + \frac{d(m+2)}{2^2} + \frac{d(m+3)}{2^3} + \ldots \right) \bmod 1.$$

What Erdős showed is that one can choose any prescribed number of succesive integers $k+1, k+2, \ldots k+K$ such that their respective divisor counts $d(k+1), d(k+2), \ldots, d(k+K)$ are respectively divisible by increasing powers $2, 2^2, 2^3, \ldots, 2^K$, and furthermore this can be done such that the subsequent terms beyond the K-th of the above series for $\{2^k E\}$ are not too large. In this way Erdős established that the binary expansion of E has arbitrarily long strings of zeros. This proves irrationality (one also argues that infinitely many 1s appear, but this is not hard). We still do not know, however, whether E is 2-dense. The primary difficulty is that the Erdős approach, which hinges on the idea that if n be divisible by j distinct primes, each to the first power, then $d(n)$ is divisible by 2^j, does not obviously generalize to the finding of arbitrary d values modulo arbitrary powers of 2. Still, this historical foreshadowing is tantalizing and

there may well be a way to establish that the E number is 2-dense.

As a computational matter, it is of interest that one can also combine the terms of E to obtain an accelerated series:

$$E = \sum_{m=1}^{\infty} \frac{1}{2^{m^2}} \frac{2^m + 1}{2^m - 1}.$$

Furthermore, the E number finds its way into complex analysis and the theory of the Riemann zeta function. For example, by applying the identity $\zeta^2(s) = \sum_{n \geq 1} d(n)/n^s$, one can derive

$$E = \frac{\gamma - \log\log 2}{\log 2} + \frac{1}{2\pi} \int_R \frac{\Gamma(s)\zeta^2(s)}{(\log 2)^s} dt,$$

where R is the Riemann critical line $s = 1/2 + it$. In this sophisticated integral formula, we note the surprise appearance of the celebrated Euler constant γ. Such machinations lead one to wonder whether γ has a place of distinction within the present context. A possibly relevant series is [Beeler et al. 72]

$$\gamma = \sum_{k=1}^{\infty} \frac{1}{2^{k+1}} \sum_{j=0}^{k-1} \binom{2^{k-j} + j}{j}^{-1}.$$

If any one of our models is to apply, it would have to take into account the fairly slow convergence of the j sum for large k. (After $k = 1$ the j-sum evidently approaches 1 from above.) Still, the explicit presence of binary powers and *rational* multipliers of said powers suggests various lines of analysis. In particular, it is not unthinkable that the j-sum above corresponds to some special dynamical map, in this way bringing the Euler constant into a more general dynamical model.

It is of interest that a certain PRNG conjecture addresses directly the character of the expansion of the Erdős-Borwein number.

Conjecture 5.3. *The sequence given by the PRNG definition*

$$x_d = \left(\sum_{k=1}^{d} \frac{2^d - 1}{2^k - 1} \right) \bmod 1 = \left(\sum_{k=1}^{d} \frac{2^{d \bmod k} - 1}{2^k - 1} \right) \bmod 1$$

is equidistributed.

Remark 5.4. One could also conjecture that the sequence in Conjecture 5.3 is merely dense, which would lead to 2-density of E.

This conjecture leads immediately, along the lines of our previous theorems pertaining to specially-constructed PRNGs, to:

Theorem 5.5. *The Erdős-Borwein number E is 2-normal iff Conjecture 5.3 holds.*

Proof: For the PRNG of Conjecture 5.3, we have

$$x_d = (2^d - 1)\left(E - \sum_{j>d}\frac{1}{2^j - 1}\right) \bmod 1,$$

so that

$$\{x_d\} = \{\{2^d E\} + \{-E - 1 + t_d\}\},$$

where $t_d \to 0$. Thus $\{2^d E\}$ is equidistributed iff (x_n) is, by Theorem 2.2(10). □

We believe that at least a weaker, density conjecture should be assailable via the kind of technique exhibited in Theorem 5.1, whereby one proceeds constructively, establishing density by forcing the indicated generator to approximate any given value in $[0, 1)$.

P. Borwein has forwarded to us an interesting observation on a possible relation between the number E and the "prime-tuples" postulates, or the more general Hypothesis H of Schinzel and Sierpinski. The idea is—and we shall be highly heuristic here—the fractional part $d(m + 1)/2 + d(m + 2)/2^2 + \cdots$ might be quite tractable if, for example, we have

$$m + 1 = p_1,$$
$$m + 2 = 2p_2,$$
$$\cdots,$$
$$m + n = np_n,$$

at least up to some $n = N$, where the $p_i > N$ are all primes that appear in an appropriate "constellation" that we generally expect to live *very* far out on the integer line. Note that in the range of these n terms we have $d(m + j) = 2d(j)$. Now if the tail sum beyond $d(m + N)/2^N$ is somehow sufficiently small, we would have a good approximation

$$\{2^m E\} \approx d(1) + d(2)/2 + \cdots = 2E.$$

But this implies in turn that some iterate $\{2^m E\}$ revisits the neighborhood of an earlier iterate, namely $\{2E\}$. It is not clear where such an argument—especially given the heuristic aspect—should lead, but it may be possible to

prove 2-density (i.e., all possible finite bitstrings appear in E) on the basis of the prime k-tuples postulate. That connection would, of course, be highly interesting. Along such lines, we do note that a result essentially of the form: "The sequence $(\{2^m E\})$ contains a near-miss (in some appropriate sense) with any given element of $(\{nE\})$" would lead to 2-density of E, because, of course, we know E is irrational and thus $(\{nE\})$ is equidistributed.

6. SPECIAL NUMBERS WITH "NONRANDOM" DIGITS

This section is a tour of side results with regard to some special numbers. We shall exhibit numbers that are b-dense but not b-normal, uncountable collections of numbers that are neither b-dense nor b-normal, and so on. One reason to provide such a tour is to dispel any belief that, because almost all numbers are absolutely normal, it should be hard to use algebra (as opposed to artificial construction) to "point to" nonnormal numbers. In fact, it is not hard to do so.

First, though, let us revisit some of the artificially constructed normal numbers, with a view to reasons why they are normal. We have mentioned the binary Champernowne, which can also be written

$$C_2 = \sum_{n=1}^{\infty} \frac{n}{2^{F(n)}}$$

where the indicated exponent is:

$$F(n) = n + \sum_{k=1}^{n} \lfloor \log_2 k \rfloor.$$

Note that the growth of the exponent $F(n)$ is slightly more than linear. It turns out that if such an exponent grows too fast, then normality can be ruined. More generally, there is the class of Erdős–Copeland numbers [Copeland and Erdős], formed by (we remind ourselves that the $(\cdot)$ notation means digits are concatenated, and here we concatenate the base-b representations)

$$\alpha = 0.(a_1)_b(a_2)_b \cdots$$

where (a_n) is *any* increasing integer sequence with $a_n = O(n^{1+\epsilon})$, any $\epsilon > 0$. An example of the class is

$$0.(2)(3)(5)(7)(11)(13)(17) \cdots_{10},$$

where primes are simply concatenated. These numbers are known to be b-normal, and they all can be written in the form $\sum G(n)/b^{F(n)}$ for appropriate numerator function G and, again, slowly diverging exponent F. We add

542   Experimental Mathematics, Vol. 11 (2002), No. 4

in passing that the generalized Mahler numbers (for any $g, b > 1$)

$$M_b(g) = 0.(g^0)_b(g^1)_b(g^2)_b \cdots$$

are known at least to be irrational [Niederreiter 86], [Zun 87], and it would be of interest to establish perturbation sums in regard to such numbers. Incidentally, it is ironic that some of the methods for establishing irrationality of the $M_b(g)$ are used by us, below, to establish *nonnormality* of certain forms.

We have promised to establish that

$$\alpha = \sum_{n \geq 1} n/2^{n^2}$$

is 2-dense but not 2-normal. Indeed, in the n^2-th binary position, we have the value n, and since for sufficiently large n, we have $n^2 - (n-1)^2 > 1 + \log_2 n$, the numerator n at bit position n^2 will not interfere (in the sense of carry) with any other numerator. One may bury a given finite binary string in some sufficiently large integer n (we say bury because a string 0000101, for example, would appear in such as $n = 10000101$), whence said string appears in the expansion. Note that the divergence of the exponent n^2 is a key to this argument that α is 2-dense. As for the lack of 2-normality, it is likewise evident that almost all bits are 0s.

Let us consider faster growing exponents, to establish a more general result, yielding a class of b-dense numbers none of which are b-normal. We start with a simple but quite useful lemma.

Lemma 6.1. *For polynomials P with nonnegative integer coefficients, $\deg P > 0$, and for any integer $b > 1$, the sequence*

$$(\{\log_b P(n)\} : n = 1, 2, 3, \dots)$$

is dense in $[0, 1)$.

Proof: For $d = \deg P$, let $P(x) = a_d x^d + \cdots + a_0$. Then $\log_b P(n) = \log_b a_d + d(\log n)/\log b + O(1/n)$. Since $\log n = 1 + 1/2 + 1/3 + \cdots + 1/n - \gamma + O(1/n^2)$ diverges with n but by vanishing increments, the sequence $(\{d(\log n)/\log b\})$ and therefore the desired $(\{\log_b P(n)\})$ are both dense by Theorem 2.2(10). $\square$

Now we consider numbers constructed via superposition of terms $P(n)/b^{Q(n)}$, with a growth condition on P, Q.

Theorem 6.2. *For polynomials P, Q with nonnegative integer coefficients, $\deg Q > \deg P > 0$, the number*

$$\alpha = \sum_{n \geq 1} \frac{P(n)}{b^{Q(n)}}$$

is b-dense, but not b-normal.

Proof: The final statement about nonnormality is easy: Almost all of the base-b digits are 0s, because $\log_b P(n) = o(Q(n) - Q(n-1))$. For the density argument, we shall show that for any $r \in (0, 1)$, there exist integers $N_0 < N_1 < \dots$ and $d_1, d_2, \dots$ with $Q(N_{j-1}) < d_j < Q(N_j)$, such that

$$\lim_{j \to \infty} \{b^{d_j} \alpha\} = r.$$

This, in turn, implies that $(\{b^d \alpha\}) : d = 1, 2, \dots\})$ is dense, hence α is b-dense. Now for any ascending chain of N_i with N_0 sufficiently large, we can assign integers d_j according to

$$Q(N_j) > d_j = Q(N_j) + \log_b r - \log_b P(N_j) + \theta_j$$
$$> Q(N_{j-1})$$

where $\theta_j \in [0, 1)$. Then

$$P(N_j)/b^{Q(N_j)-d_j} = 2^{\theta_j} r.$$

However, $(\{\log_b P(n)\})$ is dense, so we can find an ascending N_j-chain such that $\lim \theta_j = 0$. Since $d_j < Q(N_j)$ we have

$$\{b^{d_j} \alpha\} = \left(b^{\theta_j} r + \sum_{k > 0} P(N_j + k)/b^{Q(N_j+k)-d_j} \right) \bmod 1$$

and because the sum vanishes as $j \to \infty$, it follows that α is b-dense. $\square$

Consider the interesting function [Kuipers and Niederreiter 74, page 10]:

$$f(x) = \sum_{n=1}^{\infty} \frac{\lfloor nx \rfloor}{2^n}.$$

The function f is reminiscent of a degenerate case of a generalized polylogarithm form—that is why we encountered such a function during our past [Bailey and Crandall 01] and present work. Regardless of our current connections, the function and its variants have certainly been studied, especially in regard to continued fractions, [Danilov 72], [Davison 77], [Kuipers and Niederreiter 74], [Borwein and Borwein 93], [Mayer 00], [Böhmer 1926],

[Adams and Davison 77], [Bowman 95], and [Bowman 88]. If one plots the f function over the interval $x \in [0, 1)$, one sees a brand of "devil's staircase," a curve with infinitely many discontinuities, with vertical-step sizes occurring in a fractal pattern. There are so many other interesting features of f that it is efficient to give another collective theorem. Proofs of the harder parts can be found in the aforementioned references.

Theorem 6.3. (Collection.) *For the "devil's staircase" function f defined above, with the argument $x \in (0, 1)$,*

(1) f is monotone increasing.

(2) f is continuous at every irrational x, but discontinuous at every rational x.

(3) For rational $x = p/q$, lowest terms, we have

$$f(x) = \frac{1}{2^q - 1} + \sum_{m=1}^{\infty} \frac{1}{2^{\lfloor m/x \rfloor}}$$

but when x is irrational we have the same formula without the $1/(2^q - 1)$ leading term (as if to say $q \to \infty$).

(4) For irrational $x = [a_1, a_2, a_3, \ldots]$, a simple continued fraction with convergents (p_n/q_n), we have:

$$f(x) = [A_1, A_2, A_3, \ldots],$$

where the elements A_n are:

$$A_n = 2^{q_{n-2}} \frac{2^{a_n q_{n-1}} - 1}{2^{q_{n-1}} - 1}.$$

Moreover, if (P_n/Q_n) denote the convergents to $f(x)$, we have

$$Q_n = 2^{q_n} - 1.$$

(5) $f(x)$ is irrational iff x is.

(6) If x is irrational, then $f(x)$ is transcendental.

(7) $f(x)$ is never 2-dense and never 2-normal.

(8) The range $\mathcal{R} = f([0, 1))$ is a null set (measure zero).

(9) The density of 1s in the binary expansion of $f(x)$ is x itself; accordingly, f^{-1}, the inverse function on the range $\mathcal{R}$, is just 1s density.

Some commentary about this fascinating function f is in order. We see now how f can be strictly increasing,

yet manage to "completely miss" 2-dense (and hence 2-normal) values: Indeed, the discontinuities of f are dense. The notion that the range $\mathcal{R}$ is a null set is surprising, yet follows immediately from the fact that almost all x have 1s density equal to $1/2$. The beautiful continued fraction result allows extremely rapid computation of f values. The fraction form is exemplified by the following evaluation, where x is the reciprocal of the golden mean and the Fibonacci numbers are denoted F_i:

$$\begin{aligned} f(1/\tau) &= f\left(\frac{2}{1 + \sqrt{5}}\right) \\ &= [2^{F_0}, 2^{F_1}, 2^{F_2}, \ldots] \\ &= \cfrac{1}{1 + \cfrac{1}{2 + \cfrac{1}{2 + \cfrac{1}{4 + \cfrac{1}{8 + \cdots}}}}} \end{aligned}$$

It is the superexponential growth of the convergents to a typical $f(x)$ that has enabled transcendency proofs as in Theorem 6.2(6).

An interesting question is whether (or when) a companion function

$$g(x) = \sum_{n=1}^{\infty} \frac{\{nx\}}{2^n}$$

can attain 2-normal values. Evidently

$$g(x) = 2x - f(x),$$

and, given the established nonrandom behavior of the bits of $f(x)$ for any x, one should be able to establish a correlation between normality of x and normality of $g(x)$. One reason why this question is interesting is that g is constructed from "random" real values $\{nx\}$ (we know these are equidistributed) placed at unique bit positions. Still, we did look numerically at a specific irrational argument, namely

$$x = \sum_{n \geq 1} \frac{1}{2^{n(n+1)/2}}$$

and noted that $g(x)$ almost certainly is *not* 2-normal. For instance, in the first 66,420 binary digits of $g(x)$, the string '010010' occurs 3034 times, while many other length-6 strings do not occur at all.

7. CONCLUSIONS AND OPEN PROBLEMS

Finally, we give a sampling of open problems pertaining to this interdisciplinary effort:

544   Experimental Mathematics, Vol. 11 (2002), No. 4

- We have shown that for (b, c, m, n)-PRNG systems, each associated constant $\alpha_{b,p,m,n}$—under the conditions of Theorem 4.8—is b-normal. What about c-normality of such numbers for c not a rational power of b?

- The generalized Stoneham numbers (case (i) of Corollary 4.9) might be generalizable in the following way: Instead of coprimality of b, c, just specify that neither integer divides the other. Can a result on b-normality then be effected? Presumably one would need some generalization of the exponential-sum lemmas.

- We have obtained rigorous results for PRNGs that either have a certain synchronization, or have extremely small "tails." What techniques would strike at the intermediate scenario which, for better or worse, is typical for fundamental constants; e.g., the constants falling under the umbrella of Hypothesis 3.1?

- What are the fundamental connections between normality theory and automated sequences (for an excellent survey of the latter, see [Allouche and Shallit 02])? We have talked—albeit heuristically—about unnatural versus natural constructions. Perhaps there are elegant, undiscovered ways to create new normals via automatic rules.

- Does polynomial-time (in $\log n$) resolution of the n-th digit for our $\alpha_{b,c}$ and similar constants give rise to some kind of "trap-door" function, as is relevant in cryptographic applications? The idea here is that it is so very easy to find a given digit even though the digits are "random." (As in: Multiplication of n-digit numbers takes polynomial time, yet factoring into multiples is evidently very much harder.)

ACKNOWLEDGMENTS

The authors are grateful to J. Borwein, P. Borwein, D. Bowman, D. Broadhurst, J. Buhler, D. Copeland, H. Ferguson, M. Jacobsen, J. Lagarias, R. Mayer, H. Niederreiter, S. Plouffe, A. Pollington, C. Pomerance, M. Robinson, S. Wagon, T. Wieting, and S. Wolfram for theoretical and computational expertise throughout this project. We thank, in particular, J. Shallit whose acute scholarship alerted us to the deeper history of normal numbers, allowing clutch corrections to an earlier draft of our work. A referee was kind to provide key corrections as well. We would like to dedicate this work to the memory of Paul Erdős, whose ingenuity on a certain, exotic analysis dilemma—the character of the Erdős–Borwein constant E—has been a kind of guiding light for our research.

When we coauthors began—and this will surprise no one who knows of Erdős—our goals were presumed unrelated to the Erdős world. But later, his way of thinking meant a great deal. Such is the persona of genius, that it can speak to us even across rigid boundaries.

Bailey's work is supported by the Director, Office of Computational and Technology Research, Division of Mathematical, Information, and Computational Sciences of the U.S. Department of Energy, under contract number DE-AC03-76SF00098.

REFERENCES

[Adams and Davison 77] W. W. Adams and J. L. Davison. "A Remarkable Class of Continued Fractions." *Proceedings of the American Mathematical Society* 65 (1977), 194–198.

[Allouche and Shallit 02] J.-P. Allouche and J. Shallit. "Automatic Sequences Theory, Applications, Generalizations." Manuscript, 2002.

[Bailey and Rudolph 02] David H. Bailey and Daniel J. Rudolph. "An Ergodic Proof that Rational Times Normal is Normal." Manuscript, 2002. Available from World Wide Web: ⟨http://www.nersc.gov/~dhbailey/dhbpapers/ratxnormal.pdf⟩, 2002.

[Bailey et al. 97] David H. Bailey, Peter B. Borwein and Simon Plouffe. "On The Rapid Computation of Various Polylogarithmic Constants." *Mathematics of Computation* 66:218 (1997), 903–913.

[Bailey and Crandall 01] David H. Bailey and Richard E. Crandall. "On the Random Character of Fundamental Constant Expansions." *Experimental Mathematics* 10 (2001), 175–190.

[Beeler et al. 72] M. Beeler, et al. Item 120 in M. Beeler, R. W. Gosper, and R. Schroeppel, *HAKMEM*, Cambridge, MA: MIT Artificial Intelligence Laboratory, Memo AIM-239, 55, Feb. 1972.

[Böhmer 1926] P. E. Böhmer. "Über die Transzendenz gewisser dyadischer Brüche." *Mathematische Annalen* 96 (1926), 367–377; *Erratum* 96 (1926), 735.

[Borwein et al. 98] Jonathan M. Borwein, David M. Bradley, and Richard E. Crandall. "Computational Strategies for the Riemann Zeta Function." *Journal of Computational and Applied Mathematics* 121 (2000), 247–296.

[Borwein 92] Peter Borwein. "On the Irrationality of Certain Series." *Mathematical Proceedings of the Cambridge Philosophical Society* 112 (1992), 141–146.

[Borwein and Borwein 93] Jonathan Borwein and Peter Borwein. "On the Generating Function of the Integer Part of $\lfloor n\alpha + \gamma \rfloor$." *Journal of Number Theory* 43 (1993), 293–318.

[Bowman 95] Douglas Bowman. "Approximation of $\lfloor n\alpha + s \rfloor$ and the zero of $\{n\alpha + s\}$." *Journal of Number Theory* 50 (1995), 128–144.

[Bowman 88] Douglas Bowman. "A New Generalization of Davison's Theorem." *Fibonacci Quarterly* 26 (1988), 40–45.

[Broadhurst 98] David J. Broadhurst. "Polylogarithmic Ladders, Hypergeometric Series and the Ten Millionth Digits of $\zeta(3)$ and $\zeta(5)$." Preprint, 1998. Available from World Wide Web: ⟨http://xxx.lanl.gov/abs/math/9803067⟩, 1998.

[Broadhurst 00] David J. Broadhurst. "Conjecture on Integer-Base Polylogarithmic Zeros Motivated by the Cunningham Project." Manuscript, 2000.

[Cassels 57] J. W. S. Cassels. *An Introduction to Diophantine Approximations.* Cambridge: Cambridge Univ. Press, 1957.

[Champernowme 33] D. G. Champernowne. "The Construction of Decimals Normal in the Scale of Ten." *Journal of the London Mathematical Society* 8 (1933), 254–260.

[Copeland and Erdős] A. H. Copeland and P. Erdős. "Note on Normal Numbers." *Bulletin American Mathematical Society* 52 (1946), 857–860.

[Crandall 96] R. Crandall. *Topics in Advanced Scientific Computation.* Berlin: Springer-Verlag, 1996.

[Danilov 72] L. V. Danilov. "Some Classes of Transcendental Numbers." *Matematicheskie Zametki* 12 (1972), 149–154; In Russian, English translation in *Mathematical Notes of the Academy of Science of the USSR* 12 (1972), 524–527.

[Davison 77] J. L. Davison. "A Series and Its Associated Continued Fraction." *Proceedings of the American Mathematical Society* 63 (1977), 29–32.

[Devaney 95] Robert L. Devaney. *Complex Dynamical Systems: The Mathematics Behind the Mandelbrot and Julia Sets.* Providence: American Mathematical Society, 1995.

[Erdős 48] P. Erdős. "On Arithmetical Properties of Lambert Series." *Journal of the Indian Mathematical Society (N.S.)* 12 (1948), 63–66.

[Friedlander et al. 01] J. Friedlander, C. Pomerance, and I. Shparlinski. "Period of the Power Generator and Small Values of Carmichael's Function." *Mathematics of Computation* 70 (2001), 1591–1605.

[Friedlander and Shparlinski 01] J. Friedlander and I. Shparlinski. "On the Distribution of the Power Generator." *Mathematics of Computation* 70 (2001), 1575–1589.

[Friedlander and Shparlinski 02] J. Friedlander and I. Shparlinski. "Some Double Exponential Sums over $\mathbf{Z}_m$." Manuscript, 2002.

[Good 46] I. Good. "Normal Recurring Decimals." *Journal of the London Mathematical Society* 21 (1946), 167–169.

[Ferguson et al. 99] Helaman R. P. Ferguson, David H. Bailey. and Stephen Arno. "Analysis of PSLQ, An Integer Relation Finding Algorithm." *Mathematics of Computation* 68:225 (1999), 351–369.

[Hardy and Wright 79] G. H. Hardy and E. M. Wright. *An Introduction to the Theory of Numbers.* Oxford: Oxford University Press, 1979.

[Hegyvari 93] Norbert Hegyvari. "On Some Irrational Decimal Fractions." *American Mathematical Monthly* 100 (1993), 779–780.

[Jacobsen 78] Mark Jacobsen. Private communication, 2000. Jacobsen in turn references Herbert Solomon, *Geometric Probabilities.* Philadelphia: SIAM, 1978.

[Khinchin 64] A. Khinchin. *Continued Fractions.* Phoenix Books. Chicago: Univ. of Chicago Press, 1964.

[Knuth 81] Donald E. Knuth. *The Art of Computer Programming,* Volume 2, Second Edition. Menlo Park: Addison-Wesley, 1981.

[Korobov 92] N. Korobov. *Exponential Sums and their Applications.* Norwood: Kluwer Academic Publishers, 1992.

[Korobov 90] N. Korobov. "Continued Fractions of Certain Normal Numbers." *Mathematik Zametki* 47 (1990), 28–33; In Russian; English translation in *Mathematical Notes of the Academy of Science USSR* 47 (1990), 128–132.

[Korobov 72] N. Korobov. "On the Distribution of Digits in Periodic Fractions." *Matematicheskie USSR Sbornik* 18 (1972), 659–676.

[Kuipers and Niederreiter 74] L. Kuipers and H. Niederreiter. *Uniform Distribution of Sequences.* New York: Wiley-Interscience, 1974.

[Lagarias 01] J. Lagarias. "On the Normality of Fundamental Constants." *Experimental Mathematics* 10:3 (2001), 353–366.

[Levin 99] M. Levin. "On the Discrepancy Estimate of Normal Numbers." *Acta Arithmetica* LXXXVIII:2 (1999), 99–111.

[Mayer 00] R. Mayer. Private Communication, 2000.

[Mercer 94] A. McD. Mercer. "A Note on Some Irrational Decimal Fractions." *American Mathematical Monthly* 101 (1994), 567–568.

[Niederreiter 78] H. Niederreiter. "Quasi-Monte Carlo Methods and Pseudo-Random Numbers." *Bulletin of the American Mathematical Society* 84:6 (1978), 957–1041.

[Neiderreiter 92] H. Niederreiter. "Random Number Generation and Quasi-Monte Carlo Methods." In *CBMS-NSF Regional Conference Series in Applied Mathematics.* Philadelphia: SIAM, 1992.

[Niederreiter 86] H. Niederreiter. "On an Irrationality Theorem of Mahler and Bundschuh." *Journal of Number Theory* 24 (1986), 197–199.

[Niven 56] I. Niven. *Irrational Numbers,* Carus Mathematical Monographs, No. 11. New York: Wiley, 1956.

[Percival 00] C. Percival. "PiHex: A Distributed Effort to Calculate Pi." Available from World Wide Web: ⟨http://www.cecm.sfu.ca/projects/pihex/index.html⟩, 2000.

[Ribenboim 96] P. Ribenboim. *The New Book of Prime Number Records.* New York: Springer-Verlag, 1996.

546   Experimental Mathematics, Vol. 11 (2002), No. 4

[Roth 55] K. Roth. "Rational Approximations to Algebraic Numbers." *Mathematika*, 2 (1955), 1–20. Corrigendum, 168.

[Stoneham 70] R. Stoneham. "A General Arithmetic Construction of Transcendental Non-Liouville Normal Numbers from Rational Fractions." *Acta Arithmetica* 16 (1970), 239–253.

[Stoneham 73a] R. Stoneham. "On Absolute (j, ϵ)-Normality in the Rational Fractions with Applications to Normal Numbers." *Acta Arithmetica* 22 (1973), 277–286.

[Stoneham 73b] R. Stoneham. "On the Uniform Epsilon-Distribution of Residues Within the Periods of Rational Fractions with Applications to Normal Numbers." *Acta Arithmetica* 22 (1973), 371–389.

[Stoneham 74] R. Stoneham. "Some Further Results Concerning the (j, ϵ) Normality in the Rationals." *Acta Arithmetica* 26 (1974), 83–96.

[Stoneham 76] R. Stoneham. "Normal Recurring Decimals, Normal Periodic Systems, (j, ϵ)-Normality, and Normal Numbers." *Acta Arithmetica* 28 (1976), 349–361.

[Stoneham 83] R. Stoneham. "On a Sequence of (j, ϵ)-Normal Approximations to $\pi/4$ and the Brouwer Conjecture." *Acta Arithmetica* 42 (1983), 265–279.

[Weisstein 03] E. Weisstein. Available from World Wide Web: (http://www.mathworld.com), 2003.

[Zun 87] J. Zun. "A Note on Irrationality of Some Numbers." *Journal of Number Theory* 25 (1987), 211–212.

David H. Bailey, Lawrence Berkeley National Laboratory, Berkeley, CA 94720 (dhbailey@lbl.gov)

Richard E. Crandall, Center for Advanced Computation, Reed College, Portland, OR 97202 (crandall@reed.edu)

Received March 6, 2002; accepted in revised form November 14, 2002.

7. The googol-th bit of the Erdős–Borwein constant

Discussion

Who would have thought that factoring and normality theory would be linked? In this paper, it is shown how to find the googol-th bit, being the binary bit in position 10^{100} to the right-of-point, of a special constant E. (See the Discussion prior to this volume's previous paper on normality.) Factoring of numbers in the googol region is the key to success here.

The dedicated to the memory of Steve Jobs is heartfelt—he really did appreciate such abstractions, and things like huge numbers.

Challenges

Shall we ever know the googolplex-th bit of E? That would be the binary bit in position

$$10^{10^{100}}$$

to the right-of-point.

Source

Crandall, R. E., "The googol-th bit of the Erdős–Borwein constant," *INTEGERS*, **12** (2012).

#A23   INTEGERS 12 (2012)

THE GOOGOL-TH BIT OF THE ERDŐS–BORWEIN CONSTANT

Richard Crandall

Center for Advanced Computation, Reed College, Portland, Oregon
`crandall@reed.edu`

Received: 11/5/11, Accepted: 2/25/12, Published: 3/1/12

Abstract

The Erdős–Borwein constant is a sum over all Mersenne reciprocals, namely

$$E := \sum_{n \geq 1} \frac{1}{2^n - 1} = 1.1001101101010000010111111\ldots \text{ (binary)}.$$

Although E is known to be irrational, the nature of the asymptotic distribution of its binary strings remains mysterious. Herein, we invoke nontrivial properties of the classical divisor function to establish that the googol-th bit, i.e., the bit in position 10^{100} to the right of the point is a '1.' Equivalently, the integer

$$\left\lfloor 2^{10^{100}} E \right\rfloor$$

is odd. Methods are also indicated for resolving contiguously the first 2^{43} bits (one full Terabyte) in about one day on modest machinery.

– In memory of longtime comrade, Steven Paul Jobs
1955-2011

1. Preliminaries

Our primary aim is to establish specific binary bits of the Erdős–Borwein constant

$$E := \sum_{n \geq 1} \frac{1}{2^n - 1} = 1.1001101101010000010111111\ldots \text{ (binary)}.$$

This is the instance $E = EB(2)$ of the generalization for integer $t > 1$:

$$EB(t) := \sum_{n \geq 1} \frac{1}{t^n - 1}.$$

each of which real numbers having been proven irrational. Irrationality was established in 1948 by Erdős [4], and later by P. Borwein [3] via Padé approximants. There is further historical and speculative discussion in [2].

The Erdős irrationality proof is patently number-theoretical. In that same spirit we show how to resolve remote bits of E. It is striking how far "remote" can be in this context. Denoting the binary bits

$$E = e_0.e_1 e_2 e_3 \ldots,$$

we shall establish that

$$e_{10^{100}} = 1,$$

and note that this bit position is radically farther to the right-of-point than a string of say 10^{80} protons from the visible cosmos – lined up as bits – would reach.[1] It should be mentioned that in 2002 the present author and D. Bailey resolved the googol-th bit of the Stoneham number $\alpha_{2,3}$ (see [2]). Similar ideas are used here for the googol-th bit of E, but this time we need to leverage some nontrivial number theory.

2. The Erdős Paradigm

A number-theoretical approach for extraction of properties of E hinges on the observation

$$E = \sum_{k \geq 1} \frac{d(k)}{2^k}, \tag{1}$$

where $d(k)$ is the classical divisor function—the number of divisors of k, including 1 and k. It is a textbook result [5] that for any positive ϵ, $d(k) = O(k^\epsilon)$, but we shall eventually need precise, effective bounds. Erdős used a clever argument to construct finite chains of integers $n, n+1, n+2, \ldots$ such that $d(n+m)$ is divisible by 2^{m+1}. This leads—after careful bounding of carry-propagation—to the knowledge that arbitrarily long but terminating strings of '0's must appear in the binary expansion, whence E is irrational.

To modify such arguments in order to evaluate remote bits, we require effective bounds on the divisor function. It is a textbook result that $d(n) = O(n^\epsilon)$ for any $\epsilon > 0$, but we require a hard implied big-O constant. First, we have the elementary result

$$d(n) = 2 \left(\sum_{ab=n,\ a \leq b} 1 \right) - \delta_{n \in Z^2} < 2\sqrt{n}.$$

---

[1] The author hereby names this bit $e_{10^{100}} = 1$ the "Jobs bit," as it was resolved right around the time of his passing. Steven Jobs did—beyond his pragmatic eminence—always appreciate such abstract forays.

Then we have results closer to the true asymptotic character of $d(n)$, namely for $n \geq 3$,

$$d(n) \leq n^{\frac{1.5379 \log 2}{\log \log n}},$$

or alternatively

$$d(n) \leq n^{\frac{\log 2}{\log \log n}\left(1 + 1.9349 \frac{1}{\log \log n}\right)},$$

or even higher-order formulae such as

$$d(n) \leq n^{\frac{\log 2}{\log \log n}\left(1 + \frac{1}{\log \log n} + 4.7624 \frac{1}{(\log \log n)^2}\right)}.$$

Such results are due to G. Robin and J-L. Nicolas [9] [10] [11], being the fruit of a long-term study on Ramanujan's "highly composite numbers."

Careful application of these sophisticated bounding formulae results in effective bounds, as exemplified in:

Lemma 1. *The divisor function $d(n)$ satisfies $d(n) < 2n^{\delta}$, where the power δ may be interpreted as being n-dependent; for example, one may assign respective δ powers to typical intervals:*

$$n \in \left[1, 10^{10}\right) \to \delta := \frac{1}{2}, \quad n \in \left[10^{10}, 10^{26}\right) \to \delta := \frac{1}{3}, \quad n \in \left[10^{26}, 10^{56}\right) \to \delta := \frac{1}{4},$$

$$n \in \left[10^{56}, 10^{100}\right) \to \delta := \frac{1}{5}, \qquad n \in \left[10^{100}, 10^{112}\right) \to \delta := \frac{5}{29},$$

$$n \in \left[10^{112}, 10^{10^{100}}\right) \to \delta := \frac{1}{6}, \qquad n \geq 10^{10^{100}} \to \delta := \frac{1}{33}.$$

The '2'-factor in the $d(n)$ bound is just for convenience of analysis later; indeed, the prefactor '2' can itself be lowered with more work, or dropped completely for say $n \geq 10^{10}$. This lemma will be used to provide effective bounds on "tail" components of BBP-like series, as we discuss in Section 4.

3. A Variational Approach to the Divisor Function

It is sometimes lucrative to possess a bound on $d(n)$ based on knowledge of the minimum prime factor of n. For example, if n is too large to completely factor – which factoring is the only real hope of calculating $d(n)$ exactly for large n – one may still sieve n for small primes, then apply a factor-dependent bound.

In this regard it is interesting to analyze factorizations without the traditional constraint that powers of primes need be integers. For example, given that n has precisely k prime factors, say (here, $p_1 < p_2 < \cdots p_k$, but not necessarily consecutive primes)

$$n = \prod_{i=1}^{k} p_i^{t_i},$$

INTEGERS: 12 (2012)   4

and knowing that the divisor function is

$$d(n) = \prod (t_i + 1),$$

we might ask: What choice of *real-number* powers t_i maximizes this latter product? An amusing example is $n = 144$, for which the standard factorization is $144 = 2^4 \cdot 3^2$, with divisor function evaluation $d(144) = (4+1)(2+1) = 15$. But there is also the (exact, but nonstandard) factorization

$$144 = 2^{\frac{\log 216}{\log 4}} \cdot 3^{\frac{\log 96}{\log 9}} = 2^{3.877\cdots} \cdot 3^{2.077\cdots}.$$

Moreover one can show that this peculiar pair of noninteger powers atop 2,3 gives the maximum possible power-product, which maximum we shall denote $D(n)$. In our example case,

$$D(144) = \left(\frac{\log 216}{\log 4} + 1\right)\left(\frac{\log 96}{\log 9} + 1\right) = 15.0095\ldots$$

and sure enough, $d(144) = 15 < 15.0095$. We can establish a general upper bound $D \geq d(n)$ by proceeding variationally. We borrow here, from mathematical physics, the classical notion of Lagrange multipliers. The idea is to maximize (for $T_i := t_i + 1$) the product

$$\bar{D} = \prod T_i$$

subject to the constraint $n = \prod p_i^{T_i - 1}$ in the form

$$0 = C := \log n + \sum_i \log p_i \ - \sum T_i \log p_i.$$

One introduces a Lagrange multiplier λ and proceeds to render stationary (with respect to T_i variations) the unconstrained construct $S := \bar{D} + \lambda C$. Setting $\partial S / \partial T_i = 0$ for all i, we find, after backsolving for λ in the constraint, a unique maximum $D(n)$:[2]

$$d(n) \ \leq \ D(n) \ = \ \left(\frac{\log n + \sum \log p_i}{k}\right)^k \frac{1}{\prod \log p_i}. \tag{2}$$

There is a similar bounding formula in the excellent survey [7, p. 218]. Our upper bound can be put in the form

$$d(n) \ \leq \ n^{\frac{1}{A} \log \frac{2A}{G}},$$

where A is the arithmetic mean of the sequence $(t_i \log p_i)$ and G is the geometric mean of the sequence $(\log p_i)$.

---

[2]That this solution is a unique maximum over the positive-T_i manifold is not hard to prove— since the product $\bar{D}$ is monotonic in every T_i.

This "variational bound" can be useful when special information about prime factorization of n is in hand. But when the only information is a lower bound on the smallest prime factor p_1, there is a more direct route to a bound on $d(n)$, as follows.[3] Note that d is sub-multiplicative, that is $d(ab) \leq d(a)d(b)$ for all a, b (with equality when a, b are coprime). Set $\epsilon := \log 2 / \log p_1$. Then, regarding n as $\prod q_1$ with primes q_i not necessarily distinct,

$$\frac{d(n)}{n^\epsilon} \leq \prod_i \frac{d(q_i)}{q_i^\epsilon} \leq \prod_i \frac{2}{p_1^\epsilon} = \prod_i 1 = 1.$$

Thus we have

Lemma 2. *For p_1 denoting the smallest prime dividing n, we have $d(n) \leq n^{\frac{\log 2}{\log p_1}}$. In particular, if n has no divisors $< 2^q$, then $d(n) \leq n^{\frac{1}{q}}$.*

Example. If we sieve n by all primes $< 2^{32}$ without finding a factor, then we know $d(n) \leq n^{1/32}$. Such a bound can in practice be radically tighter than we glean from Lemma 1.

Another Example. For $n = 10^{50} + 1$, we find that $n = F \cdot C$, where

$$F = 101 \cdot 3541 \cdot 27961 \cdot 60101 \cdot 7019801$$

and C is composite with all prime factors $> 2^{23}$. It follows that

$$d(n) = d(F)d(C) = d(F)d(n/F) \leq 2^5 \cdot (n/F)^{1/(23)} < 549,$$

radically better than the corresponding bound from Lemma 1.

4. Generalized BBP Formula

D. Bailey, P. Borwein, and S. Plouffe invented (1995-1996) a scheme for rapidly establishing remote digits of fundamental constants (see [13]). The now celebrated "BBP" prescription presumes specific base b, then resolves (in our present nomenclature) digit position Q to the right of the point, by determining with sufficient accuracy where $\left(b^{Q-1}E \bmod 1\right)$ lies in $[0, 1)$. Application with base 2 for binary bits of E starts out with the decomposition

$$2^{Q-1}E = \sum_{k=1}^{Q-1} \frac{2^{Q-1}d(k)}{2^k} + \sum_{m=0}^{M-1} \frac{d(Q+m)}{2^{m+1}} + \sum_{m=M}^{\infty} \frac{d(Q+m)}{2^{m+1}}$$

$$= R_Q + S_{Q,M} + T_{Q,M},$$

---

[3]The present author is grateful to J-L. Nicolas for this direct, elegant argument [8].

where the $T_{Q,M}$ sum is called the "tail."

The immediate observation – one that lives at the very core of the BBP idea – is that the first sum R_Q is an integer. Most of the work involves the $S_{Q,M}$ sum, while we usually bound T with an appropriate bounding lemma. For the moment, we see that

$$\left(2^{Q-1}E\right) \bmod 1 \;=\; (S_{Q,M} \,+\, T_{Q,m}) \bmod 1,$$

so that the Q-th bit of E (to right-of-point) is '0', '1' respectively, as

$$B_Q \;:=\; S_{Q,M} \bmod 1 \;+\; T_{Q,M}$$

is in $[0,1/2), [1/2,1)$ respectively. Of course, the real trick is to use sufficiently large M to bound $T_{Q,M}$ so tightly that B_Q lies in $[0,1)$, whence a simple test of $B_Q \geq, < 1/2$ resolves the Q-th bit. To this end, we posit:

Lemma 3. *The tail is bounded by*

$$T_{Q,M} \;<\; \frac{(Q+M)^\delta}{2^M} \frac{1}{1 - \frac{1}{2}e^{\delta/(Q+M)}},$$

where δ is taken from Lemma 1 with $n := Q+M$.

Proof. We have, with $r := \frac{1}{2}e^{\delta/(Q+M)}$,

$$T_{Q,M} \;=\; \sum_{m \geq M} \frac{d(Q+m)}{2^{m+1}} \;<\; \sum_{m \geq M} \frac{(Q+m)^\delta}{2^m} \;<\; \frac{(Q+M)^\delta}{2^M}(1 + r + r^2 + \cdots). \quad \square$$

Note that this lemma's tail bound can be cut dramatically if one has a few "lucky sievings," as per Lemma 2. That is, if prime factors of $Q+M$ are sufficiently large, or in some sense sparse, one may peel off the first tail term then start a bound at index $M+1$, and so on.

5. Resolution of the Googol-th Bit

Using the apparatus of the previous section, we can find remote bits of E as in Table 1, with the count M of S-summands growing roughly logarithmically in Q. As an example table entry, the googol-th bit ($Q = 10^{100}$) involves divisor-function values as recorded in the Appendix.

All one has to do to find, say, a multibit word starting at position Q is to force the tail T to be sufficiently small. For $Q = 10^{100}$, taking $M = 94$ terms of S results in the resolution of an entire 32-bit word starting at the googol-th position. We have—from the post-googol factor tabulations in the Appendix—an exact value

$$S_{Q,94} = \frac{64355493348241192604273826566685}{1237940039285380274899124224},$$

Q	e_Q	M
10^{10}	0	21
10^{20}	0	37
10^{30}	0	39
10^{40}	1	49
10^{50}	0	49
10^{60}	0	53
10^{70}	0	61
$\vdots$	$\vdots$	$\vdots$
10^{100}	1	60

Table 1: Resolution of e_Q, the Q-th bit of E (meaning in the Q-th position to the right-of-point). The number M counts the necessary summands for the $S_{Q,M}$ term from Section 4; thus, M evaluations of the divisor function $d(Q+m)$ are used. (But Lemma 2 can accelerate this process in certain circumstances.)

while from Lemmas 1 and 3 we have $T_{Q,94} < 2 \cdot 10^{-11}$, so that in binary we know rigorously that

$$S_{Q,94} \bmod 1 + T_{Q,94} = 0.1001100001101001000001000110110010\ldots,$$

where – as claimed – the first bit to the right-of-point, being the googol-th bit itself, is '1.'

6. The Contiguous Terabyte has "Too Many '1s"

Thanks to the work of D. Mitchell, a program has been created to calculate contiguous bits from the $Q = 1$ bit (just to the right of point, the 0-th bit being the leading '1' to left-of-point) through the $Q = 2^{43}$ bit. That means we now know more than the first 8 trillion bits, and have logged populations for certain bases.

As to the method, the formula

$$E = \sum_{m \geq 1} \frac{1}{2^{m^2}} \frac{2^m + 1}{2^m - 1}$$

turns out to be efficient in a programming scenario. Rewriting,

$$E = \sum_{m \geq 1} \frac{1}{2^{m^2}} \left(1 + \frac{2}{2^m} + \frac{2}{2^{2m}} + \cdots \right),$$

we see that we can initialize each m^2-th bit to 1, than add '2's along arithmetic progressions.[4] With adroit use of threading, memory and eventual disk-writing, Mitchell was able to create the "contiguous Terabyte" in about one day.[5] As an integrity check, the remote-bits method of Sections 4, 5 was used to check the first 1 billion random bit positions in the contiguous Terabyte. In addition, several million random bit positions were likewise checked.

The discovered population counts are tabulated below. Note that in every base $2^{\beta=1,2,3,4}$ all digit populations listed add up to $2^{43}/\beta = 8796093022208/\beta$; that is, we count only nonoverlapping β-bit digits.

It is remarkable that the binary-bit counts for $0, 1$—respectively

$$4359105565638, 4436987456570$$

are so disparate. Indeed, a random coin-toss game should have, after some 8 trillion tosses, perhaps the first *four* digits in agreement. Of course, observing there are "too many '1's" in the contiguous Terabyte is merely suggestive of possible anomalies in the bit statistics of E.

Base 2 (binary)
0 : 4359105565638
1 : 4436987456570

Base 4 (quaternary)
0 : 1080561684345
1 : 1106903007230
2 : 1091079189718
3 : 1119502629811

Base 8 (octal)
0 : 359190664177
1 : 361815499829
2 : 361805068965
3 : 371183479256
4 : 361917190569
5 : 368575859354
6 : 370698715770
7 : 376844529482

---

[4]Of course, one takes every arithmetic progression beyond $Q = 2^{43}$ to guard against carry-interference. In this Terabyte case we guarded with 256 extra bits.

[5]Meaning about 24 real-time hours on an 8-core Mac Pro; still, a single desktop machine not a supercomputer warehouse! By contrast, the isolated googolth-bit resolution takes a few minutes—most of the work in that case being factorizations for post-googol integers (see Appendix).

INTEGERS: 12 (2012)   9

Base 16 (hexadecimal)
```
0  :  134567994085
1  :  132761862795
2  :  131164175459
3  :  135827739295
4  :  135994852874
5  :  139494161026
6  :  140042450578
7  :  142316547830
8  :  137111036682
9  :  135761702456
A  :  134046630124
B  :  138527844259
C  :  138566029070
D  :  141037268645
E  :  140378720036
F  :  141424240338
```

7. Open Problems

- What Erdős actually showed in 1948 was that out of the first N bits of E, there *must* be a '0'-string of length at least $c \log^\alpha N$ for some absolute constant c with $\alpha := 1/10$. Can this power α be increased? On the notion of randomness, is it legitimate to call a sequence random if it *must* contain strings of this type, for any N? (Certainly there are "coin-flip" games—real points expanded in binary—having bounded '0'-run length over the entire, infinite string.) Put another way: What is the probability measure for real numbers, in binary, that have this Erdős "string property" for a given α? (There are heuristic arguments that $\alpha = 1$ is a barrier, in the sense that we expect longest '0'-runs out of N bits to be of length $O(\log N)$ [6].) Incidentally, in the contiguous Terabyte, the first 2^{43} bits of E, the longest run of '0's has length 47, with longest run of '1's having length 41.

- What is the 10^{1000}-th bit of E? The present author knows of no way to answer this without at least *some* factorizations of 1000-decimal numbers.

- What about 2-normality of E? Can one even show that $1/2$ of the bits of E are 1's? The present author does not even know how to show that the string '11' appears infinitely often in E. In regard to 2-normality, see [2].

- What is the googolplex-th bit of E? This is equivalent to asking whether the integer $\left\lfloor 2^{10^{10^{100}}} E \right\rfloor$ is even or odd, to yield respectively '0/1.' Actually, we

INTEGERS: 12 (2012)   10

may never know this bit—certainly, factoring of numbers in the googolplex region is problematic!

- It is interesting to contemplate the true complexity of $d(n)$. Evidently, since primes p (uniquely) enjoy $d(p) = 2$, complete knowledge of $d(n)$ is at least as hard as primality proving. But what about factoring? A cryptographic RSA number $N = pq$ (product of two primes) has, trivially, $d(N) = 4$. So factoring can be quite unlike knowing $d(N)$ if there be certain extra information about N. In this regard, see [12] and references therein.

- For the constant

$$EB(10) := \sum_{k \geq 1} \frac{1}{10^k - 1}$$

there is some suspicion of statistical anomalies, at least so for the first billion decimal digits (see [1] where, as an example of statistical anomaly, it is proved that $\alpha_{2,3}$ is not 6-normal). It would be good to extend the present treatment in regard to $EB(10)$.

Acknowledgments. The author thanks D. Bailey, J. Borwein, P. Borwein, R. Brent, A. Jones, and J. Shallit for computational & mathematical insight. D. Mitchell performed the programming for storage of the contiguous Terabyte of E. Computational colleague J. Papadopoulos provided the factors of post-googol numbers. J-L. Nicolas—via messages and references—schooled the author graciously on the deeper properties of the divisor function.

References

[1] David Bailey and Jonathan Borwein, "Normal Numbers and Pseudorandom Generators," Submitted to *Proceedings of the Workshop on Computational and Analytical Mathematics in Honour of Jonathan Borwein's 60th Birthday*, Springer, (September 2011). On-line version: http://carma.newcastle.edu.au/jon/pseudo.pdf.

[2] David H. Bailey and Richard E. Crandall, "Random generators and normal numbers," *Experimental Mathematics*, vol. 11, no. 4, pg. 527–546 (2002).

[3] P. Borwein, "On the Irrationality of Certain Series," Math. Proc. Cambridge Philos. Soc. 112, 141-146, (1992).

[4] P. Erdős , "On Arithmetical Properties of Lambert Series." J. Indian Math. Soc. 12, 63-66 (1948).

[5] G. H. Hardy and E. M. Wright, *An Introduction to the Theory of Numbers*, Oxford University Press, 1979.

[6] A. Jones, private communication.

[7] J-L. Nicolas, "On highly composite numbers," *Ramanujan Revisited*, Academic Press, 215-244, (1988).

INTEGERS: 12 (2012)   11

[8] J-L. Nicolas, private communication.

[9] J-L. Nicolas and G. Robin, "Majorations explicites pour le nombre de diviseurs de N," Canad. Math. Bull. Vol 26(4), pp. 485-492 (1983).

[10] G. Robin, "Grandes valeurs de la fonction somme des diviseurs et hypothese de Riemann." J. Math. Pures Appl. 63, 187-213 (1984).

[11] G. Robin, "Majorations explicites du nombre de diviseurs d'un entier," Ph. D. thesis (1983).

[12] J. Shallit, "Number-theoretic functions which are equivalent to number of divisors," Information Processing Letters 20, 151-153, North–Holland, (1985).

[13] D. Bailey, P. Borwein and S. Plouffe, "On the Rapid Computation of Various Polylogarithmic Constants," Mathematics of Computation, vol. 66 (1997), pg. 903-913. Available at http://crd.lbl.gov/ dhbailey/dhbpapers/digits.pdf and http://en.wikipedia.org/wiki/BaileyBorweinPlouffe_formula.

8. Appendix: Post-Googol Divisors and Factors

```
d(10^100)  =  10201
(d(10^100+1), d(10^100+2),..., d(10^100+200)) =
(256, 64, 128, 96, 32, 128, 16, 768, 8, 24576, 64, 24, 48, 64, 16, 80,
64, 32, 64, 768, 32, 8, 16, 128, 12, 192, 8, 192, 32, 96, 32, 768, 4,
16, 96, 24, 16, 192, 8, 512, 8, 128, 128, 96, 32, 16, 64, 160, 64, 192,
8, 24, 24, 128, 128, 512, 256, 32, 16, 192, 64, 48, 16, 896, 512, 128,
256, 48, 64, 32, 48, 8, 256, 32, 576, 48, 8, 32, 32, 1920, 64, 32, 32,
96, 128, 64, 24, 4096, 12, 16, 32, 384, 16, 32, 64, 384, 64, 4096, 4,
288, 256, 32, 64, 64, 32, 256, 24, 48, 16, 256, 4, 80, 128, 256, 512,
144, 32, 128, 64, 256, 32, 256, 32, 24, 256, 512, 16, 128, 64, 64, 384,
48, 32, 24, 16, 96, 16, 128, 16, 192, 16, 512, 384, 320, 16, 128, 16,
96, 64, 96, 4, 640, 128, 16, 16, 24, 64, 256, 32, 192, 24, 384, 4, 96,
8, 512, 32, 64, 32, 384, 64, 48, 64, 32, 192, 160, 16, 32, 256, 48, 128,
32, 16, 256, 2048, 128, 32, 144, 8, 128, 128, 896, 16, 128, 64, 96, 48,
16, 8, 1536)
```

```
10^100+1 = 73.137.401.1201.1601.1676321.5964848081.
            1296944190290577505513857711845642744990757009476567578215372915271
            96801
10^100+2 = 2.3.4832936419.5025493293281.1061431139892014340488875721.
            6464979402011013241687574830622406864012978402059 3
10^100+3 = 7.157.769.2593.4888946572366141.2200309359940584892261 33.
            4242036622639156527880552375788044930249932162330 97
10^100+4 = 2^2.20794121.319929089.406288107529.5918277534160279189665941011889.
            1562846321861029648354357361984048909038 09
10^100+5 = 3.5.127.570527.92008683761171642627396845188769547032 0
            949797177667668839051495912373910667213989522888372 3
10^100+6 = 2.859493.2698836149.46606393157.201991350982876187.
            6930035321787863868408416051.330398011799854998020031828 31
10^100+7 = 557.294001.6908913964859.88386556167130813842350064090 5
            1132181948832801178384207854273111302870957443 689
10^100+8 = 2^3.3^2.113.593.48673.8181960160259.32930456993518040815814175510134 27.
            158048774746703862248218140371142897098668 9
10^100+9 = 3221.426362206609.72816629721289399809217825292529170113
            1895221015099208343986527943941226915328263778 1
```

INTEGERS: 12 (2012)   12

```
10^100+10 = 2.5.7.11^2.13.19.23.4093.8779.52579.599144041.7093127053.183411838171.
            141122524877886182282233539317796144938305111168717
10^100+11 = 3.7549.9604831.170862023971269597381839.159720377776884529561555201.
            168458805503671278689012389603080611 8757
10^100+12 = 2^2.43.79.735943479540771268766558728289667353547247 5
            71386517515454813070356196644097733294083014424492199
10^100+13 = 17.29^2.173.6872821583046302006287472663655 92320781.
            5882662189303137413042021569076752996489836845672351533
10^100+14 = 2.3.61.320355733.6029721445642558523.
            141445509586524630960622893455801896890879709438141927038876518056342 31
10^100+15 = 5.211.79382035150980920346405340690307261392830949801.
            11940576942571449017100623077195108726957466 6783273
10^100+16 = 2^4.241.6486935848076218529.1915250597685614658826993193.
            20873619626540002817530813703194302265479639056 2313
10^100+17 = 3^3.7.18617.25903.10971813871758633469374645226473 88
            7904955108375948552652766601255917230073634619091436160 3
10^100+18 = 2.907.59246083817.5296320795562545223259314466 30827.
            175682614675517925123451219617166344582046244425465993
10^100+19 = 419.128981.158351023.169951601411.18910843206669443.
            36358269870265130716971619469773020860088033267441390749 9
10^100+20 = 2^2.3.5.47.2988767.15506569049.1227359107267.4952827529942145668543.
            125868848880534489931181110172704915980357490 07
10^100+21 = 11.443.2916581.1104146260303595011101875420257.
            63723987021312297509098494050840802224217514357656481340308 1
10^100+22 = 2.911.54884742041712403951701427003293084522502744237
            102085620197585071350164654226125137211855104281 01
10^100+23 = 3.13.4795565371117055085672638649527185 1.
            5346820167535939222389950429483543731422341451974846366176752307
10^100+24 = 2^3.7.1109.1775532697.32354642734370330051999813271881 3.
            28029486646632139023220219672111309873253863885864462 1
10^100+25 = 5^2.53.75471698113207547169811320754716981132075471
            69811320754716981132075471698113207547169811320754717
10^100+26 = 2.3^2.31.8461.408243676198356920397439.17973559601797391007939829.
            288662616423696918682489693298266898703528917
10^100+27 = 37.7817500376745146648208006893159338109980065279.
            34572466548802211749474287766201124590387658782 40449
10^100+28 = 2^2.2543.677321.1936783.29412765531353.40204261427187057559889169.
            63373864933648884609911990883479626209574199
10^100+29 = 3.19.4673.210429649.178411330726409920627401936711 10
            3282734761915263682267134806674575739575345122794068661
10^100+30 = 2.5.17^2.48012016357931.3591932541314891.
            20064301263767150482328850050963655988361397950026433927658752114587
10^100+31 = 7.617.9084707.463548031.549808095586226854630168464 66
            0557249223439658597082353290733437006615609071798197
10^100+32 = 2^5.3.11.33521.155977777.106852828571.2120064562207281783451.
            79950517554013744522410523363155156242563037805197 21
10^100+33 = 23.43478260869565217391304347826086956521739130434782 6
            086956521739130434782608695652173913043478260871
10^100+34 = 2.7906914473.29273349974631567494719.
            216018297134610632645874688479708593987968023587206341456253683943 91
10^100+35 = 3^2.5.38851.21141503.37586080140891500526297088248071500703.
            719817481872461849061393038581795069086588340899 7
10^100+36 = 2^2.13.421.45678786771423350995797551617029051708 38
            662525123332724282843047688653389365978439612643888 17833
10^100+37 = 39640576062095087.41313840579541273.87719765535727771.
            6960925911590493224980314702905138173249716497109 7
```

INTEGERS: 12 (2012)   13

```
10^100+38 = 2.3.7^2.6763.774490190391774510933351713.121355948800841007852942196054 07.
            535101650691770679927275435877294269
10^100+39 = 628407404229083.4065739859741572599358310847655 63.
            391398465819307456740212929440975465272795781 50263391
10^100+40 = 2^3.5.41.8291.78213263.530341995529.22366776534380904027555859578 0143.
            79270101618775051135334614182509002273211
10^100+41 = 3.24032858772056765117.138698993946114804253142549 7.
            2706214173896524958912825535250360709731060328 7788191
10^100+42 = 2.29.2777.1345951.259438207754939.26383506364379192224 434030321.
            6739077161275842359144653780539104321625198617 3
10^100+43 = 11.59.1129.1346173.6787529443.6263718743647.
            238460533066503395592468182962756845218761896871 23410265282496451
10^100+44 = 2^2.3^3.32224000508338110680 7287.234579700339535290403 27927.
            1224916152689728079790911639841372159138437524825 7
10^100+45 = 5.7.409.565603.123508527311868910292026038322
            1554544661177201309727157565507831379605151377395 904945957581
10^100+46 = 2.3181.1868759255928821.84111035276901668623 81
            357532638838327724410245890759347785924047799841 46483349223
10^100+47 = 3.17.8933.5488485111634949419287863.17692570435662 499501333367419.
            22604196094974424422370620499957168355419 7
10^100+48 = 2^4.19.853.5197.58645164519523.1265296944936115806 6670
            3153017610605714547783841671102309853150725793961 093259
10^100+49 = 13.661.797.3361.23921885215780750153453456657563 65116069.
            18160715383258146933502081729547694550800118655304 1
10^100+50 = 2.3.5^2.3203.35257.70211209.8408141179126078225444 8330
            339165437517606429899225784231013549600334676728 36986553
10^100+51 = 71.2758358701518877.510611873448437102168153307109 1554638
            78745834884985513054542630430895588004592377 53
10^100+52 = 2^2.7.818118836878233844065143.
            436541540231047299693160852040621337088101414571356 81339376026806335807 2013
10^100+53 = 3^2.34981.40529381367189359.
            78370983210414966109187118581639571866860685826233 57042633484645536993928 54023
10^100+54 = 2.11.83.623401.580732441.20699532169.
            730794238533123774000776664351722659011297297969513 19227979134209000725 1
10^100+55 = 5.43.40352621.570514679894447.866599915243771.87599 044256603094139.
            266136816655836788631818935700766621650 59
10^100+56 = 2^3.3.23.2879.502509517.27060739495568903757075253.589 88880176019687972466865853.
            784450837874070274625408576045 69
10^100+57 = 31.67.2309.3821.7547.2447952814339850665 3.875961428442242224818335355 53.
            337209776583397758196966513101670840 03
10^100+58 = 2.416343878670491.19075470452091401.8046365032476149 48486627.
            782425204345425347749331027287161760530172997
10^100+59 = 3.7.43904218327282387.108461212688204976552628571826093 15
            107530646357547800416923287168905447079147237717
10^100+60 = 2^2.5.420785350832911.44186432061632719.4464888735183925 6399.
            1627023998074869019103.37018247787132672340014981 1
10^100+61 = 5689.523109.4798337.172166950433393 9.5408212763797220919797 3.
            75210343481334259612022339224193711565968339 99
10^100+62 = 2.3^2.13.2731511.1564520250331876204882024737774906820 5
            376087716667710671029713127658184441774183861988011 413
10^100+63 = 1303.17556891047924438 29.62929919145078098610248 9.
            69462561847701141837264129696659110125214783295969 01341
10^100+64 = 2^6.17.37.349.138054491513.37332834886802 41.
            3855829816411344629 3.35816618911174785100286518626464 1833810888252149
10^100+65 = 3.5.11.15073.27653.65951.1180951.14456147.1291422307404 5181
            945313691972571457369007914927796105568570648087273 6827
```

```
10^100+66 = 2.7.28409.648283.1966857343620975667.1016223574374368109205197947.
            19403910252523935366734934185104862899287773
10^100+67 = 19.47.659.2251.8641.3642168009108481.4981750128586454325416603783873O9.
            481483834172841998382407782953159063019
10^100+68 = 2^2.3.4518806497.29599291557204013994377977591524059582822681.
            6230367908453558718247933061340484046537015027
10^100+69 = 103.3823.181667.4579753.48863664957689298384943411.
            624675307793998082421640877316866659982231022483980066141
10^100+70 = 2.5.257.1439.2703996236037239436162704861514832771352782276
                        927070517517839615167255687180083445323844109209
10^100+71 = 3^5.29.315697632403487220340021.4494945222851026692801796
                        228287842072603052158559516981248119392623709533
10^100+72 = 2^3.1250000000000000000000000000000000000000000000000
                        000000000000000000000000000000000000000000000009
10^100+73 = 7.89.107.647.1889.13309.32231576674506523726164423167060419.
            286131032201005634197749446900885993031000074744501
10^100+74 = 2.3.73.137068770291678510110444307929746812672137169709.
            166566389847422325726957656884362408426022705824347
10^100+75 = 5^2.13^2.61.139.43633.825301.19224409609812811.
            403225690931051006827034621383764292357940500078338504752008096531
10^100+76 = 2^2.11.2172515904173.3515157699532874109471500687.
            29760453554670747757554886211947939814990020740179824066579
10^100+77 = 3.13469.24748187195287945158017175241913529833939663919617887
                        9897047541267602148142648550993639715890811
10^100+78 = 2.53.19041809.6198847919030513.7992358360398673669723180458516
                        94594672990286435213756257494634009520248939
10^100+79 = 23.131.482646413659.4359449947201.15773936534967843189577239482
                        70565217931108024450015566009142576951706937
10^100+80 = 2^4.3^2.5.7.307.28190375881201.43374136429285628893543.
            31713971432198155921941481.1666666666666666666666666666666667
10^100+81 = 17.41.809.4079297.166997003761959454121.
            26033028578107685036212207597755951748404897444205695108532363715081
10^100+82 = 2.24814351.44605757101.5045753779765448955569173.
            895261923721111295937622662307426417768862022444870904567
10^100+83 = 3.109.10037.140231426359.21727160459296384818152672758454825650
                        710404926974439902933184292058993245477105063
10^100+84 = 2^2.191.269.69997.111482389406555222129397018124040093.
            6235462418265224927955511580334421960539330483597943919
10^100+85 = 5.229.1823.77269.1021541371935884791.1355773132447700423.
            44767162661747939212253349597122160517899633964177303
10^100+86 = 2.3.19.311.1123.251162619206174421086488300016360733015090201789
                        573847863065738148602975744572840449800102183
10^100+87 = 7^2.11.20923687175229998149.8866924620100520579010655339523532
                        74153425142938570931294410266700242987592817
10^100+88 = 2^3.13.31.97.9241.1437967.986184733.2309103123017.854835730461250297.
            2574973404742823093663.480074153514389768077333
10^100+89 = 3^2.40156756629793913114491477.2766934395012203454514538203564
                        8988670327629874835674818966323290643352973
10^100+90 = 2.5.3079409181853103653.324737617167922950518581770926420981805685
                        233555106229023735803175419695095128853
10^100+91 = 79.6880726549933.5068013823241573808081.154972061606042703135868972981.
            23423263752533621617530706402640533
10^100+92 = 2^2.3.3571.7823.236219.2818817.6033263.7425429985798550817036897
                        202375244506903231220303001742398377095276670173
10^100+93 = 577.156192983.841220002066198331639.
            131902515215474369767608454585129809243588276452625438408233962011757
```

INTEGERS: 12 (2012)   15

```
10^100+94  = 2.7.6949.557755261667585837285371.18429178507114401861812206580516
                          44897284280159777135654220095426304719
10^100+95  = 3.5.53331697.33222956002547413.1163550228091036899011999.
                          3233701523243006231738527869087121458865612371780307
10^100+96  = 2^5.1471.14065455601.27300903463.80159555857.33984551226620654843.
                          20308118267648729400429388935023893726326361
10^100+97  = 197.4858111.180021089.1138763167.257675874383400610809009357738.
                          1978043261645793003730809769656536751933991
10^100+98  = 2.3^3.11.17.43.337.521.20233.539907194493435121.1215779969721649380466057.
                          9876297730815560076298557330035341269640291
10^100+99  = 309577.32302141308947370121165332049861585324491160519030
                          806552166343106884555377175952993923967219787
10^100+100 = 2^2.5^2.29.101.281.121499449.999999999999990000000000000009999
                          99999999990000000000000009999999999999990000000000000001
10^100+101 = 3.7.13.37.8204809363.17387650745267443.42466689592916707.
                          16340969454239796141725671994756545180514133613408902
10^100+102 = 2.23.59.499.73839646869272812389702027489023736492882596
                          382707936137566215703330002039451046529315078203571
10^100+103 = 9227.11657.18287.55294949.55184607601.
                          166612199397777827404653560556258001966697611808986746691114797388667
10^100+104 = 2^3.3.14448003448267.468728319607517981518198019211.
                          6152614833055567324841162969130771249767676707768535198531
10^100+105 = 5.19.317.2554711.1299796474040194170348435010369935170011
                          205047771396089113605109272423077187288052120235
10^100+106 = 2.4889.16223.635507.41740463.77141164072461202673.
                          120217162416660231262569387725626455263858795013700015087059
10^100+107 = 3^2.16038469.114286573037059.60617688441903879707317670409
                          780150038928091491888033559239052549191835879884137
10^100+108 = 2^2.7.42441310574479.87738667424615141969299
                          9590961514116570564622479685915208730804811601244166332403535711
10^100+109 = 11.163.57273605939085171841501.9737897151054790134802110
                          9595874772609793374591196821228602795731506592
10^100+110 = 2.3.5.149.8623.5751059.75042683729624878715019892477
                          6011431608476476919161541557817954788538561784747756144
10^100+111 = 375392363267616002656020413887213790881556622745911
                          266387944415135379724992732043044712403442990220721
10^100+112 = 2^4.347.621250308191.80708620434107.35922288913687466023021
                          78977503658087056303972667617607208685184689533131
10^100+113 = 3.1913.303846107399.1028032556303.563989491434969259411
                          286358557872369749647.3453997757912956480096879787297931
10^100+114 = 2.13.47.18289.1153007.4846883.1137995178434469963616787987502718713
                          70356452256780591392675007291015010907081871891
10^100+115 = 5.7.17.17683.1503823.44318006471.243803717223595595287573.
                          129507564403813367429407511.45166318133868365403131067161
10^100+116 = 2^2.3^2.1789.1455103152825659711.84067709962677004905937051.
                          1269299648316814823779955567405674398874917688570664991
10^100+117 = 223.2417.138337.1918067.699224053678218780258181783083375721
                          85360469781714855823268824943993565169892484175371
10^100+118 = 2.2213.5743.126893773.1787790853.17702846395518743891
                          9796019736115764349595231294751270051390061443384223492611
10^100+119 = 3.31.151.45631.3859543973142379.404337595745645929381533337809
                          097408301139398115819067748676939514714648761711
10^100+120 = 2^3.5.11.2909.423503.166531598227.1107772197296279830699572611
                          2504367658523593466746742971900512464436841223170337
10^100+121 = 113.2365541.9373933.63465865469.628823878476754072290554338460141
                          8845242739117073824649006894985312856611811
```

INTEGERS: 12 (2012)   16

```
10^100+122 = 2.3.7.41.71.750163.1280575673646290554747794213067 1063.
              85142738946645953125230980486846095525777870683767 00699
10^100+123 = 5573.5722867.10099123816709.27852851933917.
              11146639017073988804838377953892368932416775061750 2104516851701
10^100+124 = 2^2.19.67.196386488609583660644147682639434406912804 3990573448
              54673998428908091123330714846818538884524744 7
10^100+125 = 3^3.5^3.23.761.9349.18107086249193538387859961521459 4110347
              12176340040363986212378187992141847779339552 444329
10^100+126 = 2.1013.26951.66431.98419.103007.103247489.189246508793 21963.
              13917519013761920297432790919533799694481060890408 1941
10^100+127 = 13.5100877.326513248581029528021.46186067774920066381 5173080156113
              33006741894582882798247279772551954458 7
10^100+128 = 2^7.3.33246231553.62613917131.12509947185159810250399 94631488748
              05234276217057368222702720690641286050607 69
10^100+129 = 7.29.1293540926753.24971133702135855679332 1.
              30552566382184505302109553197 57.49915776522870148832884775680823
10^100+130 = 2.5.91156229.148929457013.79901404408912309.
              92188899421775182003902593678910826162513887099926 6007138529841
10^100+131 = 3.11^2.53.221957.860504803.65903001946343.1947068923 1015529163.
              21208445510441041900159338652234026444610025871 1
10^100+132 = 2^2.17.127.3249549678668340860029216104145875143102 8481.
              35633967991284747557089984670311353894151895825794927
10^100+133 = 2129.2741.61121.152924688262258047833911081812099503.
              18333578736932044222970908760327620361467020691566 7919
10^100+134 = 2.3^2.4453859921153217288588 37.1247357495275038045416 14223
              85347397807313118736226873913864177613618651311 99
10^100+135 = 5.2063.371772525345409.26076750768976332572787057452 6747249092
              35962954371366124869001267928686856316424978 1
10^100+136 = 2^3.7^2.61.17219107413143.24286977070182335135926181 44553
              72868580144157215704648794981638331592845453969 69171
10^100+137 = 3.83.1909767999449.21029068756973695958401478265718 7460987467748
              47434440074491433724265719627466876876537
10^100+138 = 2.37.137.3389.4591.883725268381227194123100448695337 5387377.
              71738382206015726952996030271965711660278544559 87
10^100+139 = 239.10181.1652120365901.24875394488527794103634024192 2104276257673
              70229325054556779841112669699324155924 21
10^100+140 = 2^2.3.5.13.167.6291975842976575437798026498410587067.
              12201181276478712966532395258880946265870116989857 425545717
10^100+141 = 43.7669.1032457.2937114203919374276971240750596119050 7216749916930
              3530746201936237680102169526804420051 39
10^100+142 = 2.11.181.1171.21521.12240944221333.298679653898882437 271.
              98863255487191253836623 7.2756918552948226009704656000 5101
10^100+143 = 3^2.7.19.889703.2406661.21025790789.38063001843713321 44457.
              48751782527356580624756215518461626501836777769332 541
10^100+144 = 2^4.293.389.304178521.93190774485577.26834573414482006 6915469.
              72088607232230102187228147800467193300591891582 9
10^100+145 = 5.7211.70358527025832438572490002956 9.
              39420103597867640422527402682359247053607336462834 5358996828991831
10^100+146 = 2.3.2282487101.10082794713401.1505384717862980247079 9.
              24042651090294165395370 1.2000919588428573457273842333 2709
10^100+147 = 73.2113.2389.271369786568829753697644684743355321495 43607172195996
              578689133646284750471722778417063182927
10^100+148 = 2^2.23.233.1299620417633.2391484801125942113.
              15009703068571887310857595413525633217692428941720 9664394284857067
10^100+149 = 3.17.12157.420190578883.667970622569.574645134365517 541604218323384
              33322537675937001111349227044891934248441
```

```
10^100+150 = 2.5^2.7.31.26891720626840241453.342729644920980089721468331055001
                        48126083732893462398747862634497615158084103
10^100+151 = 1927973835665676512899.518679238017111063918385114223719482182809088
                        6653436793207361725209210879641949
10^100+152 = 2^3.3^4.4111.6599.19819873381646179.50241036868675921087353971.
             571268029127646033945818610241224968428334573699
10^100+153 = 11.13.2791.651071.1815830747.16519134021466543611247.
             1282960047634944143273584649926925936508993671938978936979
10^100+154 = 2.227.347142871.634506238155231308460618587417993371482032510899233412
                        21112651424594420481600143970643081
10^100+155 = 3.5.17328071.3847321878278699727549977528754739443684566312468748925755
                        5942993693104481547118930125959587
10^100+156 = 2^2.27648097567.6009166703870537476739.
             15047366643178733567264408403218967581029214976607034850621132386003
10^100+157 = 7.179.193.2447.90337687157026890683257343.
             187063339483322152199522209509738366005465269643124117838952392873
10^100+158 = 2.3.29.6529.98927.1050575629.1443244473986509661734797576304534517.
             58684312183668058148562606540550519523874743
10^100+159 = 40423.61177783259.34923286955592834059.81068750046441067775940683.
             1428266113716178143065306110541400114371
10^100+160 = 2^5.5.157.743.11731.4567268213719573726027707163812296614933502171
                        5252105229120112029813174300654647891349521
10^100+161 = 3^2.47.59.400689185398886084064591096686300436751212084785831630404
                        295388067476058821172416556477140681973
10^100+162 = 2.19^2.89.5651.190360337.13443408749107.155637290760944336007726005297.
             69142958397595767222937180546465732947593
10^100+163 = 41.24390243902439024390243902439024390243902439024390243902439024
                        39024390243902439024390243902443
10^100+164 = 2^2.3.7.11.14552688897681160359740467096331204097734569.
             743677742209915030825511005107134040544950683533855319
10^100+165 = 5.589186310201415127012355311709.
             3394511999636064073600654344769386090586422336317541596961411897625237
10^100+166 = 2.13.17.7591.74873.682079.11818277.5299394027.
             931833778043405535563511073933511522393718116313116303824957664921
10^100+167 = 3.2699.109537.10903720263152091968613674605550321026159.
             1034047091409913961830214721857714914952993021468617
10^100+168 = 2^3.2826506440912283.1370368149343430083849.
             2179526410904293788782853299.148067806078017795393426744113535437
10^100+169 = 461.334993.1589939207113.12834377844581.31732771371460625718017136187
                        54517338594222020641124086447240671601
10^100+170 = 2.3^2.5.79.299177677.154000760619026940961289.1044151591430233027900199.
             29235787257428166566206783484714514363101
10^100+171 = 7.23.2671.1921363.2304215728707809514531721821395477928851.
             5252519530977697818611418843856257546375565195357
10^100+172 = 2^2.103.227496719346196361161.8759805472199688050815343.
             12179604438566406420129669983466409551161145066267547
10^100+173 = 3.116969.566045437.3802149371639624441.13406705602163997941693211196411.
             987655364635235465501493411099254497
10^100+174 = 2.57107.21480550613283265993.6221562375526746054074479.
             655142523597152438066100995052264083989761978334603
10^100+175 = 5^2.11.37.5839.331937.4044670901230701768151111939.
             125368448621667014758040059473623770992595554122825927806613
10^100+176 = 2^4.3.69931656414427.6024122118787711.14064532189166543487136136227093.
             35161378156943633740034823164308322897
10^100+177 = 5366852753.656455943061823297.65036695411041238520149.
             43643168149472786644316881718111870048527879381853
```

INTEGERS: 12 (2012)   18

```
10^100+178 = 2.7.83515028437.304534687294245804256053.
              280847501056290971775828877168561801106159488911010612113403507
10^100+179 = 3^3.13.971.1150678502221.95396635582537.1059503491609137118420461572 3.
              25228080235661492452512829808151719669569
10^100+180 = 2^2.5.107.263203937924865017242635664386495697410529.
              177539030498685830548570602454662134318629814 56495593403
10^100+181 = 19.31.13679.88735363.59052511337.163898727236840238488342835697106677.
              14451724462453436212041191526226478 36273
10^100+182 = 2.3.751.2897.766055645362812380255926934225389811245421093893755129 7
              0011526073240128875073330676652355215 1
10^100+183 = 17.331801.6961100128941734437.25468039062316830074600673014842281353
              2259000938443162808062792413329419827
10^100+184 = 2^3.43.53.65497.394169.887808639258294869.
              239299861500446984883604355572042748049829539736556391613648862 03261
10^100+185 = 3.5.7^3.97.457.121439.1265063.14671015149353017402951 93.
              57643690142089342901393656 7.3374771787410112629893662582 71
10^100+186 = 2.11.173.2243.2711.2231711918465802863807968381651158 7.
              193612900927855586506296496334190150180597675750 16721981
10^100+187 = 29.3697.9749.119659.79955273290125344645468046020744277108493604839100
              96687006714709625142839758130662428 9
10^100+188 = 2^2.3^2.379.13681.2997620694583.1787161368238253019831352803550 37
              7777980113997793330951563848538533509090401 4499
10^100+189 = 199.277.181412477550205903162019483700088892113999600892549389547013043
              557135858904437349200878036391343
10^100+190 = 2.5.4421.1045158737601119179.12678586348608424345873.
              4881771036537146264336 9.349662445189425844994521565618093
10^100+191 = 3.67.2027.469841.28431340955879.13842688194771698686788311.
              1327337919697182188672378586526157596407317941310 77
10^100+192 = 2^6.7.13.109.146527.58383439.26276153325203983.
              70078212795455149099123597642183934572642312536427680819631384963
10^100+193 = 71.4923943.210431719115074200380099604932 9.
              13593066177923892773317595096076674491769872703372318590228289
10^100+194 = 2.3.23.328591.490218640859.89897153275201034276028168 7.
              50041404289083337148686847342318334216948026508340107 1
10^100+195 = 5.327409.5014831.149251937.443634794521924801.
              18396597373804167449439806888469859915731476892432256124710393
10^100+196 = 2^2.3929.9181.5700769.84047429.14464722827296979655421424824568694064 1
              7090702391491251864859535703206568842 01
10^100+197 = 3^2.11.61.21190837.78142420483322935410220457673503256287959683002781 9
              62619503970319968732125000189445553879
10^100+198 = 2.23173.452727074623329270678427837462732608 37.
              4765969680028105997862352744190193123468416432721861558 99
10^100+199 = 7.30657788429071.4659734122298258359481810308555954077327606640900163708 00
              0818601182956158307473489476 7
10^100+200 = 2^3.3.5^2.17.19.254575361.4398346518948304621483.
              1804257591936590604367683840 19.25541233991668538051955112818103405 7
```

Part II

Analytical algorithms

"One cannot guess the real difficulties of a problem before having solved it."

—Carl Ludwig Siegel

1. Fast evaluation of multiple zeta sums

Discussion

Here the author worked out, with the aid of computationalist D. H. Bailey, an algorithm for calculating the multiple-zeta sum

$$\zeta(s_1,\ldots,s_d) := \sum_{n_1>\cdots>n_d} \frac{1}{n_1^{s_1}\cdots n_d^{s_d}}$$

is what might be called "surprising time." That is, due to certain redundancies, higher dimensions d take very little extra effort beyond the $d = 1$ cases; all of this is done with rapidly convergent series. In fact D good digits of ζ require only $O(D^{1+\epsilon})$ operations, with any dimension d dependence absorbed implicitly into the big-O constant.

After the publication of this paper, other authors found similar means of fast evaluation, as arising from a theory of multiple polylogarithms.[1]

Challenges

Even now (2012) the need for fast evaluation of multiple-zeta constructs looms important.[2] A good challenge is to explore the marriage of convergent series and quadrature, which can sometimes yield very fast evaluations.

Source

R. Crandall, "Fast evaluation of multiple zeta sums," *Math. Comp.*, **67**, 1163–1172 (1996).

---

[1]Jonathan M. Borwein, David M. Bradley, David J. Broadhurst, and Petr Lisonek, "Special Values of Multiple Polylogarithms," (1999).
http://arxiv.org/abs/math/9910045
[2]See for example, in this volume, "Unified algorithms for polylogarithm, *L*-series, and zeta variants."

MATHEMATICS OF COMPUTATION
Volume 67, Number 223, July 1998, Pages 1163–1172
S 0025-5718(98)00950-8

FAST EVALUATION OF MULTIPLE ZETA SUMS

RICHARD E. CRANDALL

ABSTRACT. We show that the multiple zeta sum:

$$\zeta(s_1, s_2, ..., s_d) = \sum_{n_1 > n_2 > ... > n_d} \frac{1}{n_1^{s_1} n_2^{s_2} ... n_d^{s_d}},$$

for positive integers s_i with $s_1 > 1$, can always be written as a finite sum of products of rapidly convergent series. Perhaps surprisingly, one may develop fast summation algorithms of such efficiency that the overall complexity can be brought down essentially to that of *one*-dimensional summation. In particular, for any dimension d one may resolve D good digits of ζ in $O(D \log D / \log \log D)$ arithmetic operations, with the implied big-O constant depending only on the set $\{s_1, ..., s_d\}$.

1. INTRODUCTION

The multiple zeta sums:

$$(1.1) \qquad \zeta(s_1, s_2, ..., s_d) = \sum_{n_1 > n_2 > ... > n_d} \frac{1}{n_1^{s_1} n_2^{s_2} ... n_d^{s_d}},$$

also called Euler/Zagier sums, have attracted considerable interest in recent times (see Bailey et al. [3], Borwein et al. [4], Borwein et al. [7], Broadhurst et al. [9], Crandall and Buhler [11], Markett [13] and Zagier [14]) (note that some previous treatments have reversed ordering of the indices n_i). The study of such sums is not only important to general zeta function theory, but also touches upon such domains as knot theory and particle physics methodology. To simplify our present analysis we shall concentrate on this scenario: each s_i is a positive integer and, to ensure convergence of the explicit sum, $s_1 > 1$. It should be noted, however, that even in absence of the integer restriction on the s_i, the ideas herein do lead to the development of convergent, albeit more intricate series.

These multiple zeta sums have been given many attractive, exact evaluations. It could be said that the research situation is quite rich, in the following sense. Whereas Euler's original evaluations, of which

$$\zeta(4, 1) = 2\zeta(5) - \zeta(2)\zeta(3)$$

is exemplary, along with more modern evaluations such as:

$$\zeta(4, 5, 1) = 2\zeta(7, 3) + \zeta(5)^2 - 17\zeta(3)\zeta(7) + \zeta(2)\zeta(5, 3) + 10\zeta(2)\zeta(3)\zeta(5) - \frac{21}{20}\zeta(10),$$

Received by the editor September 30, 1996 and, in revised form, March 3, 1997.
1991 *Mathematics Subject Classification.* Primary 11Y60, 11Y65; Secondary 11M99.

$$\zeta(6,6,6,6) = \frac{4\pi^{24}}{4326847970651925 46875},$$

are now proved (Markett [13], Borwein and Girgensohn [8], Borwein et al. [7]), suspected evaluations such as Zagier's conjectured:

$$\zeta(3,1,3,1,...,3,1) = \frac{2\pi^{4n}}{(4n+2)!}$$

(where n denotes the the number of $(3,1)$ pairs) remain elusive.* Even the two-dimensional sums $\zeta(6,2)$ and $\zeta(3,5)$ are unknown, in the sense that neither has been cast in a finite form involving "one-dimensional" sums such as Riemann zeta values. There is a growing literature on which sums are evaluable, which can be reduced to lower-dimensional forms, and which appear dimensionally irreducible and therefore "fundamental." For example, in Borwein and Girgensohn [8] it is argued that 3-dimensional sums can always be reduced (to lower-dimensional forms) if $s_1 + s_2 + s_3$ is either even or less than 10. On the other hand, $\zeta(5,3,3)$ has never been expressed as a combination of, say, two-dimensional ζ sums, "one-dimensional" Riemann sums, and fundamental constants.

To verify conjectured evaluations, indeed to *generate* reasonable conjectures, it has been important to be able to evaluate ζ numerically, to high precision (say > 100 decimal digits). The requirement of high precision arises for various reasons, such as the desire to avoid "accidental" relations and to guard against mishap when some integral coefficients of exact relations are large. The original approach of Bailey et al. [3] involved an Euler-Maclaurin scheme for estimating double sums. This approach has been effectively generalized by Broadhurst, as described in [7], to higher dimensions. A primary difficulty with an Euler-Maclaurin approach is that explicitly convergent sums do not obtain. In the special case of double sums, Crandall and Buhler [11] developed convergent series for $\zeta(r,s)$, where r, s need not even be integers. In this way, both $\zeta(6,2)$ and $\zeta(3,5)$ are now numerically resolved, each to more than 1200-digit precision (Bailey [1]). Given the present complexity claim, it should be possible to attain such extreme precision for sums of much higher dimension.

For the d-dimensional, unified approach we consider herein, the resulting convergent series will involve a single free parameter, λ, restricted to the interval $[0, 2\pi)$. One advantage of such free-parameter expansions is that one may gain substantial confidence in the numerical scheme by repeating the convergent sum for different λ choices and verifying that the numerical result is, to some satisfactory precision, invariant.

2. INTEGRAL REPRESENTATION

Our series development starts with the observation that ζ admits of certain integral representations (see Crandall [10], Borwein et al. [7], and Zagier [14]). Denote the argument vector $\vec{s} = \{s_1, s_2, ..., s_d\}$, and consider a particular d-dimensional integral representation:

$$(2.1) \qquad \zeta(\vec{s}) = \sum_{m_h \in Z^+} \int_0^\infty ... \int_0^\infty \prod_i x_i^{s_i-1} e^{-x_i \sum_{h=i}^d m_h} \frac{dx_i}{\Gamma(s_i)}$$

*Note added in proof: D. Broadhurst has recently announced a proof of this conjecture of Zagier.

in which indices i, h are understood each to run through values $1, ..., d$. This representation is proved quickly by performing the d individual (and separated) integrals, then re-indexing, with the n_i in (1.1) becoming appropriate partial sums of the m_h.

An immediate transformation of variables yields a more practical representation. Set

$$(2.2) \qquad u_k = \sum_{i=1}^{k} x_i$$

for $k = 0, 1, ..., d$, and interpret $u_0 = 0$ for convenience in what follows. Now the Jacobian of the $\vec{x} \to \vec{u}$ transformation is conveniently the identity, and we have

$$(2.3) \qquad \zeta(\vec{s}) = \sum_{m_h \in Z^+} \int_{u_d > u_{d-1} > ... > u_0} \prod_i (u_i - u_{i-1})^{s_i - 1} e^{-u_i m_i} \frac{du_i}{\Gamma(s_i)},$$

where, again, each of i, h runs through $1, ..., d$. The integrand is now somewhat more palatable, but at the expense of a more complicated domain of integration. This domain can be partitioned in the following way. Choose a positive parameter λ, and consider subdomains of integration defined for each of $k = 0, 1, ..., d$ as:

$$u_d > ... > u_{k+1} > \lambda > u_k > ... > u_0,$$

where for $k = d$ the inequality chain to the left of the λ is interpreted as empty. It is evident that the domain of integration in (2.3) can be taken to be the disjoint union of these $(d+1)$ subdomains. Now for the subdomain indexed by a $k < d$ we can expand the term $(u_{k+1} - u_k)^{s_{k+1}-1}$ in a finite binomial series, and in this way obtain a finite set of once-separated integrals. (This expansion step is, in fact, the only juncture at which the integer restriction on s_k is critical.) For the "lower" part of an integral, that is over variables $u_1, ..., u_k$, we sum formally over $m_1, ..., m_k$. In this way we obtain a kind of "factored" representation into lower and upper integrals:

$$(2.4) \qquad \begin{aligned} \zeta(\vec{s}) &= Y(\vec{s}; 0; \lambda) + Z(\vec{s}; \lambda) \\ &+ \sum_{k=1}^{d-1} \sum_{q=0}^{s_{k+1}-1} \frac{(-1)^q}{q!} Y(s_1, ..., s_k; q; \lambda) Z(s_{k+1} - q, s_{k+2}, ..., s_d; \lambda), \end{aligned}$$

where the "lower" integrals Y are defined:

$$(2.5) \qquad Y(s_1, ..., s_k; q; \lambda) = \int_{\lambda > u_k > ... > u_1 > 0} u_k^q \prod_{i=1}^{k} \frac{(u_i - u_{i-1})^{s_i - 1}}{e^{u_i} - 1} \frac{du_i}{\Gamma(s_i)},$$

while the "upper" integrals are

$$(2.6) \qquad Z(r_1, ..., r_j; \lambda) = \sum_{m_h \in Z^+} \int_{u_j > u_{j-1} > ... > u_1 > \lambda} \prod_{i=1}^{j} (u_i - u_{i-1})^{r_i - 1} e^{-u_i m_i} \frac{du_i}{\Gamma(r_i)}.$$

Note that in (2.4), when $k = d - 1$ the argument vector for Z is interpreted to be the singleton term $s_d - q$; i.e., the subsequence $s_{k+2}, ..., s_d$ is empty. Note also that evaluation of the Riemann ζ involves only the first two terms, a Y and a Z, in (2.4).

The relation (2.4) shows that, for any positive λ, every multiple zeta can be written as a finite sum of products of Y, Z functions. In the next section we show that, by way of proper constraint on the parameter λ, each of the necessary Y and Z may be cast in efficient convergent series form, whence a general multiple zeta can be calculated efficiently from (2.4).

3. OPERATOR CALCULUS AND BERNOULLI NUMBERS

For all $r_1, ..., r_j$ positive integers we can rewrite the general Z integral (2.6) by replacing terms $(u_i - u_{i-1})^{r_i-1}$ with respective operators

$$(-\partial_{m_i} + \partial_{m_{i-1}})^{r_i-1},$$

then moving these operators out of the integral. We then re-index the m_i indices according to

$$n_i = \sum_{h=i}^{d} m_h,$$

whence the transformed derivative operators ∂_{n_i} collapse conveniently to yield

$$(3.1) \qquad Z(r_1, ..., r_j; \lambda) = \frac{1}{\Gamma(r_1)} \sum_{n_1 > n_2 > ... > n_j} (-\partial_{n_1})^{r_1-1} \frac{e^{-\lambda n_1}}{n_1} \frac{1}{n_2^{r_2}...n_j^{r_j}},$$

where symbolic differentiation is understood to be performed prior to the summation. It is evident that Z looks very much like the original multiple zeta definition. Indeed Z coincides with ζ as $\lambda \to 0$, a fact also formally evident from (2.4) because said λ limit pinches all Y integrals to zero.

Now (3.1) is a patently convergent series, more efficient for larger λ. It remains to develop a converging series for the lower Y integrals. For $\lambda < 2\pi$ we can expand each term $(e^{u_i} - 1)^{-1}$ within the definition (2.5), in standard Bernoulli series. Such expansion leads to

$$Y(s_1, ..., s_k; q; \lambda) = \sum_{n_i \geq 0} \prod_{p=1}^{k} \frac{B_{n_p}}{n_p!} \int_{u_k > u_{k-1} > ... > u_0} u_k^q \prod_{i=1}^{k} (u_i - u_{i-1})^{s_i-1} \frac{du_i}{\Gamma(s_i)}.$$

This time we avoid expanding the polynomials, using instead a change of variables $u_i = v_i u_{i+1}$, with $v_0 := 0$ and, for convenience, $u := u_k$, to yield

$$Y(s_1, ..., s_k; q; \lambda)$$
$$= \sum_{n_i \geq 0} \prod_{p=1}^{k} \frac{B_{n_p}}{n_p!} \int_0^\lambda u^{S+N+q-k-1} du \prod_{i=1}^{k} \int_0^1 v_i^{S_i+N_i-i-1} (1 - v_{i-1})^{s_i-1} \frac{dv_i}{\Gamma(s_i)},$$

where we have introduced partial sum nomenclature:

$$N_p = n_1 + ... + n_p, \quad S_p = s_1 + ... + s_p$$

with $N := N_k, S := S_k$. Each v-integral is a beta function expressible in terms of gamma functions, so we immediately obtain the final result, an explicit series for Y,

$$Y(s_1, ..., s_k; q; \lambda)$$
$$(3.2) \qquad = \frac{1}{\Gamma(s_1)} \sum_{n_i \geq 0} \frac{\lambda^{N+S+q-k}}{N + S + q - k} \prod_{p=1}^{k} \frac{B_{n_p}}{n_p!} \prod_{m=1}^{k-1} \frac{\Gamma(N_m + S_m - m)}{\Gamma(N_m + S_{m+1} - m)},$$

where the indices $n_1, ..., n_k$ are restricted to non-negative integers. It turns out that the sum (3.2) for Y always converges absolutely when $\lambda < 2\pi$.

The relations (2.4), (3.1), (3.2) comprise a complete prescription for computation of multiple zeta values. It turns out that considerable reduction in the apparent complexity of the Y, Z sums is possible, as we see in the next section.

FAST EVALUATION OF MULTIPLE ZETA SUMS   1167

4. Summation collapse and Bailey acceleration

On the face of it, the Z sum (3.1) seems to require, for first index n_1 summed over the range $j \leq n_1 < L_1$, at least $O(L_1^j)$ operations; for after all, the sum is over a j-dimensional lattice region. The same basic dimension-dependent complexity would also seem to obtain for the Y sum, (3.2). However, both of these preliminary estimates turn out to be overwhelmingly pessimistic, as we presently explain.

It turns out that the Z sum can be thought of, in a computational sense to be made precise later, as *one*-dimensional. Speaking heuristically, whenever the first index n_1 in (3.1) increments, not much new information is contributed from the other indices. Specifically, the sum, which might originally be implemented as a j-nested loop, can be collapsed into just one loop. Here is a pseudocode construction for the collapsed Z sum. Denote the operator part of (3.1) by

$$f(n, r, \lambda) := (-\partial_n)^{r-1} \frac{e^{-\lambda n}}{n}.$$

Then the summation collapse algorithm starts with initialization of $(j+1)$ variables and proceeds with a singly-nested loop, like so:

```
q := t_1 := t_2 := ... := t_j := 0;
for n_1 := j to L_1
    q := q + 1;
    t_j := t_j + 1/q^{r_j};
    t_{j-1} := t_{j-1} + t_j/(q+1)^{r_{j-1}};
    t_{j-2} := t_{j-2} + t_{j-1}/(q+2)^{r_{j-2}};
    ...
    t_2 := t_2 + t_3/(q+j-2)^{r_2};
    t_1 := t_1 + f(n_1, r_1, λ) * t_2;
endfor
return t_1/Γ(r_1);
```

The final returned value, which approximates Z to order $e^{-\lambda L_1}$, is obtained thus in asymptotically jL_1 loop steps, or $O(L_1)$ arithmetic operations. It is in this precise sense that we may think of the Z sum's complexity as essentially that of a one-dimensional sum.

In a practical implementation it is clearly advantageous to pre-compute all needed values of $f(n, r, \lambda)$ in an initialization phase. To this end it should be pointed out that the multiterm expression resulting from the differentiation indicated in the definition of $f(n, r, \lambda)$ can be generated with a simple recursion. Also, for a fixed r and λ, the expression for f satisfies a recursion in n, so that repeated evaluations of the exponential function are not required. Thus an economical procedure to compute the required f values can be stated as follows. Let r_m denote largest r_k that will be encountered anywhere in the full computation, i.e. the maximum of the original s vector. Then perform the following algorithm:

```
t_0 := e^{-λ}; t_2 := 1;
for k := 1 to L_1
    t_1 := kλ; t_2 := t_0 * t_2;
    for j := 1 to r_m
        t_3 := 1; t_4 := 1;
        for i := 2 to j
            t_4 := t_4 * (j-i+1); t_3 := t_1 * t_3 + t_4;
```

```
        endfor
    f(k, j, λ) := t₂ * t₃ * k⁻ʲ;
    endfor
endfor
```

The collapse of Z sums leaves only the problematic Y sums which do not, unfortunately, enjoy an analogous collapse. But all is not lost, thanks to an acceleration first observed by Bailey [1]. He noticed that certain two-dimensional series implementations in [11] could benefit from convolution algorithms. Let us look, as a canonical example, at the problem of evaluating $\zeta(6, 2)$. In (2.4), and given knowledge of Z sum collapse, there remains only one problematic two-dimensional sum, which can be written in the following suggestive style:

$$(4.1) \quad Y(6, 2; 0; \lambda) = \frac{1}{\Gamma(6)} \sum_{N_2 \geq 0} \frac{\lambda^{N_2+6}}{N_2+6} \sum_{0 \leq N_1 \leq N_2} \frac{B_{N_1} B_{N_2-N_1}}{N_1!(N_2-N_1)!} \frac{1}{(N_1+5)(N_1+6)}.$$

We expect to perform the outer summation over a constrained range: $0 \leq N_2 \leq L_2$, to obtain an error of order $(\frac{\lambda}{2\pi})^{L_2}$. We observe that the inner sum in (4.1) is in fact the N_2-th element of an acyclic convolution. The convolution procedure is: first, find the acyclic convolution (equivalent to straightforward polynomial multiplication) of the two sequences (of coefficients):

$$x_1 := \{\frac{B_n}{n!(n+5)(n+6)}; n = 0, 1, 2, ..., L_2\},$$

$$x_2 := \{\frac{B_n}{n!}; n = 0, 1, 2, ..., L_2\}.$$

Then the Y value in (4.1) can be obtained simply from the elements of the acyclic convolution of x_1, x_2 (which elements are also the coefficients of the relevant product polynomial).

In a practical implementation it is important to efficiently compute the Bernoulli terms $B_0/0!$ through $B_{L_2}/L_2!$, even if this be done only once in a global initialization phase. Fortunately, it turns out that these Bernoulli terms can be obtained in a "parallel" fashion via a method pioneered by J. P. Buhler; namely, Newton method inversion applied to the formal generating function

$$\frac{z}{e^z - 1} = \sum_{k=0}^{\infty} \frac{B_k}{k!} z^k.$$

One uses the series expansion of $(e^z - 1)$ through a desired degree L_2 in z, then inverts the resulting polynomial, the procedure requiring $O(L_2 \log L_2)$ arithmetic operations ([10], [2]). Since the convolution inherent in (4.1) can be performed, via FFT methods, say, also in $O(L_2 \log L_2)$ arithmetic operations, it follows from all the considerations above on Y, Z that $\zeta(6, 2)$ can be obtained to D good decimal digits in $O(L_1) + O(L_2 \log L_2)$ arithmetic operations, where each of the outer summation limits L_1, L_2 has been chosen such that the respective sums are both good to the D digits.

Happily, for dimensions $d > 2$, the Y sum is still susceptible to a convolution approach. Write the double product in (3.2) abstractly, as

$$\frac{B_{N_1}}{N_1!} \frac{B_{N_2-N_1}}{(N_2-N_1)!} \cdots \frac{B_{N_k-N_{k-1}}}{(N_k-N_{k-1})!} g_1(N_1)...g_{k-1}(N_{k-1}).$$

Now observe that the sum over N_1 in the range $0 \leq N_1 \leq N_2$, which sum involving only the first two Bernoulli terms and the g_1 term, is but a certain convolution evaluated at the N_2-th element, as before. But this means that the sum over $0 \leq N_2 \leq N_3$ can then be performed via convolution of one sequence indexed by N_2 and involving g_2, with another sequence indexed by $N_3 - N_2$, and so on.

A pseudocode realization of the entire convolution procedure can be formulated as follows. Choose the outer summation limit L_2 to be one less than a power of 2 (a convenience, so that convolution lengths are themselves powers of 2). Now define sequences, each of length $L_2 + 1$, as follows:

$$B := \{\frac{B_n}{n!}; n = 0, 1, 2, ..., L_2\},$$

and for $m = 1, ..., k-1$,

$$G_m := \{\frac{\Gamma(n + S_m - m)}{\Gamma(n + S_{m+1} - m)}; n = 0, 1, 2, ..., L_2\}.$$

Define further, for two sequences X, Y each of length L, their "half-cyclic" convolution as the first L elements of the full acyclic convolution, and denote this half-cyclic by $X \times_H Y$. This half-cyclic can be obtained via these steps: zero-pad each of X, Y to length twice L, perform the cyclic convolution of the padded sequences (via standard FFT methods, say), and finally, take the first L elements of the result. Further, we denote by $X * Y$ the dyadic (elementwise) product sequence of any two equal-length sequences X, Y. Now initialize a sequence X and carry out a loop, as follows:

$X := B;$
for $m := 1$ to $k-1$
$\quad X := X * G_m;$
$\quad X := X \times_H B;$
endfor

Now the Y value from the defining series (3.2) is obtained simply as a weighted sum over elements $X[N]$ of the final half-cyclic, as

$$Y = \frac{1}{\Gamma(s_1)} \sum_{N=0}^{L_2} \frac{\lambda^{N+S+q-k}}{N+S+q-k} X[N].$$

The convolution procedure admits of interesting potential enhancements. First, if FFT method is used, the transform of the (zero-padded) B sequence may be done just once, and used repeatedly for the half-cyclic within the loop. Second, since every odd-indexed Bernoulli number vanishes except $B_1 = -1/2$, the B sequence is sparse. One may, in principle, algebraically subtract out all dependence on B_1 from (3.2), whence convolutions lengths may be effectively halved. Third, half-cyclic convolution can be done via cyclic and negacyclic convolution each of half length (*sans* zero-padding), and even done in a special, recursive fashion for lengths not a power of 2 [10]. Perhaps the most intriguing possible enhancement is to somehow invoke the negacyclic option to effect the whole pseudocode loop via one long dyadic product of various initial transforms. In other words, there may exist some means to generate "$(k-1)$-fold half-cyclic convolutions" in such a maximally efficient manner.

We have thus realized the considerable advantage of the Bailey acceleration: if proper precision is maintained for each convolution, we see that the total operation cost is still that of the $(k = 1)$-dimensional case; namely $O(L_2 \log L_2)$.

5. COMPLEXITY

It follows from all the above complexity estimates that, if the respective outer summation limits L_1, L_2 be equal, then the multiple zeta with pure-integer argument vector $\vec{s}$ can be resolved to D digits in $O(D \log D)$ arithmetic operations. But we can optimize a little further. Since the complexity for the Z sum is slightly lower (i.e. linear in L_1), the optimal parameter λ can be slightly lowered (with $L_1 \sim L_2 \log D$) to yield an overall complexity of $O(D \log D / \log \log D)$. The problem of what should be the overall big-O constant is interesting. Perhaps it is reasonable to conjecture that the constant can be taken to be some simple function only of the dimension d and the "weight" $\sum_{i=1}^{d} s_i$. (In this regard, note Bailey's empirical discovery of quadratic growth (in dimension d) of this constant for a particular class of zetas, as we discuss in the next section.) Incidentally, many existing schemes for evaluation of, say, an arbitrary Riemann $\zeta(n), n \in Z$, require $O(D \log D)$ operations or slightly better than this [10]; although there are elegant, Apery-like formulae involving $O(D)$ operations and small implied big-O constants [6]; and a complicated but asymptotically efficient scheme involving only $O(M(D) \log^2 D)$ *bit* operations [12], where $M(D)$ is the bit-complexity of multipication of two D-digit numbers. It may also be possible, for our present multiple zeta algorithms or the cited Riemann zeta ones, to effect a hypergeometric acceleration, as in [5] for further reduction in operation complexity, to say $O(D^a)$ for some $a < 1$. In any event it is clear that the expedient of summation collapse combined with Bailey acceleration is efficient indeed: our d-dimensional sums reduce, up to a big-O constant, to our own $(d = 1)$-dimensional case. If an implementation be sufficiently adroit, the penalty for higher dimensions should be correspondingly light.

6. EXPERIMENT AND PRACTICAL CONSIDERATIONS

Even if the acceleration of section four is implemented, the computation of the Y function is generally more expensive than the Z function. If the acceleration is not employed, the disparity between these two costs is even more pronounced. Fortunately, these costs can be brought more into balance, with an overall savings in computation time, by adjusting the value of λ. Smaller values of λ tend to lower the cost of the Y function, but increase the cost of the Z function. To be precise, the two limits L_1 (the limit for the Z function) and L_2 (the limit for the Y function) are given by

$$L_1 = \frac{-\log \epsilon}{\lambda}, \qquad L_2 = \frac{\log \epsilon}{\log(\lambda/(2\pi))},$$

where ϵ is the desired accuracy of the result. Using these formulae, an optimal value of λ can be determined.

The author employed this asymmetrical technique to evaluate the four-dimensional sum $\zeta(3, 1, 3, 1)$, using $\lambda = 1/20$, $L_1 = 12000$ and $L_2 = 120$, so that the collapsed loop for Z is about as fast as the straightforward (i.e. without acceleration) evaluation of Y using (3.2). The numerical result agrees with Zagier's conjectured $\zeta(3, 1, 3, 1) = 2\pi^8/10!$ to more than 250 decimal digits.

At the time of this writing, the present author has been informed by David H. Bailey that the latter has succeeded implementing the complete procedure described herein, including the convolution acceleration for the Y function. Bailey employed $\lambda = 5/32 = 0.15625$, $L_1 = 12000$ and $L_2 = 512$, which yields more than 800 decimal digit accuracy in the result. The resulting program verified $\zeta(3,1,3,1) = 2\pi^8/10!$, to 800 decimal place accuracy, in only four minutes run time on an IBM RS6000/590 workstation.

Bailey has further verified the Zagier conjecture, as it is stated in our introduction, for $1 \leq n \leq 10$, again to 800 decimal place accuracy for each number n of argument pairs $\{3,1\}$. The entire run over the stated range of n required just 4.3 hours on an IBM RS6000/590 workstation. Bailey found that the CPU time required for each instance n (not including global initialization, such as computing the Bernoulli coefficients used for all dimensions), accurately satisfies an empirical quadratic formula in n. These results suggest that evaluations of multiple zetas of very high dimensions (perhaps dimensions into the hundreds) can be feasibly computed using the techniques described herein.

There are many useful, sharp tests of any numerical algorithm for mutiple zetas. The compendium [7] contains many fascinating exact evaluations, as well as various well-motivated conjectures. One way to test higher-dimensional implementations is to exploit the complete reducibility of any multiple zeta of the form $\zeta(s,s,...,s)$; i.e. with all argument components equal. Denote such a ζ having exactly N appearances of s by $\zeta(\{s\}_N)$. Then, via a combinatorical argument, one may establish the recursion

$$\zeta(\{s\}_N) = \frac{1}{N}\sum_{j=1}^{N}(-1)^{j-1}\zeta(js)\zeta(\{s\}_{N-j}),$$

where we also define the empty case $\zeta(\phi) := 1$, revealing immediately that any $\zeta(\{s\}_N)$ can always be reduced to "one-dimensional" (Riemann zeta) sums. The exact evaluation of $\zeta(6,6,6,6)$ exhibited in the Introduction may be effected in this way. Another rigorous test for any implementation is the sum rule [7]:

$$\sum_{n_j > \delta_{j,1}; \sum n_j = N} \zeta(n_1,...,n_d) = \zeta(N).$$

Finally, we address briefly the scenario in which at least one of the argument components is not an integer. It turns out that the Z integral defined by (2.6) can always be cast as a sum of the type (3.1), except that for r_1 non-integral we use the correspondence

$$(-\partial_{n_1})^{r_1-1}\frac{e^{-\lambda n_1}}{n_1} \leftrightarrow \frac{\Gamma(r_1, \lambda n_1)}{n_1^{r_1}},$$

where Γ here is the incomplete gamma function. The Y sum (3.2) is, as it stands, a correct representation of the defining form (2.5) for any set of real $s_1,...,s_k$, as long as the sum converges. Therefore, the only thing preventing a complete series development for arbitrary real argument components is that the factored representation (2.4) involves explicit summation indices $s_{k+1} - 1$ which must be integers. Thus more work must be put into the expansion of the integrand of (2.3) in order to handle arbitrary real argument vector components. However, as s_1 in particular does not appear as a summation index, we can certainly evaluate, via the tools already in hand, such *a priori* poorly convergent entities as

$$\zeta(\frac{3}{2},1)$$
$$= 4.68181441155622703569221027933719959407392140996102081308586936...,$$

$$\zeta(\frac{3}{2},1,1)$$
$$= 9.12448455911887955642845079650575695833536290219085798530250 3...,$$

$$\zeta(\frac{3}{2},1,1,1)$$
$$= 18.1147001562818106030733870815411024229762662312206606418 5...,$$

each obtained, via the aforementioned asymmetrical series approach with $L_1 = 2L_2, 20L_2,$ and $50L_2$ respectively, in a few minutes on a typical workstation.

ACKNOWLEDGMENTS

The author is grateful to V. Adamchik, D. Bailey, D. Bradley, J. Buhler, J. Borwein, and D. Broadhurst for their theoretical and computational ideas pertaining to this work.

REFERENCES

1. D. Bailey, private communication, 1994.
2. D. Bailey, J. Borwein, and R. Crandall, "On the Khintchine Constant," *Math. Comp.*, **66** (1997), 417–431. MR **97c**:11119
3. D. Bailey, J. Borwein and R. Girgensohn, "Experimental evaluation of Euler sums," *Experimental Mathematics* **3**, (1994), 17–30. MR **96e**:11168
4. D. Borwein, J. Borwein and R. Girgensohn, "Explicit evaluation of Euler sums," *Proc. Edinburgh Math. Soc.* **38**, (1995), 277–294. MR **96f**:11106
5. J. M. Borwein and P. B. Borwein, *Pi and the AGM*, John Wiley & Sons, 1987. MR **89a**:11134
6. J. Borwein and D. Bradley, "Empirically Determined Apery-like Formulae for $\zeta(4n+3)$," manuscript, 1996.
7. J. M. Borwein, D. M. Bradley and D. J. Broadhurst, "Evaluations of k-fold Euler/Zagier sums: a compendium of results for arbitrary k," manuscript, 1996.
8. J. M. Borwein and R. Girgensohn, "Evaluation of Triple Euler Sums," manuscript, 1995.
9. D. J. Broadhurst, R. Delbourgo and D. Kreimer, *Phys. Lett.* **B366** (1996), 421.
10. R. E. Crandall, *Topics in Advanced Scientific Computation*, TELOS/Springer-Verlag, New York, 1996. MR **97g**:65005
11. R. E. Crandall and J. P. Buhler, "On the Evaluation of Euler Sums," *Experimental Mathematics* **3**, (1994), 275-285. MR **96e**:11113
12. E. A. Karatsuba, "Fast Calculation of the Riemann Zeta Function $\zeta(s)$ for Integer Values of the Argument s," *Probs. Information Trans.* **31** (1995), 353-355. MR **96k**:11155
13. C. Markett, "Triple Sums and the Riemann Zeta Function," *J. Number Theory*, 48 (1994), 113-132. MR **95f**:11067
14. D. Zagier, "Values of zeta functions and their applications," preprint, Max-Planck Institut, 1994.

CENTER FOR ADVANCED COMPUTATION, REED COLLEGE, PORTLAND, OREGON 97202
E-mail address: crandall@reed.edu

2. On the Khintchine Constant

Discussion

I still remember where I was sitting—in a number-theory class given by J. Roberts in 1969—and the way the sunbeams poured into the room from a (too rare) Portland, Oregon sun, when I learned that the integer elements a_k of a typical continued fraction

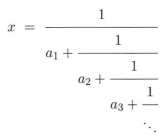

$$x = \cfrac{1}{a_1 + \cfrac{1}{a_2 + \cfrac{1}{a_3 + \cfrac{1}{\ddots}}}}$$

have the astounding property that for almost all $x \in (0,1)$ the geometric mean

$$K := \lim_{n \to \infty} (a_1 a_2 a_3 \cdots a_n)^{1/n}$$

exists as a limit, and moreover—this is the focus of my remembrance—said limit is almost always the *same constant*, called the Khintchine constant K_0. The constant is about 2.685, which right away suggests that typical real x have "lots of 1s, 2s, and 3s" in the element list.

Challenges

In the paper, various "artificially constructed" continued fractions are exhibited that either provably have or are conjectured to have the "correct" mean K_0. A challenge is to find a fundamental number whose mean is *provably* K_0. As with normal numbers, no non-artificial examples are known. Then there is the reflexive dilemma: Does K_0 itself have geometric mean K_0? (It is not even known whether K_0 is algebraic or transcendental.)

Source

Bailey, D., Borwein, J. and Crandall, R.E., "On the Khintchine Constant," *Math. Comp.*, **66**, 417–431 (1997).

On the Khintchine Constant

David H. Bailey[*]

Jonathan M. Borwein[†]

Richard E. Crandall[‡]

Abstract

We present rapidly converging series for the Khintchine constant and for general "Khintchine means" of continued fractions. We show that each of these constants can be cast in terms of an efficient free-parameter series, each of which involves only values of the Riemann zeta function, rationals, and logarithms of rationals. We provide an alternative, polylogarithm series for the Khintchine constant and indicate means to accelerate such series. We discuss properties of some explicit continued fractions, constructing specific fractions that have limiting geometric mean equal to the Khintchine constant. We report numerical evaluations of such special numbers and of various Khintchine means. In particular, we used an optimized series and a collection of fast algorithms to evaluate the Khintchine constant to more than 7000 decimal places.

Keywords and phrases. Khintchine constant, continued fractions, geometric mean, harmonic mean, computational number theory, zeta functions, polylogarithms.

1991 Mathematics subject classification. Primary 11Y60, 11Y65; Secondary 11M99.

[*]NASA Ames Research Center, Mail Stop T27A-1, Moffett Field, CA 94035-1000, USA. dbailey@nas.nasa.gov

[†]Centre for Experimental and Constructive Mathematics, Simon Fraser University, Burnaby, BC V5A 1S6, Canada. jborwein@cecm.sfu.ca. (Research supported by the Shrum Endowment at Simon Fraser University and NSERC.)

[‡]Center for Advanced Computation, Reed College, Portland, OR 97202. crandall@reed.edu.

2 BAILEY, BORWEIN AND CRANDALL

1. INTRODUCTION

The Khintchine constant arises in the measure theory of continued fractions. Every positive irrational number can be written uniquely as a simple continued fraction $[a_0; a_1, a_2, \cdots, a_n, \cdots]$, i.e., with a_0 a nonnegative integer and all other a_i positive integers. The *Gauss-Kuz'min distribution* [13] predicts that the density of occurrence of some chosen positive integer k in the fraction of a random real number is given by

$$\mathrm{Prob}(a_n = k) = -\log_2 \left[1 - \frac{1}{(k+1)^2} \right].$$

In his celebrated text, Khintchine [13] uses the Gauss-Kuz'min distribution to show that for almost all positive irrationals the limiting geometric mean of the positive elements a_i of the relevant continued fraction exists and equals

$$K_0 := \prod_{k=1}^{\infty} \left[1 + \frac{1}{k(k+2)} \right]^{\log_2 k} = \prod_{k=1}^{\infty} k^{\log_2 \left[1 + \frac{1}{k(k+2)} \right]}.$$

The fundamental constant K_0 is the *Khintchine constant*. Ever since Khintchine's elegant discovery there has been a keen interest in the numerical evaluation of K_0 [14, 20, 26, 27, 24]. It is known that this constant can be cast in terms of various converging series, the following example of which having been used by Shanks and Wrench to provide the first high-precision numerical values for K_0:

$$\log(K_0)\log(2) = \sum_{s=1}^{\infty} \frac{\zeta(2s) - 1}{s} \left(1 - \frac{1}{2} + \frac{1}{3} - \cdots + \frac{1}{2s - 1} \right). \qquad (1)$$

This series can be rendered even more computationally efficient via the introduction of a free integer parameter. We used a carefully optimized free-parameter series to resolve K_0 to 7350 decimal places ($K_0 = 2.68545200106\ldots$ see §5).

The Khintchine constant can be thought of as a member of a certain class of constants we shall call *Khintchine means* K_p, for real numbers $p < 1$. The *Hölder mean* of order p of the continued fraction elements, namely $\lim_k [(a_1^p + a_2^p + \cdots + a_k^p)/k]^{1/p}$, also exists with probability one and again with probability one equals the constant:

$$K_p := \left\{ \sum_{k=1}^{\infty} -k^p \log_2 \left[1 - \frac{1}{(k+1)^2} \right] \right\}^{1/p}.$$

(See the final section of Khintchine's book [13] for a proof for $p < \frac{1}{2}$, or more modern references on ergodic theory for a proof for $p < 1$ [19].) We may interpret K_0 as the limiting instance of the K_p definition as $p \to 0$. We shall show in

Theorem 6 below that for any negative integer p the Khintchine mean of order p satisfies an identity

$$(K_p)^p \log(2) = \sum_{s=2}^{\infty} (\zeta(s-p) - 1) Q_{sp}, \tag{2}$$

where each coefficient Q_{sp} is rational. Again, there is a free-parameter generalization, which we employed to resolve the *harmonic mean* K_{-1} also to over 7000 decimal places ($K_{-1} = 1.74540566240\ldots$ see §5). It is of interest that, evidently, only K_0 can be written as a series involving exclusively even zeta arguments. The computational implications of this unique property of K_0 are discussed in §5. We should mention that aside from the series (1) for K_0 there are other previously known formulae for Khintchine means, some of which involve derivatives of the zeta function [24].

In the next section we establish the series forms (1) and (2) for K_0 and the general Khintchine means K_p, respectively. Actually, (1) and (2) can be thought of as degenerate cases of free-parameter forms in which an integer parameter can be optimized for numerical efficiency. Then in §3 we present polylogarithm series and a certain zeta-like function whose evaluations can be used to accelerate the polylogarithm series. In §4 we discuss explicit continued fractions with a view toward determining whether Hölder means exist and coincide with Khintchine means. In particular, some numbers known to have geometric mean (zeroth Hölder mean) equal to K_0 are presented. Finally, in §5 we discuss numerical details relevant to very-high-precision evaluation of Khintchine means.

2. Fundamental Identities

This section is devoted to presenting the basic identities. We begin with a list of preliminary, largely elementary, results needed in the paper. The first lemma amounts to a set of observations due to Wrench and Shanks [27].

Lemma 1. (a) *We have*

$$-\log(1-x)\log(1+x) = \sum_{k=1}^{\infty} \frac{A_k}{k} x^{2k},$$

where $A_s := \sum_{m=1}^{2s-1} (-1)^{m-1}/m$.
(b) *Further,*

$$\sum_{k=2}^{N} \log(1 - \frac{1}{k}) \log(1 + \frac{1}{k}) - \sum_{k=2}^{N} \log(k-1) \log(1 - \frac{1}{k^2}) = -\log(N) \log(1 + \frac{1}{N}).$$

(c) *Thus,*

$$\sum_{k=2}^{\infty} \log(1 - \frac{1}{k}) \log(1 + \frac{1}{k}) = -\log(K_0) \log(2).$$

4   BAILEY, BORWEIN AND CRANDALL

Proof. Part (a) is most easily seen by differentiating both sides. The left–hand side becomes $f(x) - f(-x)$ where $f(x) := \log(1+x)/(1-x)$. Using the standard relationship

$$\sum_{k=1}^{\infty} \frac{a_k}{1-x} x^k = \sum_{k=1}^{\infty} \{\sum_{j=1}^{k} a_j\} x^k$$

produces (a).

Part(b) is easily established inductively after expanding the left–hand side.

Part (c) follows on taking limits and noting that

$$-\sum_{k=2}^{N} \log(k-1) \log(1 - \frac{1}{k^2}) = \log(K_0) \log(2),$$

as follows from the definition of K_0. $\square$

We shall find it convenient to use the *Hurwitz zeta function* defined by

$$\zeta(s, N) := \sum_{n=1}^{\infty} \frac{1}{(n+N)^s},$$

so that $\zeta(s) = \zeta(s,0)$ and so that for N a nonnegative integer

$$\zeta(s, N) = \zeta(s) - \sum_{n=1}^{N} \frac{1}{n^s}.$$

With this notation we have:

Lemma 2. (a) *For N a positive integer,*

$$\sum_{n=2}^{\infty} \zeta(n, N) = \frac{1}{N}.$$

(b) *For N a positive integer,*

$$\sum_{n=1}^{\infty} \frac{\zeta(2n, N)}{n} = \log(\frac{N+1}{N}).$$

(c) *The integral*

$$\int_0^1 \frac{\log(1-t^2)}{t(1+t)} dt = -\log^2(2).$$

Proof. The proofs of the first two identities are similar and rely on expanding the zeta terms, rearranging the order of summation and re-evaluating. In both cases, the result telescopes to the desired conclusion.

Part (c) is less immediate. Actually, the indefinite integral is evaluable with the aid of the theory of the dilogarithm [15]. The integral

$$\int_0^t \frac{\log(1-x^2)}{x(1+x)}\,dx$$

equals the log terms

$$-\frac{\log^2(1+t)}{2} - \log^2(2) + \log(2)\log(1-t) - \log(1+t)\log(1-t) + \log(t)\log(1-t)$$

plus the dilog terms

$$\mathrm{dilog}(t) - \mathrm{dilog}(1+t) - \mathrm{dilog}(\frac{1+t}{2}),$$

and the limit as $t \to 1$ yields the desired result $-\log^2(2)$. $\square$

We are now in a position to establish a general Shanks-Wrench identity [20] for K_0.

Theorem 3. *For any positive integer N,*

$$\log(K_0)\log(2) = \sum_{s=1}^{\infty} \zeta(2s, N)\frac{A_s}{s} - \sum_{k=2}^{N} \log(1-\frac{1}{k})\log(1+\frac{1}{k}), \qquad (3)$$

where $A_s := \sum_{m=1}^{2s-1}(-1)^{m-1}/m$.

Remark. The integer N is a free parameter that can be optimized in actual computations to significantly reduce the number of zeta evaluations required. Variation of this parameter also provides a kind of error check, for whatever the choice of positive integer N, one expects the same result for the left-hand side. Note that in the case $N = 1$ the second summation is empty, and we recover precisely the K_0 identity (1) of §1.

Proof. Let $f(N)$ denote the right-hand side of (1). Then

$$f(N-1) - f(N) = \sum_{s=1}^{\infty} \frac{A_s}{s}N^{-2s} + \log(1-\frac{1}{N})\log(1+\frac{1}{N})$$

which equals zero by Lemma 1(a). Thus, since $\zeta(2s, N) \to 0$ sufficiently rapidly,

$$f(1) = f(N) = \lim_{N\to\infty} f(N)$$

$$= -\sum_{k=2}^{\infty} \log(1-\frac{1}{k})\log(1+\frac{1}{k}).$$

By Lemma 1(c), this sum agrees with $\log(K_0)\log(2)$. $\square$

As a companion relation to the identity of Theorem 3, we can establish an elegant integral representation for the left-hand side. There is a powerful generalization of Lemma 2(b) in the form of a generating function based on Euler's product for $\sin(\pi t)/(\pi t)$ (see [23, p. 249]). For real t in $[0,1)$ define $g(t)$ by

$$g(t) := \sum_{s=1}^{\infty} \frac{\zeta(2s)-1}{s} t^{2s} = -\log\left(\frac{\sin(\pi t)}{\pi t}\right) + \log(1-t^2), \qquad (4)$$

and define also the limiting case $g(1) := \log 2$. We only need observe now that, on the basis of Theorem 3 with parameter $N = 1$,

$$\log(K_0)\log(2) = \int_0^1 \frac{\log(2) + g(t)/t}{1+t} dt,$$

and with the help of the previous dilogarithm integral evaluation we thus arrive at an integral representation. (Reference [20] contains an equivalent integral identity.)

Corollary 4. *The following integral representation holds for K_0:*

$$\int_0^1 \frac{\log[\sin(\pi t)/(\pi t)]}{t(1+t)} dt = -\log(K_0)\log(2).$$

It is amusing to observe that Lemma 1(c) may also be turned into an analogous integral form:

$$\log(K_0)\log(2) = \int_1^{\infty} \frac{\log(\lfloor t \rfloor)}{t(1+t)} dt = \int_0^1 \frac{\log(\lfloor 1/t \rfloor)}{1+t} dt.$$

This was observed from a very different starting point by Robert Corless [11] but follows immediately on breaking the first integral up at integer points.

We now derive new, corresponding identities for the higher–order Khintchine means. They are in some sense simpler, since one logarithmic term is replaced by a negative integral power. There is an observation that leads directly to a zeta function expansion for these general Khintchine means. Note that a sum of terms $k^p \log(1-(k+1)^{-2})$ can be expressed, via expansion of the logarithm, in terms of sums of the form (note p is assumed to be a negative integer):

$$\sum_{n=2}^{\infty} \frac{1}{n^{2s-p}(1-1/n)^{-p}}.$$

Upon expansion of the term

$$1/(1-1/n)^{-p}$$

ON THE KHINTCHINE CONSTANT   7

in powers of $1/n$, we obtain an identity for the pth power of K_p as a series of zeta functions. The result, after the same manner of free-parameter manipulation we used for K_0, is a new series that can be thought of as a companion identity to the Shanks-Wrench expansion of Theorem 3.

Theorem 6. *For negative integer p and positive integer N we have*

$$K_p^p \log(2) = \sum_{n=1}^{\infty} \frac{\sum_{j=0}^{\infty} \binom{j-p-1}{-p-1} \zeta(2n+j-p, N)}{n} - \sum_{k=2}^{N} \log(1 - \frac{1}{k^2})(k-1)^p.$$

Remark. Note that for $N = 1$ the final sum is empty, the coefficient of any given $\zeta(s)$ is an easily computed rational, and we immediately establish a general series with rational coefficients, (2) of §1.

Corollary 7. *The constant K_{-1} satisfies, for any integer $N > 0$,*

$$\frac{\log(2)}{K_{-1}} = \sum_{n=1}^{\infty} \frac{\frac{1}{N} - \sum_{k=2}^{2n} \zeta(k, N)}{n} - \sum_{k=2}^{N} \frac{\log(1 - k^{-2})}{k-1}.$$

Proof. It suffices to show that for every positive integer n

$$\sum_{j=0}^{\infty} \zeta(2n+j+1, N) + \sum_{k=2}^{2n} \zeta(k, N) = \frac{1}{N}.$$

This follows immediately from Lemma 2(a). □

3. POLYLOGARITHM SERIES

There exist some interesting identities for the Khintchine constant in terms of polylogarithm evaluations. One particularly interesting polylogarithm identity is obtained by resolving the integral representation of Corollary 4 in polylogarithm terms [28]. One may employ the Euler product for $\sin z/z$ to write the integral as a sum of logarithmic integrals, each in turn expressible in terms of polylogarithms. This procedure leads to the series

$$\log(K_0) \log(2) = \log^2(2) + \mathrm{Li}_2(-\frac{1}{2}) + \frac{1}{2} \sum_{n=2}^{\infty} (-1)^n \mathrm{Li}_2(\frac{4}{n^2}),$$

where, $\mathrm{Li}_m(z)$ is the *polylogarithm* function: $\mathrm{Li}_m(z) := \sum_{k=1}^{\infty} z^k k^{-m}$.

A somewhat different application of polylogarithms is to use the classical Abel identity [15]

$$\log(1-x) \log(1-y) = \mathrm{Li}_2(\frac{x}{1-y}) + \mathrm{Li}_2(\frac{y}{1-x}) - \mathrm{Li}_2(x) - \mathrm{Li}_2(y) - \mathrm{Li}_2(\frac{xy}{(1-x)(1-y)})$$

8   BAILEY, BORWEIN AND CRANDALL

together with Lemma 1(c), setting $x := 1/n, y := -1/n$ to obtain

$$\log(K_0)\log(2) = \frac{\pi^2}{6} - \frac{1}{2}\log^2(2) + \sum_{n=2}^{\infty}\mathrm{Li}_2\left(\frac{-1}{n^2-1}\right).$$

An interesting line of analysis starting from this last polylogarithm series is to "peel off" parts of the Li_2 function, casting the corrections in closed form. Such a procedure gives polylogarithm-based analogues of Theorem 3. For example, one can replace the last Li_2 summand above with a more rapidly decaying term

$$\mathrm{Li}_2\left(\frac{-1}{n^2-1}\right) - \frac{-1}{n^2-1} - \frac{1}{4}\frac{1}{(n^2-1)^2}$$

and add back a correction

$$-\Omega(1) + \frac{1}{4}\Omega(2) = \frac{\pi^2}{48} - \frac{59}{64},$$

where Ω is the zeta-like function

$$\Omega(m) := \sum_{n=2}^{\infty}\frac{1}{(n^2-1)^m}.$$

Careful Eulerian partial fraction decomposition (as detailed in [7]) can be used to produce a closed-form evaluation of $\Omega(m)$ for any positive integer m. In this way one may accelerate the convergence of relevant polylogarithm sums.

4. EXPLICIT CONTINUED FRACTIONS

It is remarkable that, even though a random fraction's limiting geometric mean exists and furthermore equals the Khintchine constant with probability one, not a single explicit real number (e.g., a real number cast in terms of fundamental constants) has been demonstrated to have elements whose geometric mean equals K_0. Likewise, for any negative integer p, not a single explicit real number has been shown to have elements whose Hölder mean equals K_p. In any event, it is worthwhile to mention some classical continued fractions with respect to this theoretical impasse.

The continued fraction for e is

$$e = [2; 1, 2, 1, 1, 4, 1, 1, 6, 1, 1, 8, 1, 1, 10, 1, 1, 12, \ldots].$$

The elements are eventually comprised of a meshing of two arithmetic progressions, one of which has zero common difference while the other has difference two and diverges. Thus the meshing has diverging geometric mean. Thus, e does not possess geometric mean K_0. The harmonic mean for e does exist, but equals $3/2$, which is not K_{-1}. It turns out that any fraction with elements lying in a single arithmetic progression can be evaluated in terms of special functions. Explicitly, for any positive integers a, d we have [2, eq. 9.1.73]

$$[a; a + d, a + 2d, a + 3d, ...] = \frac{I_{a/d-1}(\frac{2}{d})}{I_{a/d}(\frac{2}{d})},$$

where I_ν is the *modified Bessel function* of order ν. These arithmetic progression fractions are certainly interesting, and not beyond deep analysis. It was known, for example, to C. L. Siegel that these fractions are transcendental [22]. But each such fraction has diverging geometric mean and indeed diverging Hölder means. Note that the means are monotone nondecreasing in p, and so a fraction with lim inf of its elements infinite has infinite means.

Another example of interest is π, whose continued fraction expansion is

$$\pi = [3; 7, 15, 1, 292, 1, 1, 1, 2, 1, 3, \ldots].$$

The continued fraction elements do not appear to follow any pattern and are widely suspected to be in some sense random. Based on the first 17,001,303 continued fraction elements, the geometric mean (of the fraction elements yielding the same precision) is 2.686393 and the harmonic mean is 1.745882 [12]. These values are reasonably close to K_0 and K_{-1}, but of course no conclusion can be drawn beyond this.

It is a well-known theorem of Lagrange that the elements of a simple continued fraction form an eventually periodic sequence if and only if the fraction is an irrational quadratic surd. All Hölder means for $p = 0, -1, -2, \ldots$ then exist, and are completely determined by one period of elements. Hence, each Hölder mean of a quadratic surd is an algebraic number. Clearly, for any algebraic number $c = a^{1/b}$ formed from integers a, b, one can write down a quadratic surd having geometric mean c. Along these lines, it is not hard to show that if there exists an integer $m > 2$ such that

$$\frac{\log(K_0/m)}{\log(2/m)}$$

is rational, then there exists a quadratic surd with geometric mean K_0. Thus the issue of transcendence for K_0 and related numbers is an interesting one, and one we return to in the next section.

Even though no explicit real number is known to have elements whose geometric mean is K_0, it is still possible to fabricate explicit lists of *elements* whose geometric mean does equal K_0. If one were in possession of some representation of K_0 to arbitrary accuracy, one could of course construct a fraction having geometric mean K_0 by appending a "2" (respectively, "3") to the element list whenever the current geometric mean were above (below) K_0. There seems to be no way to determine a *priori* the value of, say, the nth element. Thus, such a constructed fraction is not explicit.

But it is possible to give an explicit list of elements having the desired property. One successful construction has been given by [1], as follows. Consider the naturally ordered rationals in (0,1); that is, consider

$$1/2, 1/3, 2/3, 1/4, 2/4, 3/4, 1/5, 2/5, 3/5, 4/5, \ldots,$$

where for each successive denominator $d = 2, 3, 4, \ldots$ we employ in increasing order all numerators between 1 and $d - 1$ inclusive. Now consider the (finite) set of fraction elements for each rational in the list. (We also demand the caveat that no such terminating fraction is allowed to end with element 1, so for example 2/3 is the fraction $[1, 2]$ rather than $[1, 1, 1]$.) If we concatenate the elements from all the terminating fractions, the infinite chain of elements has limiting geometric mean equal to K_0. The resulting sequence of elements starts out:

$$A = [2; 3, 1, 2, 4, 2, 1, 3, 5, 2, 2, 1, 1, 2, 1, 4, 6, 3, 2, 1, 2, 1, 5, 7, 3, 2, \ldots].$$

The geometric mean of the first 15,000 elements of A is 2.35821 ..., which appears low but note that as the denominator d increases during construction of the elements, larger and larger elements (such as d itself) appear.

But one may construct elements whose geometric mean converges much more rapidly to K_0. One such construction is based on a deterministic stochastic sampling of the Gauss-Kuz'min density, and proceeds as follows. First, for non-negative integer n define the *van der Corput discrepancy sequence* [16] to be a set of the base-2 numbers

$$d(n) = 0.b_0 b_1 b_2 \ldots,$$

where the b_i are the binary bits of n, with b_0 being least significant. As n runs through positive integers, the sequence of $d(n)$ is confined to (0,1) and has appealing pseudorandom properties. The construction of the number we shall call Z_2 then starts with $a_0 := 0$, and loops as follows:

For $n = 1$ to ∞, set $a_n := \lfloor 1/(2^{d(n)} - 1) \rfloor$.

The continued fraction elements a_n thus determined start out

$$Z_2 = [0; 2, 5, 1, 11, 1, 3, 1, 22, 2, 4, 1, 7, 1, 2, 1, 45, 2, 4, 1, 8, 1, 3, 1, 14, 1, \ldots].$$

On the idea that the discrepancy sequence is in a certain sense equidistributed, we are moved to posit:

Conjecture. *The geometric mean of the number Z_2 is in fact the Khintchine constant K_0. Furthermore, every limiting p-th Hölder mean of Z_2 for $p = -1, -2, \ldots$ is the respective Khintchine mean K_p.*

With regard to the above conjecture, S. Plouffe [18] has reported a computation of the geometric and harmonic means through 5206016 continued fraction elements of Z_2. His results are 2.6854823207 and 1.7454074435, respectively, which are remarkably close to the expected theoretical values. The authors have also been informed by T. Wieting that he has an unpublished proof of the basic conjecture, i.e., that the limiting Hölder means of Z_2 exist for $p = 0, -1, -2, \ldots$ and furthermore equal the corresponding Khintchine means [25]. In the wake of such positive results, one can ask in addition for, say, the rate of convergence to K_0. Again on the basis of the distribution of discrepancy values we think it reasonable to conjecture that the geometric mean G_n of the first n elements of Z_2 satisfies

$$|G_n - K_0| < c/n^q$$

for some absolute constant c and some $q > 1/2$.

Yet a third, and novel construction runs as follows. First, for the correct Gauss-Kuz'min density of 1's, namely $p_1 = \log_2(4/3)$, generate a list of 0's and $1's$ by assigning element values $a_n = k(\lceil np_1 \rceil - \lceil (n-1)p_1 \rceil)$ for $n > 0$ and $k = 1$. Now replace the remaining 0's in the sequence with 2's, by replacing p_1 with $p_2/(1 - p_1)$ and incrementing k to 2, in which case the index n on a_n refers, naturally, to the nth zero of the preceding list. Continuation of this construction for $k = 3, 4, \cdots$ gives a real number:

$$R = [0; 1, 2, 1, 3, 1, 4, 2, 1, 5, 6, 2, 1, 3, 1, 7, 1, 8, 2, 1, 4, 1, 3, 2, 1, 9, 1, \ldots].$$

Through 100,000 elements the geometric and harmonic means for R work out to be 2.6753 and 1.7454, respectively. Although we have not done so, it should be possible to prove for example that the limiting geometric mean of these elements is indeed K_0.

5. COMPUTATION OF KHINTCHINE MEANS

As intimated in our introduction, calculation of K_0 has occupied the attention of various researchers. For example, Gosper recently computed K_0 to 2217 digits [12]. Our 7350-digit value is in complete agreement with Gosper's 2217 digits. There is also the interesting problem of computing fraction elements from decimal representations of certain real numbers, a task that one may wish to do in, say, statistical experiments involving Khintchine means. We mention that in [21] an interesting algorithm is presented for computation of fraction elements without recourse to decimal input. Instead, differential properties of an appropriate function are used. For example, Shiu resolved 10,000 elements of the fraction for e^π using properties of the function $f(t) = \sin(\log(t))$.

By exploiting various modern algorithms to be described presently, the present authors have explicitly computedcdots K_0 and K_{-1} to more than 7350 decimal digit accuracy. These computations were performed with the aid of the

MPFUN multiprecision software [4, 5], which was found to be significantly faster for our purposes than other available multiprecision facilities. One utilizes this software by writing ordinary Fortran-90 code, with multiprecision variables declared to be of type `mp_integer`, `mp_real` or `mp_complex`. In the computations described below, the level of precision was sufficiently high that the "advanced" routines of the Fortran-90 MPFUN library were employed. These routines employ special algorithms, including fast Fourier transform (FFT) multiplication, which are efficient for extra-high levels of precision.

The constant K_0 was computed using the formula given above in Theorem 3, with the free integer parameter $N = 100$, and with $N = 120$ as a check. The implementation of this formula was straightforward except for the computation of the Riemann zeta function. To obtain 7350-digit accuracy in the final result, 2048 terms of the indicated series were evaluated, which requires $\{\zeta(2k),\ 0 \le k \le 2048\}$ to be computed. One approach to compute these zeta function values is to apply formulas due to P. Borwein [8]. These formulas are very efficient for computing one or a few zeta function values, but when many values are required as in this case, another approach was found to be more efficient. This method is based on an observation that has previously been used in numerical approaches to Fermat's "Last Theorem" [9, 10], namely

$$
\begin{aligned}
\coth(\pi x) &= \frac{-2}{\pi x} \sum_{k=0}^{\infty} \zeta(2k)(-1)^k x^{2k} \\
&= \cosh(\pi x)/\sinh(\pi x) \\
&= \frac{1}{\pi x} \cdot \frac{1 + (\pi x)^2/2! + (\pi x)^4/4! + (\pi x)^6/6! + \cdots}{1 + (\pi x)^2/3! + (\pi x)^4/5! + (\pi x)^6/7! + \cdots}.
\end{aligned}
$$

Let $N(x)$ and $D(x)$ be the numerator and denominator polynomials obtained by truncating these two series to n terms. Then the approximate reciprocal $Q(x)$ of $D(x)$ can be obtained by applying the Newton iteration

$$
Q_{k+1}(x) := Q_k(x) + [1 - D(x)Q_k(x)]Q_k(x).
$$

Once $Q(x)$ has been computed to sufficient accuracy, the quotient polynomial is simply the product $N(x)Q(x)$. The required values $\zeta(2k)$ can then be obtained from the coefficients of this polynomial.

Computation time for the Newton iteration procedure can be reduced by starting with a modest polynomial length and precision level, iterating to convergence, doubling each, etc., until the final length and precision targets are achieved. Computation time can be further economized by performing the two polynomial multiplications indicated in the above formula using a FFT-based convolution scheme. In our implementation, FFTs were actually performed at two levels of this computation: (i) to multiply pairs of polynomials, where the data elements to be transformed are the multiprecision polynomial coefficients,

ON THE KHINTCHINE CONSTANT   13

and (ii) to multiply pairs of multiprecision numbers, where the data elements to be transformed are integers representing successive sections of the binary representations of the two multiprecision numbers.

The constant K_{-1} was computed by applying the formula in Corollary 7. Again, the challenge here is to precompute values of the Riemann zeta function for integer values. But in this case both odd and even values are required. The odd values can be economically computed by applying the following two formulas, the first given by Ramanujan, but simplified slightly and known earlier; the second derived by differentiating a companion identity of Ramanujan [6, ch. 14]:

$$\zeta(4N+3) = -2\sum_{k=1}^{\infty}\frac{1}{k^{4N+3}(\exp(2k\pi)-1)}$$
$$-\pi(2\pi)^{4N+2}\sum_{k=0}^{2N+2}(-1)^k\frac{B_{2k}B_{4N+4-2k}}{(2k)!(4N+4-2k)!},$$
$$\zeta(4N+1) = -\frac{1}{N}\sum_{k=1}^{\infty}\frac{(2\pi k+2N)\exp(2\pi k)-2N}{k^{4N+1}(\exp(2k\pi)-1)^2}$$
$$-\frac{1}{2N}\pi(2\pi)^{4N}\sum_{k=1}^{2N+1}(-1)^k\frac{B_{2k}B_{4N+2-2k}}{(2k-1)!(4N+2-2k)!}.$$

Here B_{2k} is as always the $2k$th Bernoulli number.

Alternatively, the formulas can be written in terms of the even zetas as

$$\zeta(4N+3) = -2\sum_{k=1}^{\infty}\frac{1}{k^{4N+3}(\exp(2k\pi)-1)}$$
$$+\frac{1}{\pi}\{\frac{(4N+7)}{2}\zeta(4N+4)-\sum_{k=1}^{N}2\zeta(4k)\zeta(4N+4-4k)\},$$
$$\zeta(4N+1) = -\frac{1}{N}\sum_{k=1}^{\infty}\frac{(2\pi k+2N)\exp(2\pi k)-2N}{k^{4N+1}(\exp(2k\pi)-1)^2}$$
$$+\frac{1}{2N\pi}\{\sum_{k=1}^{2N}(-1)^k2k\zeta(2k)\zeta(4N+2-2k)+(2N+1)\zeta(4N+2)\}.$$

These two formulas are not very economical for computing a single odd value or just a few odd values of $\zeta(k)$ — again, the formulas in [8] are more efficient for such purposes. But these Ramanujan formulas are quite efficient when a large number of odd zetas are required. Note that the infinite series in the two formulas can be inexpensively evaluated for many N simultaneously, since the expensive parts of these expressions do not involve N. Further, the evaluation

Constant	Cont. Frac. Elements	Geometric mean	Harmonic mean
K_0	7182	2.660716	1.745541
K_{-1}	7052	2.722471	1.746871
A	15000	2.358251	1.745395
R	100000	2.6753	1.7454
Z_2	5206016	2.685482	1.7454074
π	17001303	2.686393	1.745882

Table 1: Continued Fraction Statistics

of the infinite series can be cut off once terms for a given N are smaller than the "epsilon" of the numeric precision level being used. Happily, convergence here is fairly rapid for large N.

At first glance, the latter summations in these two formulas may appear quite expensive to evaluate. But note that each is merely the polynomial product of two vectors consisting principally of even zeta values. Thus, both sets of summation results can be computed using multiprecision FFT-based convolutions.

Computation of K_0 to 7350 digit precision required 2.5 hours on an IBM RS6000/ 590 workstation, and computation of K_{-1} also to 7350 digits required some 12 hours. Excerpts of the resulting decimal expansions for each are included in the appendix. The complete expansions are available from the authors.

One intriguing question that was raised decades ago [27] is whether the continued fraction elements of K_0 themselves enjoy a limiting geometric mean K_0. We can of course ask more generally whether, for the fraction elements of any Khintchine mean K_p, the limiting Hölder mean of order q is in fact K_q. During the task of computing from a given decimal representation a Hölder mean of some order, the issue of where to terminate the list of continued fraction elements is an interesting one. We employed a simple criterion: if x is known numerically, to D decimals to the right of the decimal point, generate continued fraction elements for x until a convergent p/q has $2q^2 > 10^D$. The motivation for choosing this simple criterion is the theorem that at least one of any two successive convergents must satisfy

$$|\frac{p}{q} - x| < \frac{1}{2q^2}$$

and conversely, any reduced ratio p/q satisfying this inequality must be a convergent of x [13].

On the Khintchine Constant 15

Our results are shown in Table 1, together with results for the constants K_{-1}, A, R, Z_2 (which were defined above), and π. To give statistical perspective to our results for K_0 and K_{-1}, we computed the geometric and harmonic means of the first 7000 fraction elements for each of 100 pseudorandom multiprecision numbers of the same precision, namely 7350 decimal digits. The average and standard deviation of their geometric means were 2.683740 and 0.030124, respectively. The same statistics for their harmonic means were 1.745309 and 0.011148, respectively. Note that these two averages are in good agreement with the theoretical values K_0 and K_{-1}. In any event, it appears that the geometric and harmonic means for the first 7182 elements of our 7350-digit K_0 are within reasonable statistical limits of the expected theoretical values.

A question implicitly asked in the previous section is whether K_0 or K_{-1} is algebraic. This question can be numerically explored by means of integer relation algorithms. A vector of real numbers $(x_1, x_2, \ldots, x_n)$ is said to possess an *integer relation* if there exist integers a_k such that $a_1 x_1 + a_2 x_2 + \cdots + a_n x_n = 0$. It can easily be seen that a real number α is algebraic of degree $n-1$ if and only if the vector $(1, \alpha, \alpha^2, \cdots, \alpha^{n-1})$ possesses an integer relation. Even if α is not algebraic, integer relation algorithms produce bounds that allow one to exclude relations within a region.

We employed the "PSLQ" algorithm developed by Ferguson and one of the authors, a simplified version of which is given in [3]. This algorithm, when applied to power vectors generated from our computed values of K_0 or K_{-1}, found no relations for either. On the contrary, we obtained the following result: if K_0 satisfies a polynomial of the form

$$0 = a_0 + a_1 \alpha + a_2 \alpha^2 + a_3 \alpha^3 + \cdots + a_{50} \alpha^{50}$$

in the variable α, then the magnitude of some integer coefficient a_k exceeds 10^{70}. The same was found to be true for K_{-1}.

In a second experiment, we explored the possibility that K_0 or K_{-1} is given by a multiplicative formula involving powers of primes and some well-known mathematical constants. To that end, let p_k denote the kth prime. We established, using PSLQ, that neither K_0 nor K_{-1} satisfies a relation of the form

$$0 = a_0 \log \alpha + \sum_{k=1}^{15} a_k \log p_k$$
$$+ a_{16} \log \pi + a_{17} \log e + a_{18} \log \gamma + a_{19} \log \zeta(3) + a_{20} \log \log 2$$

with integer coefficients a_k of absolute value 10^{20} or less. By exponentiating this expression, it follows that neither K_0 nor K_{-1} satisfies a corresponding multiplicative formula with exponents of absolute value 10^{20} or less.

16   BAILEY, BORWEIN AND CRANDALL

There are many other tests that might be applied. For example, further work might be to rule out the possibility that $\log K_0$, $(\log K_0)(\log 2)$, or one of many other forms involving K_0 be an algebraic number of low degree.

Acknowledgment. Thanks are due to Robert Corless, Greg Fee, Thomas Wieting, Simon Plouffe, and Joe Buhler for many helpful discussions. We are grateful to a referee for suggesting the construction of the candidate number R at the end of §4.

ON THE KHINTCHINE CONSTANT   17

References

[1] R. ADLER, M. KEANE AND M. SMORODINSKY, "A construction of a normal number for the continued fraction transformation," *J. of Number Theory* **13** (1981), 95-105.

[2] M. ABRAMOWITZ AND I. A. STEGUN, *Handbook of mathematical functions*, Dover Publications, Inc., New York, 1972.

[3] D. H. BAILEY, J. M. BORWEIN, AND R. GIRGENSOHN, "Experimental evaluation of Euler sums," *Experimental Math.* **3** (1994), 17–30.

[4] D. H. BAILEY, "A Fortran-90 based multiprecision system," *ACM Trans. on Math. Software* **21** (1995), 379–387.

[5] D. H. BAILEY, "Multiprecision translation and execution of Fortran programs," *ACM Trans. on Math. Software* **19** (1993), 288–319.

This software and documentation, as well as that described in [4], may be obtained by sending electronic mail to mp-request@nas.nasa.gov, or by using Mosaic at address http://www.nas.nasa.gov.

[6] B.C. BERNDT, *Ramanujan's notebooks, Part III*, Springer Verlag, New York, 1991.

[7] D. BORWEIN, J.M. BORWEIN, AND R. GIRGENSOHN, "Explicit evaluation of Euler sums," *Proc. Edinburgh Math. Soc.* **38** (1995), 277–294.

[8] P. BORWEIN, "An efficient algorithm for the Riemann zeta function," CECM preprint, Dept. of Math. and Statistics, Simon Fraser University, Burnaby BC V5A 1S6, Canada.

[9] J. BUHLER, R. CRANDALL, AND R. SOMPOLSKI, "Irregular primes to one million," *Math. Comp.* **59** (1992), 717–722.

[10] J. BUHLER, R. CRANDALL, R. ERNVALL, AND T. METSÄNKYLÄ, "Irregular primes to four million," *Math. Comp.* **61** (1993), 151–153.

[11] R. CORLESS, personal communication.

[12] R. W. GOSPER, personal communication.

[13] A. KHINTCHINE, *Continued fractions,* University of Chicago Press, Chicago, 1964.

[14] D. LEHMER, "Note on an absolute constant of Khintchine," *Amer. Math. Monthly* **46** (1939), 148-152.

[15] L. LEWIN, *Polylogarithms and associated functions*, North Holland, New York, 1981.

18   BAILEY, BORWEIN AND CRANDALL

[16] H. NIEDERREITER, *Random number generation and quasi-Monte-Carlo methods*, CBMS–NSF Regional Conf. Ser. Appl. Math., vol. 63, 1992.

[17] N. NIELSEN, *Die Gammafunktion*, Princeton University Press, 1949.

[18] S. PLOUFFE, personal communication.

[19] C. RYLL-NARDZEWSKI, "On the ergodic theorems (I,II)," *Studia Math.* **12** (1951) 65-79.

[20] D. SHANKS AND J. W. WRENCH, "Khintchine's constant," *Amer. Math. Monthly* **6** (1959), 276-279.

[21] P. SHIU, "Computation of continued fractions without input values," *Math. Comp.* **64** (1995), 1307-1317.

[22] C.L. SIEGEL, *Transcendental numbers*, Chelsea, New York, 1965.

[23] K. R. STROMBERG, *An introduction to classical real analysis*, Wadsworth, Belmont, CA, 1981.

[24] I. VARDI, *Computational recreations in mathematica*, Addison-Wesley, Redwood City, CA, 1991.

[25] T. WIETING, personal communication.

[26] J. W. WRENCH, "Further evaluation of Khintchine's constant," *Math. Comp.* **14** (1960), 370-371.

[27] J. W. WRENCH AND D. SHANKS, "Questions concerning Khintchine's constant and the efficient computation of regular continued fractions," *Math. Comp.* **20** (1966), 444-448.

[28] D. ZAGIER, personal communication.

On the Khintchine Constant   19

Appendix: The Khintchine Constant K_0 to 7,350 Digits

2.
6854520010653064453097148354817956938203822939944 6
29530511523455572188595371520028011411749318476979
95153465905288090082897677716410963051792533483259
66838185231542133211949962603932852204481940961806
86641664289308477880620360737053501033672633577289
04990427070272345170262523702354581068631850103237
46558037750264425248528694682341899491573066189872
07994137235500057935736698933950879021244642075289
74145914769301844905060179349938522547040420337798
56398310157090222339100002207725096513324604444391
. . . . .
36909874406573435125594396103980583983755664559601

The Khintchine Harmonic Mean K_{-1} to 7,350 Digits

1.
74540566240734686349459630968366106729493661877798
42565950137735160785752208734256520578864567832424
20977343982577985596531102601834294460206578713176
15026238960612981165718728271638949622593992929776
06160830078357479801549029312671643067241248453710
96077711207484391474195803753220015690822609477078
44894635568203493582068440202422591615018316479048
29229656977733143662210991806388842581650599997697
61391683577259217628635718712601565066754443340174
00283376465305136584406098398017126202832041200630
. . . . .
78553128249666473680304034761497467330708479436280

Khintchine Means K_p to 50 Digits for Various Negative p

p	K_p
-2	1.45034032849563040605298307668069788140829997960 5904...
-3	1.31350707868798576671733944707278682815812986148 4792...
-4	1.23696180942373005262622724445342256742024113154 8937...
-5	1.18900392646551315406236373277140339738609251263 9671...
-6	1.15655237442151442315260599874341004684021307071 8761...
-7	1.13332336395086579491028969490886836359909828241 1797...
-8	1.11596440897871669061915641934534969576949118223 0400...
-9	1.10254313667072801383609340252256835102222128414 9318...
-10	1.09187704120961267827611097947763825649327265142 9656...

3. On the dynamics of certain recurrence relations

Discussion

This paper can be viewed either as a strenuous exercise in continued fraction theory, or a profound foray into advanced matrix algebra—said foray mostly and brilliantly by our coauthor R. Mayer. The paper brings closure on many aspects of a Ramanujan problem outlined in previous works.[1]

Challenges

Section 8 of this paper discusses some open problems, for example there is a suggestion on how to move to 3×3 matrix theory.

Source

D. Borwein, J. Borwein, R. Crandall, and R. Mayer, "On the dynamics of certain recurrence relations," *Ramanujan J.*, 13:63-101 (2007).

---

[1] J. Borwein, R. Crandall and G. Fee, "On the Ramanujan AGM fraction. Part I: The real-parameter case," *Experimental Mathematics* 13, 275-286 (2004).
J. Borwein and R. Crandall, "On the Ramanujan AGM fraction. Part II: The complex-parameter case," *Experimental Mathematics* 13, 287-296 (2004).

Ramanujan J (2007) 13:63–101
DOI 10.1007/s11139-006-0243-3

On the dynamics of certain recurrence relations

D. Borwein · J. Borwein · R. Crandall · R. Mayer

Dedicated respectfully to Richard Askey on the occasion of his 70th Birthday.
Received: 15 January 2004 / Accepted: 15 July 2004
© Springer Science + Business Media, LLC 2007

Abstract In recent analyses [3, 4] the remarkable AGM continued fraction of Ramanujan—denoted $\mathcal{R}_1(a, b)$—was proven to converge for almost all complex parameter pairs (a, b). It was conjectured that $\mathcal{R}_1$ diverges if and only if $(0 \neq a = be^{i\phi}$ with $\cos^2 \phi \neq 1)$ or $(a^2 = b^2 \in (-\infty, 0))$. In the present treatment we resolve this conjecture to the positive, thus establishing the precise convergence domain for $\mathcal{R}_1$. This is accomplished by analyzing, using various special functions, the dynamics of sequences such as (t_n) satisfying a recurrence

$$ t_n = (t_{n-1} + (n - 1)\kappa_{n-1} t_{n-2})/n, $$

where $\kappa_n := a^2, b^2$ as n be even, odd respectively.

Research supported by NSERC.

Research supported by NSERC, the Canada Foundation for Innovation and the Canada Research Chair Program.

D. Borwein (✉)
Department of Mathematics, University of Western Ontario, London Ontario Canada N6A 5B7
e-mail: dborwein@uwo. ca

J. Borwein
Faculty of Computer Science Dalhousie University, Halifax NS Canada B3H 1W5
e-mail: jborwein@cs.dal.ca

R. Crandall
Center for Advanced Computation, Reed College, Portland, Oregon 97202
e-mail: crandall@reed.edu

R. Mayer
Department of Mathematics, Reed College, Portland Oregon 97202
e-mail: mayer@reed.edu

As a byproduct, we are able to give, in some cases, exact expressions for the n-th convergent to the fraction $\mathcal{R}_1$, thus establishing some precise convergence rates. It is of interest that this final resolution of convergence depends on rather intricate theorems for complex-matrix products, which theorems evidently being extensible to more general continued fractions.

Keywords Complex continued fractions · Dynamical systems · Arithmetic-geometric mean · Matrix analysis · Stability theory

2000 Mathematics Subject Classification Primary—11J70, 11Y65, 40A15

1 Nomenclature

In companion treatments [3, 4] we considered the Ramanujan AGM fraction

$$\mathcal{R}_1(a, b) = \cfrac{a}{1 + \mathcal{S}(a, b)} \cfrac{a}{1 + \cfrac{b^2}{1 + \cfrac{4a^2}{1 + \cfrac{9b^2}{1 + \ddots}}}} \tag{1.1}$$

one of whose attractive properties being a formal AGM relation—known to be true at least for positive real a, b—

$$\mathcal{R}_1\left(\frac{a + b}{2}, \sqrt{ab}\right) = \frac{\mathcal{R}_1(a, b) + \mathcal{R}_1(b, a)}{2},$$

but of dubious validity for general complex parameters [3]. The work [4] focused on the convergence domain

$$\mathcal{D}_0 := \{(a, b) \in \mathcal{C} \times \mathcal{C} : \mathcal{R}_1(a, b) \text{ converges on } \hat{\mathcal{C}}\},$$

where $\hat{\mathcal{C}} := \mathcal{C} \cup \{\infty\}$ denotes the extended complex field. It was proved therein that if we define

$$\mathcal{D}_2 := \{(a, b) \in \mathcal{C} \times \mathcal{C} : |a| \neq |b|\},$$
$$\mathcal{D}_3 := \{(a, b) \in \mathcal{C} \times \mathcal{C} : a^2 = b^2 \notin (-\infty, 0)\},$$
$$\mathcal{D}_1 := \mathcal{D}_2 \cup \mathcal{D}_3,$$

then

$$\mathcal{D}_1 \subseteq \mathcal{D}_0,$$

so that the Ramanujan fraction converges for almost all complex pairs (a, b). There is a conjecture [4, Conjecture 5.4], effectively saying that in fact

$$\mathcal{D}_1 = \mathcal{D}_0.$$

Equivalently: The fraction $\mathcal{R}_1$ diverges whenever $(0 \neq a = be^{i\phi}$ with $\cos^2 \phi \neq 1)$ or $(a^2 = b^2 \in (-\infty, 0))$. This divergence behavior can be understood, as we do presently, in terms of the special dynamics of certain recurrence relations.

In what follows, we consider classical convergents p_n/q_n to the fraction $\mathcal{S}$ from (1.1), but renormalize to obtain certain sequences. Specifically, to evaluate $\mathcal{S}$ we use initial values $(p_{-1}, p_0, q_{-1}, q_0) = (1, 0, 0, 1)$ and a recurrence (also satisfied by the p_n)

$$q_n = q_{n-1} + n^2 \kappa_n q_{n-2},$$

where $\kappa_n = a^2, b^2$ as n be even, odd respectively. Whether this classical procedure has p_n/q_n approaching a limit depends in a delicate way, as we have intimated, on the parameters a, b.

For the theoretical treatments to follow, we now establish some sequence renormalizations, each theoretically interesting in its own right. We shall consider renormalized sequences denoted (t_n), (r_n), (v_n) in what follows. The first renormalization is

$$t_n := \frac{q_{n-1}}{n!},$$

so that

$$t_n = \frac{t_{n-1} + (n-1)\kappa_{n-1}t_{n-2}}{n}, \tag{1.2}$$

as intimated in our Abstract. Another interesting renormalization is

$$r_n := \frac{q_n}{a^n \Gamma(n + 3/2)}, \tag{1.3}$$

with recurrence

$$r_n = \frac{1}{a(n + 1/2)} r_{n-1} + \frac{n^2}{n^2 - 1/4} r_{n-2}, \quad n \text{ even,}$$

$$r_n = \frac{1}{a(n + 1/2)} r_{n-1} + \frac{n^2 e^{-2i\phi}}{n^2 - 1/4} r_{n-2}, \quad n \text{ odd,}$$

as is used in Conjecture 5.4 of [4]. A slight variation on (r_n) turns out to be optimal in some ways:

$$v_n := \frac{q_n}{\Gamma(n + 3/2)} \frac{1}{\kappa_n^{(n+1)/2}}. \tag{1.4}$$

The recurrence relation for this (v_n) sequence will later be put into a convenient matrix form. Now the classical separation of convergents to S can be written, for n even, as

$$\frac{p_n}{q_n} - \frac{p_{n-1}}{q_{n-1}} = -\frac{b^n a^n n!^2}{q_n q_{n-1}}.$$

Given our various renormalized sequences (t_n), (r_n), (v_n), we can also write (again, for n even)

$$\frac{p_n}{q_n} - \frac{p_{n-1}}{q_{n-1}} = -\frac{b^n a^n}{t_{n+1} t_n (n+1)},$$

$$= -\frac{a(b/a)^n}{r_n r_{n-1}} \left\{ 1 + O\left(\frac{1}{n}\right) \right\}, \tag{1.5}$$

$$= -\frac{1}{a v_n v_{n-1}} \left\{ 1 + O\left(\frac{1}{n}\right) \right\}.$$

If we assume $a = be^{i\phi} \neq 0$, the S fraction—and hence $\mathcal{R}_1(a, b)$—diverges if

$$t_n = O(|a|^n/\sqrt{n}), \tag{1.6}$$

or

$$(r_n) \text{ is bounded,}$$

or

$$(v_n) \text{ is bounded,}$$

since any one of such growth conditions implies eventual separation of the convergents. Our program for finally resolving the convergence domain of $\mathcal{R}_1(a, b)$ is now evident: We only need show that one of these growth conditions is true, for $a = be^{i\phi} \neq 0$ and $\cos^2 \phi \neq 1$.

But first we offer in the next section a digression, an analysis of the (t_n) recurrence in the cases $\cos^2 \phi = 1$. Note that [4] has established convergence/divergence theory for these cases $a = \pm b$; yet, the present analysis of sequence dynamics yields substantially more information on convergence rates.

2 The instance $a = \pm b$

Assume $a = \pm b$, so that $\mathcal{R}_1(a, a)$ is to be studied. For $a = b$ the recurrence (1.2) is blind to the parity of n (since $\kappa_n := a^2$, always), and the manipulations leading to the bound (1.5) are especially elegant and tractable. We note that divergence of $\mathcal{R}_1(a, a)$ for any pure-imaginary a (i.e. $a^2 \in (-\infty, 0)$) is established via a different approach in [4].

Let us assume $t_0 := 1$ so that fixing t_1 completely determines the entire sequence (t_n). An exponential generating function can be established as follows: Let $s_n := t_n/n!$

and consider

$$y(x) := \sum_{n=0}^{\infty} s_n x^n. \tag{2.1}$$

Using

$$\sum_{n=2}^{\infty} n^2 s_n x^n = \sum_{n=2}^{\infty} s_{n-1} x^n + a^2 \sum_{n=2}^{\infty} s_{n-2} x^n$$

we obtain

$$\sum_{n=2}^{\infty} n^2 s_n x^n = (x + a^2 x^2) \sum_{n=0}^{\infty} s_n x^n - t_0 x,$$

and hence that

$$\sum_{n=0}^{\infty} n^2 s_n x^n = (x + a^2 x^2) \sum_{n=0}^{\infty} s_n x^n + (t_1 - 1)x.$$

Thus

$$x \frac{d}{dx}(xy'(x)) = \sum_{n=0}^{\infty} n^2 s_n x^n = x(1 + a^2 x)y(x) + x(t_1 - 1).$$

Therefore $y := y(x)$ satisfies the differential equation

$$xy'' + y' - (1 + a^2 x)y = (t_1 - 1), \quad y(0) = 1, \, y'(0) = t_1.$$

This, for $t_1 = 1$ and general a with the help of *Maple*, has the solution

$$y(x) = e^{-ax} {}_1 F_1 \left(\frac{a+1}{2a}; 1; 2ax \right) = e^{-ax} \sum_{n \geq 0} \left(\frac{a+1}{2a} \right)_n \frac{(2ax)^n}{n!^2},$$

where $(a)_n := a(a+1) \cdots (a+n-1)$ is the Pochhammer symbol.

On equating coefficients we find that (here, by $t_n(t_0, t_1)$ is meant the recurrence solution starting with given t_0, t_1):

$$t_n(1, 1) = a^n \sum_{k=0}^{n} \binom{n}{k} \frac{(-2)^k}{k!} \omega_k = a^n {}_2 F_1(-n, \omega; 1; 2), \tag{2.2}$$

where

$$\omega = \omega(a) := \frac{1 - 1/a}{2}$$

turns out to be an ubiquitous entity in our analysis.

Though (2.2) gives a finite form for t_n, it is not clear how to deduce other sequences (say, starting with $t_1 \neq 1$), moreover the large-n asymptotic behavior is nontrivial. One way to proceed on asymptotics is to use the rather intricate hypergeometric theory of J. Fields and Y. Luke [10, pp. 247–254], wherein the focus is upon expansions in the *parameters* rather than the argument. However, for our particular hypergeometric cases, a certain approach as outlined below is more specific and direct, less technical. Incidentally, and remarkably, by using the program ct, (described in [14, p. 112], and available on the home page of that book) one may actually *derive and prove* the recurrence (1.2) (with such assignments as $\kappa_n := -1$, all n) from the hypergeometric form (2.2), and likewise for (2.3) below.

Remarks. We have not been able to achieve similar success for $a \neq b$ by like methods. However

1. For general a, b the corresponding exponential generating function z can be neatly placed in the following coupled form:

$$t \frac{d^2}{dt^2} y(t) + \frac{d}{dt} y(t) - a^2 t y(t) = x(t), \quad t \frac{d^2}{dt^2} x(t) + \frac{d}{dt} x(t) - b^2 t x(t) = y(t),$$

   with $x(0) = 0$, $y(0) = 1$ and where y and x are the even and odd terms of z. Adding the odd and even terms when $a = b$ recovers the differential equation above.
   Note also that for $a = 0$ or $b = 0$ the underlying recursion is easy to solve explicitly.
2. Additionally, for $a = 1$, $b = i$, in terms of Bessel functions, we have a functional equation for the even part of the ordinary generating function:

$$y(t) = I_0(t) + \frac{\pi}{2} \int_0^t K(w, t) y(w) dw$$

   where the kernel is

$$K(w, t) := \int_w^t \{I_0(t) K_0(z) - K_0(t) I_0(z)\} \{J_0(z) Y_0(w) + Y_0(z) J_0(w)\} dz.$$

   The odd part has a similar equation. Note that K_0 and Y_0 have logarithmic singularities at 0 while $I_0(0) = J_0(0) = 1$.
3. Correspondingly, we can construct functional equations for general a and b by considering the appropriate differential equations

$$D_a(y) = x, \quad D_b(x) = y.$$

   Let

$$D_a(y_a) = 0, \quad y_a(0) = 1 \quad \text{and} \quad D_b(x_b) = 0, \quad x_b(0) = 0,$$

   independently; then the corresponding Green's functions lead to the desired functional equation.

Inter alia, differential-equation techniques applied instead to the standard generating function can also yield similar results. One finds such oddities as the following, for the case $a = i$:

$$\sum_{n \geq 0} t_n(1, 1)x^n = \frac{e^{\arctan x}}{\sqrt{1 + x^2}}$$

$$= \frac{1}{(1 - ix)^{-\omega(i)^*}(1 + ix)^{\omega(i)}}.$$

But still, the asymptotic behavior of the coefficients in such expansions is unclear. As we shall show, said coefficients can exhibit quite peculiar oscillations.

Armed with an exact form (2.2) for $t_n(1, 1)$, we require an independent sequence in order to forge a general solution to recurrence (1.2). To this end, we found that a hypergeometric form related to that in (2.2) is

$$F_n(a) := a^n 2^{1-\omega} \frac{\Gamma(n + 1)}{\Gamma(n + 1 + \omega)\Gamma(1 - \omega)} \, _2F_1\left(\omega, \omega; n + 1 + \omega; \frac{1}{2}\right). \quad (2.3)$$

which satisfies the recursion (1.2), for any parameter choice $a = b$ (for which, we recall, $\kappa_n := a^2$, always)—as can be checked in a computer algebra system. But here is an important observation: The same recurrence is also satisfied by $F_n(-a)$, since, after all, a^2 is the only a-dependent component of (1.2). This means that any solution of (1.2) is a superposition, such as

$$t_n(0, 1) = \alpha F_n(a) + \beta F_n(-a), \quad (2.4)$$

$$t_n(1, 0) = \gamma F_n(a) + \delta F_n(-a),$$

where the constants $\alpha, \beta, \gamma, \delta$ can be written always in terms of the four constants $F_0(\pm a)$, $F_1(\pm a)$. We may also recast $t_n(1, 1)$ as a superposition:

$$t_n(1, 1) = \frac{1}{2}F_n(a) + \frac{1}{2}F_n(-a). \quad (2.5)$$

amounting to a hypergeometric identity involving the F_n and the right-hand side of (2.2). Such an identity, once known, can be derived from known transformation formulae [1, (2.3.12), (2.2.7) p. 68].

As to the asymptotic character of F_n for large n, denote

$$\Gamma_n(x) := \frac{n! n^x}{x(x + 1) \cdots (x + n)} = \Gamma(x)\left\{1 + O\left(\frac{1}{n}\right)\right\},$$

and also

$$\omega := \omega(a), \quad \omega' := \omega(-a).$$

Then

$$F_n(a) = \frac{a^n 2^{\omega'}}{n^\omega} \frac{\Gamma_n(\omega)}{\Gamma(\omega)\Gamma(\omega')} \, {}_2F_1\left(\omega, \omega; n+1+\omega; \frac{1}{2}\right).$$

Thus, the large-n behavior of (2.3) is

$$F_n(a) \sim \frac{2^{\omega'}}{\Gamma(\omega')} \frac{a^n}{n^\omega} \left\{1 + O\left(\frac{1}{n}\right)\right\}. \tag{2.6}$$

We now know from (2.4) the general asymptotic for arbitrary a, in the form

$$|t_n| = O\left(\frac{|a|^n}{n^{\mathrm{Re}(\omega)}} + \frac{|a|^n}{n^{\mathrm{Re}(\omega')}}\right) \tag{2.7}$$

a bound valid for *any* initial values t_0, t_1. This amounts to an analytic proof of [4, Theorem 5.1], that the Ramanujan fraction $\mathcal{R}_1(a, a)$ diverges for any pure-imaginary a; indeed, for such a we have $\mathrm{Re}(\omega) = \mathrm{Re}(\omega') = 1/2$, and the argument following (1.4) goes through.

We observe that the asymptotic relation (2.6) explains, finally, the interesting oscillation of a typical t_n sequence. For example, when $a = i$ we see that the t_n typically exhibit a fourfold quasi-oscillation, as (large) n runs through values modulo 4. (That is, when plotted versus n, the (real) sequence $t_n(1, 1)$ exhibits the "snaking" of four separate "necklaces.") In fact, for $a = i$ the detailed asymptotic is

$$t_n(1, 1) = \sqrt{\frac{2}{\pi}} \cosh\frac{\pi}{2} \frac{1}{\sqrt{n}} \left(1 + O\left(\frac{1}{n}\right)\right)$$

$$\times \begin{cases} (-1)^{n/2} \cos(\theta - \log(2n)/2) & \text{if } n \text{ is even} \\ (-1)^{(n+1)/2} \sin(\theta - \log(2n)/2) & \text{if } n \text{ is odd} \end{cases}$$

where $\theta := \arg \Gamma((1 + i)/2)$. This behavior is certainly difficult to infer directly from the recurrence (1.3), or even from our first hypergeometric form (2.2) for $t_1(1, 1)$.

For $\mathcal{S}(ri, ri)$ and any $r \neq 0$ we see a similar oscillatory behavior; indeed it is exactly in this case that (2.6) and (2.7) have precisely $\sqrt{n}$ growth.

3 The fraction convergents for $\mathcal{R}_1(a, a)$

The recurrence solutions (2.4) lead immediately to expressions for the convergents p_n/q_n to the $\mathcal{S}(a, a)$ fraction—and hence to the convergents, say P_n/Q_n of the original

$\mathcal{R}_1$ fraction, since $\mathcal{R}_1 =: 1/(1 + \mathcal{S})$). In fact,

$$\frac{p_{n-1}}{q_{n-1}} = \frac{n!\,t_n(1,0)}{n!\,t_n(0,1)} = \frac{t_n(1,0)}{t_n(0,1)}$$

$$= \frac{F_1^- F_n^+ - F_1^+ F_n^-}{-F_0^- F_n^+ + F_0^+ F_n^-}, \tag{3.1}$$

where we denote $F_n^\pm := F_n(\pm a)$.

In an obvious sense we therefore have a closed form for the general convergent. Two interesting cases are $a = i$ and $a = 1$. In the former case the fraction diverges; in the latter case $\mathcal{R}_1 = \log 2$, as discussed in [3]. For the case $a = i$, our previous frustration in tracking the convergents numerically is now explained. In fact, on the basis of (3.1) and (2.6) the convergents to $\mathcal{S}(i, i)$ have asymptotic behavior, for n even,

$$\frac{p_{n-1}}{q_{n-1}} \sim -\frac{\cos \phi_0}{\cos \phi_1} \frac{\cos(\theta - \phi_1 - \log(2n)/2)}{\cos(\theta - \phi_0 - \log(2n)/2)},$$

where $\phi_{0,1}$ are certain (unequal) constant angles; while for n odd the second ratio of cosines is to be changed to a ratio of sines, of the same respective arguments. Also, in this asymptotic, it must be understood that $O(1/n)$ additive terms exist in both numerator and denominator.

More generally, (3.1) and (2.6) can be employed in this fashion to show that for nonzero real r, *the convergents p_n/q_n to the fraction $\mathcal{S}(ir, ir)$ are dense on the real axis*. It is possible, for example, to work out a (large) explicit integer n for which the n-th convergent to $\mathcal{S}(i, i)$ exceeds a googol (10^{100}). (Or for that matter, one can locate an n for which p_n/q_n is less than minus a googol.) Incidentally all this means that neither the even nor odd part of the fraction converges—a unique situation since for other, diverging cases of $|a| = |b|$ one still has separate, even/odd convergence [4].

The case $a = 1$, via relation (3.1) (and proper limit-taking, as $\omega' = \omega(-1) = 1$ and the ratio (3.1) needs be taken delicately) enjoys a striking, exact form for the general convergent of $\mathcal{R}_1$, namely

$$\frac{P_n}{Q_n} = \log 2 - \frac{(-1)^n}{2n + 2} \, {}_2F_1\left(1, 1; n + 2, \frac{1}{2}\right).$$

We deduce that, remarkably, the right-hand side here is always rational. Moreover, we now know the precise convergence rate, namely $P_n/Q_n = \log 2 - (-1)^n/(2n + 2) + O(1/n^2)$, and such an observation may be a clue to acceleration algorithms for continued fractions of the Ramanujan type. Notice also how very much stronger this convergence knowledge is, over such as [3, Theorem 7.3].

For general a for which $\mathcal{R}_1(a, a)$ converges, let us with impunity force Re $(a) > 0$ (knowing that $\mathcal{R}_1(-a, -a) = \mathcal{R}_1(a, a)$), for which we have Re$(\omega' := \omega(-a)) >$

$\text{Re}(\omega(a))$ and therefore, from (3.1) and some manipulation,

$$\frac{P_n}{Q_n} = \frac{a}{1 + \frac{a}{1+\omega'}\frac{{}_2F_1(\omega';\omega';2+\omega';1/2)}{{}_2F_1(\omega';\omega';1+\omega';1/2)}} + O\left(\frac{1}{n^{\text{Re}(1/a)}}\right) = \mathcal{R}_1(a,a) + O\left(\frac{1}{n^{\text{Re}(1/a)}}\right),$$

$$(3.2)$$

so that the compound fraction involving the hypergeometric functions is the actual fraction value. It is interesting to compare said compound fraction with a previous exact evaluation from [3], namely

$$\mathcal{R}_1(a,a) = \frac{1}{\omega'}\, {}_2F_1\left(\omega', 1; 1+\omega'; -1\right).$$

There is an interesting check of formula (3.2), namely we take $a = \infty$ so $\omega = \omega' = 1/2$ and

$$\frac{{}_2F_1(1/2, 1/2; 3/2; 1/2)}{{}_2F_1(1/2, 1/2; 5/2; 1/2)} = \frac{\pi/\sqrt{8}}{3/\sqrt{8}},$$

and sure enough, as explained in [3],

$$\mathcal{R}_1(\infty, \infty) = \frac{\pi}{2}.$$

It also follows from [3] that such hypergeometric ratios as appear in (3.2) can be put in closed form for any rational ω'.

4 Some initial matrix analysis

As precursor to what follows, let us do some elementary matrix analysis, focusing on the particular sequence (v_n) from (1.4). Consider the relevant two-step matrix iteration, with here, $\omega := \kappa_1/\kappa_0 = b^2/a^2$,

$$\begin{bmatrix} v_{2n} \\ v_{2n-1} \end{bmatrix} = Y_n \begin{bmatrix} v_{2n-2} \\ v_{2n-3} \end{bmatrix},$$

$$(4.1)$$

where the Y matrix can be worked out to be

$$Y_n := \begin{bmatrix} \frac{4n^2+1/a^2}{4n^2-1/4} & \frac{\omega^n(2n-1)^2}{a(2n+1/2)((2n-1)^2-1/4)} \\ \frac{\omega^{-n}}{a(2n-1/2)} & \frac{(2n-1)^2}{(2n-1)^2-1/4} \end{bmatrix} \left(= I + \frac{1}{2an}\begin{bmatrix} 0 & \omega^n \\ \omega^{-n} & 0 \end{bmatrix} + O\left(\frac{1}{n^2}\right)\right).$$

But this means that

$$\begin{bmatrix} v_{2n} \\ v_{2n-1} \end{bmatrix} = Z_n \begin{bmatrix} v_0 \\ v_{-1} \end{bmatrix},$$

with

$$Z_n := \prod_{k=1}^{n} Y_k. \tag{4.2}$$

(Here and elsewhere, such a matrix product is interpreted as "left-handed," in that the $k = 1$ matrix is on the far right (see Remark to Theorem 6.1).)

　　Now

$$\det(Y_n)^{-1} = \left(1 - \frac{1}{(4n)^2}\right)\left(1 - \frac{1}{(4n-2)^2}\right)$$

is independent of (nonzero) (a, b) and so by a Wallis formula

$$\lim_{n \to \infty} \det(Z_n) = \prod_{n=1}^{\infty} \left(1 - \frac{1}{(2n)^2}\right)^{-1} = \frac{\pi}{2}.$$

In spite of this easily derived determinantal limit—actually easier to see from the one-step iteration—we cannot yet assert that the matrices Z_n converge to a finite matrix, say Z_∞ Assume, though, that there *is* convergence, with

$$Z_\infty = \begin{bmatrix} A & C \\ B & D \end{bmatrix}, \qquad A, B, C, D \in \mathcal{C}. \tag{4.3}$$

Now let (u_n) denote the sequence (v_n) with $q_n \to p_n$ in (1.4). From the standard initial conditions for the $\mathcal{S}(a, b)$ fraction, we have

$$(u_{-1}, u_0, v_{-1}, v_0) = (1/\sqrt{\pi}, 0, 0, 2/(a\sqrt{\pi})), \tag{4.4}$$

implying that

$$v_{2n} \sim 2A/(a\sqrt{\pi}), \quad u_{2n} \sim B/\sqrt{\pi},$$
$$v_{2n-1} \sim 2C/(a\sqrt{\pi}), \quad u_{2n-1} \sim D/\sqrt{\pi}.$$

Such analysis leads immediately to

Theorem 4.1. *For nonzero complex parameters (a, b) with $\omega := b^2/a^2$, assume that the matrix Z_n converges to Z_∞ as in (4.3). Then the even/odd parts of $\mathcal{S}(a, b)$ both converge, respectively to*

$$\mathcal{S}^{(even)} = \frac{aB}{2A}, \quad \mathcal{S}^{(odd)} = \frac{aD}{2C},$$

which limits cannot be equal; in fact we have an explicit separation

$$\mathcal{S}^{(even)} - \mathcal{S}^{(odd)} = -\frac{a\pi}{4AC}.$$

🐾 Springer

The import of Theorem 4.1 is as follows: Whenever we can show that the infinite product Z_∞ exists, we have divergence of the $\mathcal{S}$ and perforce of $\mathcal{R}_1$, albeit with even/odd parts separately converging. (It was noted in [4] on the basis of established theory that the separate even/odd convergence holds for $a = be^{i\phi}$ with $\cos^2 \phi \neq 1$, so any proof that Z_∞ exists repeats, at least, that result.)

5 Operator algebra theory

Our next foray into matrix analysis involves the matrix or sup-norm defined as follows. First, the norm of a complex number z will be the usual absolute value $|z|$, while the norm of a vector x of complex elements will be $|x| := (\sum |x_j|^2)^{1/2}$. For a complex $k \times k$ matrix M, we recall that the matrix or *operator norm* is defined as the "sup-norm" over all unit-norm vectors as

$$|M| := \sup_{|x|=1} |Mx|.$$

This norm is sometimes denoted $||M||_2$ [7], and has the requisite properties

$$
\begin{aligned}
|cM| &= |c||M|, \quad \text{for scalar } c, \\
|M + N| &\leq |M| + |N|, \\
|MN| &\leq |M||N|, \\
|M| &\leq k \max|M_{ij}|.
\end{aligned}
\tag{5.1}
$$

An important additional property is that

$$|M|^2 = \text{maximum eigenvalue of } M^\dagger M, \tag{5.2}$$

where $\dagger$ denotes adjoint (conjugate-transpose), and one may equally well use $MM^\dagger$.

We may apply sup-norm theory to establish divergence theorems for certain (a, b). Consider the (t_n) sequence, from (1.2). We define a vector τ_n,

$$\tau_n := \begin{bmatrix} t_{2n} \\ t_{2n-1} \end{bmatrix} = \frac{2n-2}{2n-1} E_n \begin{bmatrix} t_{2n-2} \\ t_{2n-3} \end{bmatrix}, \tag{5.3}$$

where the E_n matrix is

$$E_n := \begin{bmatrix} \dfrac{1 + b^2(2n-1)^2}{4n(n-1)} & \dfrac{a^2}{2n} \\ \dfrac{1}{2n-2} & a^2 \end{bmatrix}.$$

In turn, we have

$$E_n = F_n + O(1/n^2),$$

where

$$F_n = \begin{bmatrix} b^2 & \frac{a^2}{2n} \\ \frac{1}{2n} & a^2 \end{bmatrix}.$$

Being as, via the Wallis/Stirling formula,

$$\prod_{n=2}^{N} \frac{2n-2}{2n-1} \sim \sqrt{\frac{\pi}{4N}},$$

we have for a constant c

$$|\tau_N| \le \frac{c}{\sqrt{N}} \prod_{n=1}^{N} \left(|F_n| + O\left(\frac{1}{n^2}\right) \right). \tag{5.4}$$

However, from (5.2) we know $|F_n|$ is determined by the largest eigenvalue of $F_n F_n^\dagger$:

$$2|F_n|^2 = |a|^4 + |b|^4 + O\left(\frac{1}{n^2}\right) + \sqrt{\left(|b|^4 - |a|^4 + \frac{|a|^4 - 1}{(4n^2)} \right)^2 + \frac{|b^2 + |a|^4|^2}{n^2}}.$$

In this way the sup-norm theory applies to two significant cases. When $|a| \ne |b|$ we have

$$|F_n| = \max(|a|^2, |b|^2) + O\left(\frac{1}{n^2}\right),$$

leading via (5.4) and (1.5) to

Theorem 5.1. *If $|a| \ne |b|$ then any solution of recurrence (1.2) has*

$$|t_n| = O\left(\frac{\max(|a|, |b|)^n}{\sqrt{n}} \right),$$

and the convergents to $\mathcal{S}(a, b)$ have, for constant c,

$$\left| \frac{p_{2n}}{q_{2n}} - \frac{p_{2n-1}}{q_{2n-1}} \right| > c \min\left(\left|\frac{a}{b}\right|, \left|\frac{b}{a}\right| \right)^{2n}.$$

Remark. It is already known [4] that for $|a| \ne |b|$ the $\mathcal{S}$ fraction converges on $\hat{\mathcal{C}}$. The sup-norm theory here bounds the convergence; this bound is consistent with the convergence rates for $|a| \ne |b|$ in [3, 4].

The second significant application of this sup-norm theory is the case $|a| = 1, b = i$, for which the eigenvalue calculation reduces to

$$|F_n| = 1 + O\left(\frac{1}{n^2}\right),$$

so that, from (5.4) and (1.5) we have

Theorem 5.2. *If* $|a| = 1, b = i$ *then any solution of recurrence (1.2) has*

$$|t_n| = O\left(\frac{1}{\sqrt{n}}\right),$$

and $\mathcal{S}(a, b)$, *perforce* $\mathcal{R}_1(a, b)$ *diverges.*

Remark. We have already taken care of the instance $(a, b) = (i, i)$ in our present Section 2, and in [4] we also handled $(a, b) = (i, i)$. Note that the present analysis also applies to sequences (t_n) for $b = i$ but with a chosen *randomly* from the unit circle.

A third application of sup-norm methods gives *a lower* bound on the accuracy of convergents. If we go back to the (v_n) sequence and observe that in (4.2), with $a \neq 0, |\omega| = 1$,

$$|Y_n| \leq \left|I + \frac{1}{2an}\Omega_n\right| + O\left(\frac{1}{n^2}\right).$$

where

$$\Omega_n := \begin{bmatrix} 0 & \omega^n \\ \omega^{-n} & 0 \end{bmatrix}, \tag{5.5}$$

then a similar eigenvalue analysis on $(I + \Omega_n/(2an))^{\dagger}(I + \Omega_n/(2an))$ yields, for constants c_1, c_2,

$$|Z_N| \leq c_1 \sqrt{\prod_{n=1}^{N}\left(1 + \frac{1}{n}\left|\mathrm{Re}\left(\frac{1}{a}\right)\right|\right)} < c_2 N^{|\mathrm{Re}(1/a)|/2}.$$

Therefore $|v_n| < c_2 n^{|\mathrm{Re}(1/a)|/2}$, so by the last relation of (1.5) we obtain

Theorem 5.3. *If* $|a| = |b|$ *then the convergents to* $\mathcal{S}(a, b)$ *have, for positive constant* c

$$\left|\frac{p_{2n}}{q_{2n}} - \frac{p_{2n-1}}{q_{2n-1}}\right| \geq \frac{c}{n^{|\mathrm{Re}(1/a)|}}.$$

Remark. This result is consistent with our exact, hypergeometric analysis of Section 2 when $a = b$. It is also a corollary of Theorem 5.3 that—as we have seen several times already—fractions $\mathcal{S}, \mathcal{R}_1$ diverge for $a^2 = b^2 \in (-\infty, 0)$. Indeed, for such cases the exponent of n in the bounding right-hand side is zero, so the fraction cannot converge. In spite of the allure of this sup-norm theory, it appears difficult to generalize Theorem 5.2 to more general cases $|a| = |b|$ using sup-norm methods alone. Evidently, deeper results are required to resolve finally the convergence problem for $\mathcal{R}_1$. Accordingly, with a view to Theorem 4.1, our next foray will address not just sup-norm bounds on, but actual convergence of matrix products.

6 Some deeper matrix analysis

We know now that $\mathcal{S}$ and therefore $\mathcal{R}_1$ diverges if the product matrix Z_n in (4.2) converges to a finite complex matrix. To this end, we shall now prove a general "perturbation" theorem which will turn out to have application even beyond our present analysis.

Theorem 6.1. *Let (a_n), (b_n) be sequences of $k \times k$ complex matrices. Suppose that $\prod_{j=1}^{n} a_j$ converges to L as $n \to \infty$ where L is invertible, and that $\sum_{j=1}^{\infty} |b_j| < \infty$. Then*

$$\prod_{j=1}^{n}(a_j + b_j)$$

converges to a finite complex matrix as $n \to \infty$.

Remark. Again, matrix products involving symbology $\prod_{k \geq 1}$ are interpreted as having the ($k = 1$) matrix at the far right. However, oppositely defined products can easily be handled via transposition of whole products.

Note that in one dimension if the limit is singular (i.e., zero) one talks about divergence to zero.

We shall require a series of lemmas, whose development could be stream-lined by applying more functional analytic ideas—for example, Lemma 6.2 is a consequence of the Banach open mapping theorem along with the principle of uniform boundedness, or of the stability of the condition number—but we opt to leave everything explicit.

Lemma 6.2. *Let (l_n) be a sequence in the set of $k \times k$ complex matrices $M_{k \times k}$. Suppose that $(l_n) \to L$, where L is an invertible matrix. Then there exists an $N \in Z^+$ and a $K \in R^+$ such that for all $n \in Z^+$, $n \geq N$ implies*

$$|bl_n| \geq K|b| \text{ for all } b \in M_{k \times k}.$$

Proof: Since the invertible elements form an open set in $M_{k \times k}$, there is an $\varepsilon > 0$ such that the set $B_\varepsilon = \{x \in M_{k \times k} : |x - L| \leq \varepsilon\}$ consists entirely of invertible elements.

Springer

Let $U = \{x \in M_{k \times k} : |x| = 1\}$. Define $\Phi : U \times B_\varepsilon \to R$ by

$$\Phi(u, x) = |ux| \text{ for all } (u, x) \in U \times B_\varepsilon.$$

Then Φ is a continuous function on the compact set $U \times B_\varepsilon$, so it has a minimum value K. Since x is invertible, and $u \neq 0$ implies $|ux| > 0$, we see that $K > 0$. We have

$$\left| \frac{b}{|b|} \cdot x \right| \geq K \text{ for all } b \in (M_{k \times k} \backslash \{0\}), x \in B_\varepsilon,$$

and hence

$$|bx| \geq K|b| \text{ for all } b \in M_{k \times k}, x \in B_\varepsilon.$$

Since $(l_n) \to L$, there is an $N \in Z^+$ such that $l_n \in B_\varepsilon$ for $n \geq N$. Then $n \geq N$ implies $|bl_n| \geq K|b|$ for all $b \in M_{k \times k}$. $\qquad \square$

Let us say that a matrix product is *tail-Cauchy* if $\prod_{j=p+1}^{q} a_j \to I$ as $q > p \to \infty$ (i.e., both tend to infinity). We note for future use that every tail-Cauchy sequence is bounded.

Lemma 6.3. *Let (a_n) be a sequence in $M_{k \times k}$.*

a. Every tail-Cauchy product converges;
b. Conversely, suppose $(\prod_{j=1}^{n} a_j) \to L$, where L is invertible, then (a_n) is tail-Cauchy.

Proof:

(a) As observed every tail-Cauchy product is bounded. Indeed, there is an $M \in Z^+$ such that

$$q > M \implies \left| \prod_{j=M+1}^{q} a_j - I \right| < 1 \implies \left| \prod_{j=1}^{q} a_j \right| < 2 \left| \prod_{j=1}^{M} a_j \right|.$$

Let $R := 2 |\prod_{j=1}^{M} a_j|$ (or any upper bound), and let $\varepsilon \in R^+$. By the tail-Cauchy property, we can choose $P > M$ such that

$$q > p \geq P \implies \left| \prod_{j=p+1}^{q} a_j - I \right| < \frac{\varepsilon}{R}.$$

Then

$$q > p \geq P \implies \left| \prod_{j=1}^{q} a_j - \prod_{j=1}^{p} a_j \right| \leq \left| \prod_{j=p+1}^{q} a_j - I \right| \left| \prod_{j=1}^{p} a_j \right| < \frac{\varepsilon}{R} \cdot R = \varepsilon.$$

Thus $(\prod_{j=1}^{n} a_j)$ is a Cauchy sequence, and hence it converges.

(b) Conversely, suppose the product converges to an invertible limit. By Lemma 6.2 there is an $N \in Z^+$ and $K \in R^+$ such that

$$p \geq N \implies \left| x \prod_{j=1}^{p} a_j \right| \geq K|z| \text{ for all } x \in M_{k \times k}.$$

Since $(\prod_{j=1}^{n} a_j)$ converges, it is Cauchy and so there is an $M \in Z^+$ such that

$$q > p \geq M \implies \left| \prod_{j=1}^{q} a_j - \prod_{j=1}^{p} a_j \right| < \varepsilon K \implies \left| \left(\prod_{j=p+1}^{q} a_j - I \right) \prod_{j=1}^{p} a_j \right| < \varepsilon K.$$

Let $P = \max\{M, N\}$. Then, as required

$$q > p \geq P \implies \left| \prod_{j=p+1}^{q} a_j - I \right| \leq \left| \left(\prod_{j=p+1}^{q} a_j - I \right) \prod_{j=1}^{p} a_j \right| \Big/ K < \varepsilon.$$

$\square$

Lemma 6.4 (Key inequality). *Let (a_n), (b_n) be sequences of $k \times k$ complex matrices. Suppose p in Z^+ is given such that*

$$q > r \geq p \longrightarrow \left| \prod_{j=r+1}^{q} a_j \right| \leq M < \infty. \tag{6.1}$$

Then for positive n,

$$\left| \prod_{j=p+1}^{p+n} (a_j + b_j) - \prod_{j=p+1}^{p+n} a_j \right| \leq M \left(\prod_{j=p+1}^{p+n} (1 + M|b_j|) - 1 \right).$$

Proof: For $n \in Z^+$, let $Z_n = \{1, 2, \ldots, n\}$, and for $p, n \in Z^+$ and $S \subset Z_n$ let

$$\pi_S^{p,n} := x_n x_{n-1} \ldots x_2 x_1$$

where

$$x_j := b_{p+j} \text{ if } j \in S \text{ but } x_j := a_{p+j} \text{ if } j \notin S.$$

Then

$$\prod_{j=p+1}^{p+n} (a_j + b_j) - \prod_{j=p+1}^{p+n} a_j = \sum_{S \subset Z_n, S \neq \emptyset} \pi_S^{p,n}.$$

Let $|S|$ denote the number of elements in S. By hypothesis (6.1) we have $M > 1$ such that

$$q > r \geq p \implies \left| \prod_{j=r+1}^{q} a_j \right| \leq M.$$

If $n \in Z^+$ and $S \subset Z_n$, then since there are at most $|S| + 1$ 'runs' of a_{p+j}'s

$$\left| \pi_S^{p,n} \right| \leq M^{|S|+1} \prod_{j \in S} |b_{p+j}|,$$

Hence,

$$\left| \prod_{j=p+1}^{p+n} (a_j + b_j) - \prod_{j=p+1}^{p+n} a_j \right| \leq \sum_{k=1}^{n} \sum_{S \subset Z_n, |S|=k} \left| \pi_S^{p,n} \right| \leq \sum_{k=1}^{n} \sum_{S \subset Z_n, |S|=k} M^{k+1} \prod_{j \in S} |b_{p+j}|$$

$$= M \left(\sum_{k=1}^{n} \sum_{S \subset Z_n, |S|=k} \prod_{j \in S} (M|b_{j+p}|) \right)$$

$$= M \sum_{S \subset Z_n, S \neq \emptyset} \prod_{j \in S} (M|b_{j+p}|)$$

$$= M \left(\prod_{k=1}^{n} (1 + M|b_{k+p}|) - 1 \right)$$

$$= M \left(\prod_{j=p+1}^{p+n} (1 + M|b_j|) - 1 \right).$$

$\square$

We may now establish the perturbation result

Proof of Theorem 6.1: By Lemma 6.3(b), (a_n) has a tail-Cauchy product and so hypothesis (6.1) holds for large p. Also, since $\sum |b_j|$ is convergent, $\prod (1 + M|b_j|)$ converges. Hence Lemma 6.4, shows that.

$$\left| \prod_{j=p+1}^{q} (a_j + b_j) - \prod_{j=p+1}^{q} a_j \right| \to 0$$

when $q > p \to \infty$. By the triangle inequality, $(a_n + b_n)$ has a tail-Cauchy product. An appeal to Lemma 6.3(a) concludes the proof of Theorem 6.1. $\square$

The importance of the perturbation Theorem 6.1 is evident: When analyzing the dynamics of recurrence sequences, we may—with impunity—discard certain troublesome terms in the overall matrix analysis. An immediate application of this idea appears in the next section.

Examples. It is instructive to note counterexamples to broader variants of Theorem 6.1, or Lemma 6.3(b), for example the invertibility of L is mandatory. For $n \in Z^+$, define

$$a_n := \begin{bmatrix} 1 & n^3 \\ 0 & \frac{1}{(n+1)^3} \end{bmatrix} \qquad b_n := \begin{bmatrix} 0 & 0 \\ \frac{1}{n^2} & 0 \end{bmatrix}.$$

Then we see by induction that

$$\prod_{j=1}^{n} a_j = \begin{bmatrix} 1 & \sum_{j=0}^{n-1} \frac{1}{(j!)^3} \\ 0 & \frac{1}{((n+1)!)^3} \end{bmatrix},$$

so $(\prod_{j=1}^{n} a_j)$ converges to a singular matrix, although the finite products are all nonsingular. However $\prod_{j=1}^{n}(a_j + b_j)$ *diverges* since $\det \prod_{j=1}^{n}(a_j + b_j) = \prod_{j=1}^{n}(\frac{1}{(j+1)^3} - j)$ which clearly diverges. Also (a_n) is clearly divergent.

Another example (from the scalar $(k = 1)$ domain) is $a_n := 2^n$, for n even and $a_n := 4^{-n}$ for n odd, yield convergent (to zero) products that are not tail-Cauchy and such that adding $b_n := 2^{-n}$ provides a summable perturbation for which $\prod_{n>0}(a_n + b_n)$ oscillate unboundedly. Indeed, for $P_n := \prod_{j=1}^{n}(a_j + b_j)$ we have $P_{2n} \to \infty$ while $P_{2n+1} \to 0$.

Note also that $a_n := 1$ and $b_n := -x^2/n^2$ provides a summable perturbation for which $\prod_{n>0}(a_n + b_n) = \sin(\pi x)/(\pi x)$ has zeros.

We conclude the section by recording the operator theoretic analogue of Theorem 6.1.

Corollary 6.5. *Theorem 6.1 holds in any Banach algebra with unit, and so in particular, for the bounded linear operators endowed with the uniform norm.*

Proof: We replace Lemma 6.2 by the standard fact [12, p. 310] that the mapping $x \mapsto x^{-1}$ is continuous at any invertible element. $\qquad\square$

7 Exponential-sum analysis

In view of the perturbation Theorem 6.1, when $|\omega| = 1$ we may write our key matrix product (4.2) relevant to the (v_n) sequence as

$$Z_N = \prod_{n=1}^{N} \left(I + \frac{1}{2an} \Omega_n + O\left(\frac{1}{n^2}\right) \right),$$

where

$$\Omega_n := \begin{bmatrix} 0 & \omega^n \\ \omega^{-n} & 0 \end{bmatrix}.$$

In this way, Theorem 6.1 essentially allows simplification of the Y_n matrix, because $Y_n \sim I + \Omega_n/(2an)$. As soon as we can show

$$U_N := \prod_{n=1}^{N}\left(I + \frac{1}{2an}\Omega_n\right)$$

converges for $a \neq 0$, $\omega := b^2/a^2 \neq 1$, $|\omega| = 1$, then the analysis has been brought full-circle, with $\mathcal{R}_1(a, b)$ diverging for such parameters via Theorem 4.1, so that the convergence region for $\mathcal{R}_1$ is finally resolved. (Note that, in view of the perturbation Theorem 6.1, if some U_N is singular we may take the product from some sufficiently large $n = n_0$ to ensure convergence to an invertible U_∞.)

With a view to calculation of the U_N products, we introduce certain exponential sums—which might be called *multiple-Lerch sums*—via

$$T_n(N, \omega) := \sum_{N \geq j_n > j_{n-1} > \cdots > j_1 \geq 1} \frac{\omega^{j_n - j_{n-1} + \cdots}}{j_n j_{n-1} \cdots j_1}, \qquad (7.1)$$

where the indices are alternating in the exponent of ω, and the sum is considered empty (0) if $n > N$, with also the assignment $T_0(N, \omega) := 1$. These sums are generally very difficult to cast into closed form, even for the full sum $T_n(\infty, \omega)$; however, we shall be able to establish useful upper bounds—indeed, we shall see that in our preferred scenario $|\omega| = 1$, $\omega \neq 1$, the sum $T_n(\infty, \omega)$ decays very rapidly in n.

Evidently,

$$U_N = \begin{bmatrix} \alpha_N(\omega) & \beta_N(\omega) \\ \beta_N(\omega^{-1}) & \alpha_N(\omega^{-1}) \end{bmatrix},$$

where

$$\alpha_N(\omega) := \sum_{n=0}^{\infty} (2a)^{-2n} T_{2n}(N, \omega),$$

$$\beta_N(\omega) := \sum_{n=0}^{\infty} (2a)^{-(2n+1)} T_{2n+1}(N, \omega). \qquad (7.2)$$

Note that these sums are manifestly finite, since $T_m(N, \omega)$ vanishes for $m > N$. Our goal is to show that both α_N, β_N converge to finite complex values as $N \to \infty$, regardless of complex $a \neq 0$ with $|\omega| = 1$, $\omega \neq 1$.

Examples. More explicitly let

$$L_\psi(\omega, n) := T_n(\infty, \omega) = \sum_{j_n > j_{n-1} > \cdots > j_1 \geq 1} \frac{\omega^{j_n - j_{n-1} + \cdots}}{j_n j_{n-1} \cdots j_1}.$$

Observe that $L_\psi(\omega, 1) = -\log(1 - \omega)$ for any $\omega \neq 1$, $|\omega| \leq 1$. For the case $\omega = -1$, $L_\psi(-1, n)$ is evaluated in [2] and elsewhere via $L_\psi(-1, n) = \zeta(-1, -1, \ldots, -1)$ (repeated n times) which is the coefficient of t^n in

$$A(t) := \frac{\sqrt{\pi}}{\Gamma\left(1 + \frac{t}{2}\right)\Gamma\left(\frac{1}{2} - \frac{t}{2}\right)}.$$

Our subsequent analysis will reveal that said coefficient decays very rapidly as $n \to \infty$.

Analyzing more general L_ψ seems substantially more difficult. For example, taking $\omega := i$ we obtain

$$L_\psi(i, 1) = i \int_0^1 \frac{dx}{1 - ix} = -\log(1 - i) = \frac{\pi}{4} + \frac{i}{2}\ln(2) = i\,\mathrm{Li}_1\left(\frac{1}{2} - \frac{i}{2}\right),$$

$$L_\psi(i, 2) = \sigma(1) = \frac{5\pi^2}{96} - \frac{1}{8}\ln^2(2) + i\left(\frac{\pi}{8}\ln(2) - \mathrm{G}\right) = -\mathrm{Li}_2\left(\frac{1}{2} - \frac{i}{2}\right),$$

where $\mathrm{G} := \sum_{n \geq 0}(-1)^n/(2n + 1)^{-2}$ is the Catalan constant, and Li_n is the standard polylogarithm of order n. This last relation follows from

$$\sigma(t) := \int_0^t \int_0^x \frac{1}{(y + i)(x + i)x}\,dy\,dx$$

$$= \frac{\zeta(2)}{4} + \frac{1}{2}\mathrm{Li}_2\left(\frac{2t}{t + i}\right) - \frac{1}{2}\mathrm{Li}_2\left(\frac{1 + it}{2}\right) - \frac{1}{4}\mathrm{Li}_2(-t^2)$$

$$+ \frac{1}{4}\ln(2)\ln\left(\frac{t^2 + 1}{2}\right) - \frac{3}{16}\ln^2(t^2 + 1)$$

$$+ \arctan(t)\left\{\frac{i}{4}\ln(4t^2 + 4) - \frac{1}{4}\arctan(t)\right\}.$$

With significant symbolic computational help we can derive

$$L_\psi(i, 3) = -\frac{1}{4}\int_0^1 \frac{\ln^2\left(\frac{t^2 + 1}{2}\right)}{1 - t}dt + \frac{1}{16}\ln^3(2) - \frac{7}{32}\zeta(3) - \frac{5\pi^2}{192}\ln(2)$$

$$- i\left\{\frac{7\pi^3}{128} - \frac{\mathrm{G}}{2}\ln(2) + \frac{\pi}{32}\ln^2(2) + 2\,\mathrm{Im}\,\mathrm{Li}_3\left(\frac{1 - i}{2}\right)\right\}.$$

The real part is also expressible as:

$$\frac{5\pi^2}{192}\ln(2) - \frac{1}{4}\zeta(3) - \frac{1}{48}\ln^3(2) - \frac{1}{4}\sum_{n=1}^{\infty} \frac{2^n\left(\sum_{k=1}^{n-1}\frac{1}{k}\right)}{n^2\binom{2n}{n}},$$

and one may use

$$\int_0^1 \frac{\ln^2\left(\frac{t^2+1}{2}\right)}{1-t}\,dt = \frac{1}{2}\ln^3(2) - \frac{5\pi^2}{24}\ln(2) + 2\sum_{n=2}^{\infty}\frac{(-1)^n}{n}\sum_{j=1}^{n-1}\frac{1}{j}\sum_{k=1}^{2n}\frac{1}{k}.$$

Consider now a general, truncated, constrained exponential sum

$$W_n(N; \rho, \ldots, \rho_n) := \sum_{N \ge j_n > j_{n-1} > \cdots > j_1 \ge 1} \frac{\rho_n^{j_n} \cdots \rho_1^{j_1}}{j_n j_{n-1} \cdots j_1}$$

$$= \left(\prod_{k=1}^{n}\rho_k\right)\int_0^1 \cdots \int_0^1 dx_1 \ldots dx_n \, S_n(N; \rho_1 x_1, \ldots, \rho_n x_n), \quad (7.3)$$

(when the integral exists), where we define

$$S_n(N; z_1, \ldots, z_n) := \sum_{N > k_n > \cdots k_1 \ge 0} z_n^{k_n} z_{n-1}^{k_{n-1}} \cdots z_1^{k_1}.$$

Note that the full constrained exponential sum is interpreted as $W_n(\infty, \rho_1, \ldots, \rho_n)$ which is the integral (7.3) with $S_n(\infty; \rho_1 x_1, \ldots, \rho_n x_n)$ used on the right side. Note now, the combinatorial matter,

$$S_n(\infty; z_1, \ldots, z_n) = \frac{z_n^{n-1}}{1 - z_n} \frac{z_{n-1}^{n-2}}{1 - z_n z_{n-1}} \cdots \frac{1}{1 - z_n z_{n-1} \cdots z_1}. \quad (7.4)$$

Our exponential sum (7.1) now takes the form

$$T_n(N, \omega) = W_n(N; \omega^{\pm 1}, \ldots, \omega^{-1}, \omega),$$

so that the full multiple-Lerch sum is

$$T_n(\infty, \omega) = \sum_{j_n > j_{n-1} > \cdots > j_i \ge 1} \frac{\omega^{j_n - j_{n-1} + \cdots}}{j_n j_{n-1} \cdots j_1}$$

$$= \omega^{n \bmod 2}\int_0^1 \cdots \int_0^1 dx_1 \ldots dx_n \, S_n(\infty; \omega^{2(n \bmod 2)-1} x_1, \ldots, \omega^{-1} x_{n-1}, \omega x_n)$$

$$= \omega^{\lfloor (n+1)/2 \rfloor}\int_0^1 \cdots \int_0^1 \frac{x_n^{n-1}}{1 - \omega x_n} \frac{x_{n-1}^{n-2}}{1 - x_n x_{n-1}} \frac{x_{n-2}^{n-3}}{1 - \omega x_n x_{n-1} x_{n-2}}$$

$$\cdots dx_1 \ldots dx_n. \quad (7.5)$$

We now establish four lemmas, with a view to bounding both $T_n(\infty, \omega)$ and $T_n(\infty, \omega) - T_n(N, \omega)$ for finite N:

Lemma 7.1. *The truncation error term*

$$E_n(N; z_1, \ldots, z_n) := S_n(\infty, z_1 \ldots, z_n) - S_n(N, z_1, \ldots, z_n)$$

is given by ($S_0 := 1$, and assuming all $|z_k| < 1$)

$$E_n(N) = \sum_{j=1}^{n} (-1)^{j-1} \frac{S_{n-j}(\infty, z_1, \ldots, z_{n-j})(z_n \ldots z_{n-j+1})^N}{(1 - z_{n-j+1})(1 - z_{n-j+1} z_{n-j+2}) \cdots (1 - z_{n-j+1} \ldots z_n)}.$$

Proof: Suppose that $|z_k| < 1$ for all $k \geq 1$. Let $S_n := S_n(\infty, z_1, \ldots, z_n)$, and

$$c_{k,n} := \frac{(-1)^{k+1} S_{k-1}}{\prod_{j=k}^{n}(1 - z_k z_{k+1} \ldots z_j)}.$$

The required conclusion is then

$$E_n(N) = \sum_{k=1}^{n} c_{k,n} \left(\prod_{j=k}^{n} z_j \right)^N. \tag{7.6}$$

Recursively, for $n > 1$, we have

$$E_n(N) = \frac{z_n^N}{1 - z_n} S_{n-1} - z_n^N \sum_{m=0}^{\infty} z_n^m E_{n-1}(m + N)$$

and

$$E_1(N) = \frac{z_1^N}{1 - z_1}.$$

We now prove (7.6) by induction. Direct calculation shows that (7.6) holds for $n = 2$ and $n = 3$. Suppose it holds with $n - 1$ in place of n, so that

$$E_{n-1}(N) = \sum_{k=1}^{n-1} c_{k,n-1} \left(\prod_{j=k}^{n-1} z_j \right)^N,$$

and hence

$$z_n^N \sum_{m=0}^{\infty} z_n^m E_{n-1}(m + N) = z_n^N \sum_{k=1}^{n-1} c_{k,n-1} \sum_{m=0}^{\infty} z_n^m \left(\prod_{j=k}^{n-1} z_j \right)^{N+m}$$

$$= \sum_{k=1}^{n-1} c_{k,n-1} \frac{\left(\prod_{j=k}^{n} z_j \right)^N}{1 - \prod_{j=k}^{n} z_j} = -\sum_{k=1}^{n-1} c_{k,n} \left(\prod_{j=k}^{n} z_j \right)^N.$$

Thus, by the recursion,

$$E_n(N) = \frac{z_n^N}{1-z_n} S_{n-1} + \sum_{k=1}^{n-1} c_{k,n} \left(\prod_{j=k}^n z_j\right)^N = \sum_{k=1}^{n} c_{k,n} \left(\prod_{j=k}^n z_j\right)^N,$$

as desired. □

We shall eventually use the idea that the truncation error in the multiple-Lerch sum itself has a representation

$$T_n(\infty, \omega) - T_n(N, \omega) = \omega^{n \bmod 2} \int_0^1 \cdots \int_0^1 dx_1 \ldots dx_n\, E_n(N; \omega^{2(n \bmod 2)-1}x_1,$$
$$\ldots, \omega^{-1}x_{n-1}, \omega x_n). \tag{7.7}$$

This integral can be given in a more explicit form such as (7.5), using Lemma 7.1 and the symbolic form (7.4).

Now we define for ω on the unit complex circle,

$$\sigma(\omega) := \sin(\min(\pi/2, |\arg(\omega)|)).$$

Lemma 7.2. *For positive integer p, with $q = p$ or $p-1$, and $|\omega| = 1$, real $\lambda \in [0, 1]$ define*

$$I(p, q, \omega, \lambda) := \int_0^1 \int_0^1 \frac{x^p y^q}{(1-\omega\lambda x)(1-\lambda xy)} dx dy.$$

Then

$$|I| \le \frac{1}{\sigma(\omega)p}.$$

Proof: We have

$$|I| \le \sup_{0 \le \rho \le 1} \frac{1}{|1-\omega\rho|} \sum_{m=1}^{\infty} \frac{1}{p+m}\frac{1}{q+m}.$$

Note first that $|1-\omega\rho|^2 = 1 - 2\rho \cos \arg(\omega) + \rho^2 \ge \sigma(\omega)^2$. The final sum is easily estimated by

$$\sum_{m=1}^{\infty} \frac{1}{(p+m)^2} < \sum_{m=1}^{\infty} \frac{1}{(p+m)(p-1+m)} = \frac{1}{p}.$$ □

We also have

Lemma 7.3. *For* $|\omega| = 1$*, the truncated multiple-Lerch sum has*

$$|T_n(N, \omega)| < \frac{2^N}{n!}.$$

Proof: Clearly, by considering the possible indices j_m in (7.1),

$$|T_n(N, \omega)| \leq \frac{\binom{N}{n}}{n!}$$

and the result follows. $\square$

These observations lead to sufficiently strong exponential-sum bounds:

Lemma 7.4. *For* $|\omega| = 1$*,* $\omega \neq 1$ *the multiple-Lerch sum* (7.1) *has*

$$|T_n(\infty, \omega)| \leq \frac{1}{\sigma^{\lfloor (n+1)/2 \rfloor}} \frac{1}{(n-1)!!},$$

and for $n < N$,

$$|T_n(\infty, \omega) - T_n(N, \omega)| \leq \frac{1}{\sigma^{n/2+1}} \frac{1}{(n-1)!!} \frac{2n}{N-n},$$

where $\sigma := \sigma(\omega)$ *and the double-factorial is defined by* $q!! := q(q-2)\cdots$
$((q \bmod 2) + 1)$ *for positive integers* q*, with* $0!! = (-1)!! := 1$.

Proof: We think of the integral representation (7.5) as a chain of $\lfloor n/2 \rfloor$ double integrals of the I-type of Lemma 7.2, with possibly one integral (over x_1) left over. Lemma 7.2 thus gives immediately

$$|T_n(\infty, \omega)| \leq 1/\sigma \cdot 1/(\sigma \cdot (n-1) \cdot \sigma \cdot (n-3)\cdots),$$

which is the desired bound for $N = \infty$. The integral (7.7) is more intricate, but happily the result of Lemma 7.1 is that E_n becomes separated with respect to integration variables $x_1, \ldots, x_{n-j}$ (via the S_{n-j} terms) and the variables $x_{n-j+1}, \ldots, x_n$. So the integral (7.7) separates to give, again via Lemma 7.2,

$$|T_n(\infty, \omega) - T_n(N, \omega)| \leq \sum_{j=1}^{n} \frac{1}{\sigma^{\lfloor (n-j+1)/2 \rfloor}} \frac{1}{(n-j-1)!!} \frac{1}{\sigma^{\lfloor (j+1)/2 \rfloor}} \frac{1}{N^{\lfloor (j+1)/2 \rfloor}}$$

$$\leq \frac{1}{\sigma^{n/2+1}} \left(\frac{1}{N} \left(\frac{1}{(n-2)!!} + \frac{1}{(n-3)!!} \right) \right.$$

$$\left. + \frac{1}{N^2} \left(\frac{1}{(n-4)!!} + \frac{1}{(n-5)!!} \right) + \cdots \right)$$

         Springer

$$\leq \frac{1}{\sigma^{n/2+1}} \frac{1}{(n-1)!!} \left(\frac{2n}{N} + \frac{2n^2}{N^2} + \cdots \right)$$

$$\leq \frac{1}{\sigma^{n/2+1}} \frac{1}{(n-1)!!} \frac{2n}{N} \frac{1}{1-n/N},$$

which proves the second bound of the theorem.   □

Next we define functions (recall $0!! = (-1)!! := 1$)

$$F(z) := \sum_{n=0}^{\infty} \frac{z^{2n}}{(2n-1)!!} = 1 + z e^{z^2/2} \int_0^z e^{-t^2/2} dt,$$

$$G(z) := \sum_{n=0}^{\infty} \frac{z^{2n}}{(2n)!!} = e^{z^2/2}.$$

The sums α_N, β_N in (7.2) may now be addressed. We shall say that a matrix A *dominates* B if $|A_{jk}| \geq |B_{jk}|$ for all index pairs j, k.

Theorem 7.5. *For $a \neq 0$, $|\omega| = 1$, $\omega \neq 1$ the matrix*

$$U_N := \prod_{n=1}^{N} \left(I + \frac{1}{2an} \Omega_n \right),$$

where

$$\Omega_n := \begin{bmatrix} 0 & \omega^n \\ \omega^{-n} & 0 \end{bmatrix},$$

converges as $N \to \infty$. Moreover, U_∞ is dominated by the matrix (with $\sigma := \sigma(\omega)$, $\rho := 1/|2a\sqrt{\sigma}|$):

$$\begin{bmatrix} F(\rho) & \rho G(\rho)/\sqrt{\sigma} \\ \rho G(\rho)/\sqrt{\sigma} & F(\rho) \end{bmatrix}.$$

Proof: Consider for the moment the α_N element from (7.2), assume integer $M \in [1, N/2)$, and decompose (recalling that α_N, β_N are actually finite sums, as $T_m(N, \omega) := 0$ for $m > N$):

$$\alpha_N(\omega) = \sum_{n=0}^{\lfloor N/2 \rfloor} \frac{1}{(2a)^{2n}} T_{2n}(N, \omega)$$

$$= \left(\sum_{0 \le n \le M} + \sum_{n=M+1}^{\lfloor N/2 \rfloor} \right) \frac{1}{(2a)^{2n}} T_{2n}(N, \omega)$$

$$= \sum_{0 \le n \le M} \frac{1}{(2a)^{2n}} T_{2n}(\infty, \omega) + \sum_{0 \le n \le M} \frac{1}{(2a)^{2n}} (T_{2n}(N, \omega) - T_{2n}(\infty, \omega))$$

$$+ O \left(\sum_{n=M+1}^{\lfloor N/2 \rfloor} \frac{1}{(2a)^{2n}} \frac{2^N}{(2n)!} \right).$$

It is evident from the first bound of Lemma 7.4 that the first sum here converges absolutely as $M \to \infty$. The second sum, by the second bound of Lemma 7.4, is bounded by the same F factor times $4\sqrt{\sigma} M/(N - 2M)$, and the big-$O$ term follows from Lemma 7.3. Now, for positive real x and the choice $M = \lceil \frac{1}{2} \frac{N}{\log N} - 1 \rceil$ with also $2M^2 > x^2$,

$$2^N \sum_{n>M} \frac{x^{2n}}{(2n)!} < 2^N \frac{x^{2M+2}}{(2M+2)!} \frac{1}{1 - x^2/(4M^2)} < 2 \left(\frac{3x \log N}{N^{1-\log 2}} \right)^{N/\log N},$$

where the last inequality arises from the Sterling bound $k! > (k/3)^k$. For our current choice of M we thus have

$$\alpha_N(\omega) = \sum_{0 \le n \le M} \frac{1}{(2a)^{2n}} T_{2n}(\infty, \omega) + O \left(\frac{1}{\log N} \right).$$

This means that the $\alpha_N(\omega)$ matrix element converges to the value

$$\alpha_\infty(\omega) := \sum_{n \ge 0} \frac{1}{(2a)^{2n}} T_{2n}(\infty, \omega),$$

which is in turn bounded in magnitude by $F(\rho)$. The same analysis applies to $\alpha_N(\omega^{-1})$ (note $\sigma(\omega) = \sigma(\omega^{-1})$). The same overall procedure works for β_N sums together with use of the G function, and the theorem is in this way proved. □

Corollary 7.6. *For $a \ne 0$, $\omega := b^2/a^2 \ne 1$, $|\omega| = 1$ the Ramanujan fraction $\mathcal{R}_1(a, b)$ diverges. In particular, the even/odd parts of $\mathcal{R}_1$ both converge, but to distinct limits.*

Proof: If some U_N from Theorem 7.5 is singular, then redefine the U_N product to start from some sufficiently large $n = n_0$. By the perturbation Theorem 6.1, and Theorem 7.5, the product Z_N of (4.2) thus converges to a finite matrix, and Theorem 4.1 applies. □

Via Corollary 7.6—and previous results for the $a^2 = b^2$ cases—we finally resolve the convergence domain for the Ramanujan AGM fraction, in the sense that the set equality $\mathcal{D}_1 = \mathcal{D}_0$ indeed holds. Note that, in terms of recurrence relations, we have hereby shown that (v_n) as defined in (1.4) converges to finite, bifurcated values (as n be even/odd) for the a, b parameters of Corollary 7.6, and in this way we also have immediate, precise knowledge of the asymptotic—and sometimes peculiar—behavior of sequences (t_n), (r_n), (q_n) for such parameter choices.

8 Matrix analysis toward a generalization of $\mathcal{R}_1$

Our exponential-sum resolution of the convergence domain was seen to be rather intricate, but it did, after all, provide explicit bounds on matrix elements, and did reveal some interesting aspects of multiple-Lerch sums. But there is a way to proceed more generally, to assail more general recurrence schemes than those attendant on the Ramanujan fraction. We hereby establish

Theorem 8.1. *Denote*

$$\Omega_n = \begin{bmatrix} 0 & \omega^n \\ \omega^{-n} & 0 \end{bmatrix}$$

where $|\omega| = 1$, $\omega \neq 1$. If z is an arbitrary complex number and a real, decreasing sequence (m_j) has $\sum_{j>0} m_j^2 < \infty$, then the matrix product

$$\prod_{j=1}^{n}(I + zm_j\Omega_j)$$

converges to a finite matrix as $n \to \infty$

Remark. Note that Theorem 8.1 implies the convergence in Theorem 7.5, since $\sum 1/n^2$ is finite. We have already intimated, though, as to the benefits of our quantitative multiple-Lerch sum analysis that led to Theorem 7.5.

In order to prove Theorem 8.1, we first establish some nomenclature and lemmas. Let (a_n) be a sequence of complex numbers, and let (ε_n) be a sequence of positive numbers. Denote the limit of (a_n) by a_∞. We write

$$(a_n) \prec (\varepsilon_n)$$

to mean that (a_n) converges with the caveat

$$|a_k - a_\infty| = O(\varepsilon_k).$$

Lemma 8.2. *Let (a_n) and (b_n) be complex sequences, let (ε_n) be a positive sequence, and let c be a complex number. Suppose that*

$$(a_n) \prec (\varepsilon_n) \text{ and } (b_n) \prec (\varepsilon_n).$$

Then

$$(a_n + b_n) \prec (\varepsilon_n), \quad (a_n b_n) \prec (\varepsilon_n), \quad (c a_n) \prec (\varepsilon_n).$$

Proof: The proof is straightforward. □

Lemma 8.3. *Let* (p_n) *be a decreasing sequence of non-negative real numbers that converges to* 0. *Then*

$$\left(\sum_{j=1}^{n} p_j w^j \right) \prec (p_n),$$

and

$$\left(\sum_{j=1}^{n} p_j \omega^{-j} \right) \prec (p_n).$$

Proof: By [16, Theorem 2.2 chapter 1] it follows that

$$\left| \sum_{j=n}^{\infty} p_j \omega^j \right| \leq p_n \max_{k \geq n} \left| \sum_{j=n}^{k} \omega^j \right| \leq \frac{2 p_n}{|1 - \omega|}.$$

□

Lemma 8.4. *The product*

$$\prod_{j=1}^{n} (1 + z m_j \omega^j)$$

converges as $n \to \infty$.

Proof: It follows from [9, p. 225] that if (a_n) is sequence such that $\sum_{j=1}^{n} a_j$ converges as $n \to \infty$, and $\sum_{j=1}^{n} |a_j|^2$ likewise converges, then $\prod_{j=1}^{n} (1 + a_j)$ also converges. Hence Lemma 8.4 follows from Lemma 8.3. □

Lemma 8.5. *Let*

$$U = \begin{bmatrix} 0 & 1 \\ 0 & 0 \end{bmatrix}, \quad L = \begin{bmatrix} 0 & 0 \\ 1 & 0 \end{bmatrix}.$$

For all $n \in \mathbb{Z}^+$, *let*

$$\Pi_U^n = \prod_{j=1}^{n} (I + z m_j \omega^j U) = \prod_{j=1}^{n} \begin{bmatrix} 1 & z m_j \omega^j \\ 0 & 1 \end{bmatrix},$$

$$\Pi_L^n = \prod_{j=1}^n (I + zm_j\omega^{-j}L) = \prod_{j=1}^n \begin{bmatrix} 1 & 0 \\ zm_j\omega^{-j} & 1 \end{bmatrix},$$

$$\Sigma_U^n = z\sum_{j=1}^n m_j\omega^j, \quad \Sigma_L^n = z\sum_{j=1}^n m_j\omega^{-j}.$$

Then (Π_U^n) and (Π_L^n) converge, and

$$\left(\Sigma_U^n\right) \prec (m_n), \quad \left(\Sigma_L^n\right) \prec (m_n). \tag{8.1}$$

Proof: We see by induction that

$$\Pi_U^n = \begin{bmatrix} 1 & \Sigma_U^n \\ 0 & 1 \end{bmatrix} \text{ and } \Pi_L^n = \begin{bmatrix} 1 & 0 \\ \Sigma_L^n & 1 \end{bmatrix}.$$

Conditions (8.1) hold by Lemmas 8.2 and 8.3, so (Π_L^n) and (Π_U^n) converge. □

Lemma 8.6. *For all $n \in Z^+$, let*

$$\Pi_{UL}^n = \prod_{j=1}^n ((I + zm_j\omega^j U)(I + zm_j\omega^{-j}L)).$$

Then

$$\Pi_{UL}^n = \Pi_U^n \Pi_L^n \prod_{j=1}^n (I + R_j),$$

where

$$R_n = zm_n\omega^{-n} \begin{bmatrix} -\Sigma_U^{n-1} - \left(\Sigma_U^{n-1}\right)^2\Sigma_L^{n-1} & -\left(\Sigma_U^{n-1}\right)^2 \\ \Sigma_U^{n-1}\Sigma_L^n + \Sigma_U^{n-1}\Sigma_L^{n-1} + \Sigma_L^{n-1}\Sigma_L^n\left(\Sigma_U^{n-1}\right)^2 & \Sigma_U^{n-1} + \left(\Sigma_U^{n-1}\right)^2 - \Sigma_L^n \end{bmatrix}.$$

Here we interpret Σ_U^0 and Σ_L^0 to be zero, so $R_1 := 0$.

Proof: The proof is by induction on n. For $n = 1$ the result is clear. Suppose the result holds for n. Then

$$\Pi_{UL}^{n+1} = (I + zm_{n+1}\omega^{n+1}U)(I + zm_{n+1}\omega^{-(n+1)}L)\Pi_{UL}^n$$

$$= (I + zm_{n+1}\omega^{n+1}U)(I + zm_{n+1}\omega^{-(n+1)}L)\Pi_U^n\Pi_L^n\prod_{j=1}^n(I + R_j).$$

We calculate the commutator

$$
\begin{aligned}
K &= \left[\left(I + z m_{n+1} \omega^{-(n+1)} L \right), \Pi_U^n \right] \\
&= \left[z m_{n+1} \omega^{-(n+1)} L, \, I + \Sigma_U^n U \right] = z m_{n+1} \omega^{-(n+1)} \left[L, \Sigma_U^n U \right] \\
&= z m_{n+1} \omega^{-(n+1)} \Sigma_U^n [L, U] \\
&= z m_{n+1} \omega^{-(n+1)} \begin{bmatrix} -\Sigma_U^n & 0 \\ 0 & \Sigma_U^n \end{bmatrix}.
\end{aligned}
$$

We have used the fact that $[L, U] = \begin{bmatrix} -1 & 0 \\ 0 & 1 \end{bmatrix}$. Since $XY = YX + [X, Y]$ for any 2×2 matrices X, Y, we have

$$
\Pi_{UL}^{n+1} = \left(I + z m_{n+1} \omega^{n+1} U \right) \left(\Pi_U^n \left(I + z m_{n+1} \omega^{-(n+1)} L \right) + K \right) \Pi_L^n \prod_{j=1}^{n} (I + R_j)
$$

$$
= \Pi_U^{n+1} \Pi_L^{n+1} (I + V_n) \prod_{j=1}^{n} (I + R_j)
$$

where

$$
\begin{aligned}
V_n &= \left(\Pi_L^{n+1} \right)^{-1} \left(\Pi_U^n \right)^{-1} K \Pi_L^n \\
&= z m_{n+1} \omega^{-(n+1)} \begin{bmatrix} 1 & 0 \\ -\Sigma_L^{n+1} & 1 \end{bmatrix} \begin{bmatrix} 1 & -\Sigma_U^n \\ 0 & 1 \end{bmatrix} \begin{bmatrix} -\Sigma_U^n & 0 \\ 0 & \Sigma_U^n \end{bmatrix} \begin{bmatrix} 1 & 0 \\ \Sigma_L^n & 1 \end{bmatrix} \\
&= z m_{n+1} \omega^{-(n+1)} \begin{bmatrix} -\Sigma_U^n - \left(\Sigma_U^n\right)^2 \Sigma_L^n & -\left(\Sigma_U^n\right)^2 \\ \Sigma_U^n \Sigma_L^{n+1} + \Sigma_U^n \Sigma_L^n + \Sigma_L^n \Sigma_L^{n+1} \left(\Sigma_U^n\right)^2 & \Sigma_U^n + \left(\Sigma_U^n\right)^2 \Sigma_L^{n+1} \end{bmatrix} \\
&= R_{n+1}.
\end{aligned}
$$

It follows that

$$
\Pi_{UL}^{n+1} = \Pi_U^{n+1} \Pi_L^{n+1} (I + R_{n+1}) \prod_{j=1}^{n} (1 + R_j) = \Pi_U^{n+1} \Pi_L^{n+1} \prod_{j=1}^{n+1} (I + R_j).
$$

This completes the proof of Lemma 8.6.   □

Lemma 8.7. *The sequence (Π_{UL}^n) converges.*

Proof: By Lemmas 8.5 and 8.6, it is sufficient to show that $\prod_{j=1}^{n}(I + R_j)$ converges. From the definition of R_n and Lemmas 8.2 and 8.5, we see that

$$R_n = zm_n\omega^{-n} \begin{bmatrix} q_{11}^n & q_{12}^n \\ q_{21}^n & q_{22}^n \end{bmatrix}$$

where

$$\left(q_{ij}^n\right) \prec (m_n) \text{ for all } i \text{ and } j.$$

Thus

$$R_n = P_n + zm_n\omega^{-n}T$$

where

$$P_n := zm_n\omega^{-n} \begin{bmatrix} q_{11}^n - q_{11}^\infty & q_{12}^n - q_{12}^\infty \\ q_{21}^n - q_{21}^\infty & q_{22}^n - q_{22}^\infty \end{bmatrix} \quad \text{and} \quad T := \begin{bmatrix} q_{11}^\infty & q_{12}^\infty \\ q_{21}^\infty & q_{22}^\infty \end{bmatrix}.$$

We have a matrix norm relation

$$|P_n| = O\left(m_n^2\right), \quad \text{and} \quad \text{hence} \quad \sum_{n=1}^{\infty} |P_n| \text{ converges.}$$

By our perturbation Theorem 6.1, if we show that $\prod_{j=1}^{n}(I + zm_j\omega^{-j}T)$ converges, it will follow that $\prod_{j=1}^{n}(I + R_j)$ converges. From the development for V_n above, we see that

$$\det(T) = -\left(\Sigma_U^\infty\right)^2 \quad \text{and that} \quad \text{trace}(T) = 0.$$

Hence the characteristic polynomial of T is given by

$$p(x) = x^2 - \left(\Sigma_U^\infty\right)^2.$$

If $\Sigma_U^\infty \neq 0$, then T has distinct eigenvalues $\pm\lambda$, so T is diagonalizable and there is a matrix M such that

$$I + zm_n\omega^{-n}T = M \begin{bmatrix} (1 + \lambda zm_n\omega^{-n}) & 0 \\ 0 & (1 - \lambda zm_n\omega^{-n}) \end{bmatrix} M^{-1},$$

and hence

$$\prod_{j=1}^{n}(I + zm_j\omega^{-j}T) = M \begin{bmatrix} \prod_{j=1}^{n}(1 + \lambda zm_j\omega^{-j}) & 0 \\ 0 & \prod_{j=1}^{n}(1 - \lambda zm_j\omega^{-j}) \end{bmatrix} M^{-1}.$$

This product converges by Lemma 8.4. If $\Sigma_U^\infty = 0$, then $T = 0$, so $\prod_{j=1}^n (I + zm_j\,\omega^{-j}T)$ certainly converges. $\qquad\square$

We may now move to

Proof of the Theorem 8.1: We have

$$I + zm_j\Omega^j = (I + zm_j\omega^j U)(I + zm_j\omega^{-j}L) - z^2 m_n^2 U L.$$

By Lemma 8.7, we know that $\prod_{j=1}^n (I + zm_j\omega^j U)(I + zm_j\omega^{-j}L)$ converges, and since $\sum_{j=1}^n |z^{2j}m_j^2 U L| < \infty$, it follows from our perturbation Theorem 6.1 that $\prod_{j=1}^n (I + zm_j\Omega^j)$ converges. $\qquad\square$

Let us now generalize the Ramanujan AGM fraction in the following manner. Define for an extra parameter c the continued fraction (which may or may not converge)

$$\mathcal{Q}(a, b, c) := \cfrac{1^c b^2}{1 + \cfrac{2^c a^2}{1 + \cfrac{3^c b^2}{1 + \cfrac{4^c a^2}{1 + \cfrac{}{\ddots}}}}}$$

Note that $\mathcal{S}(a, b) = \mathcal{Q}(a, b, 2)$ so that the Ramanujan fraction (1.1) itself is the specific assignment

$$\mathcal{R}_1(a, b) = \frac{a}{1 + \mathcal{Q}(a, b, 2)}.$$

Thus we know, for example, that

$$\mathcal{Q}(1, 1, 2) = \frac{1}{\log 2} - 1.$$

With a view to our general Theorem 8.1, we may define a sequence $(v_n^{(c)})$ as a modification of our (v_n) in (1.4):

$$v_n^{(c)} := \frac{q_n}{\Gamma^{c/2}(n + 3/2)\kappa_n^{(n+1)/2}},$$

where as usual $\kappa_n := a^2, b^2$ as n be even, odd respectively. Now the algebra that opened Section 4 reads

$$\begin{bmatrix} v_{2n}^{(c)} \\ v_{2n-1}^{(c)} \end{bmatrix} = Y_n^{(c)} \begin{bmatrix} v_{2n-2}^{(c)} \\ v_{2n-3}^{(c)} \end{bmatrix},$$

          Springer

where the Y matrix here is

$$
Y_n^{(c)} := \begin{bmatrix} \frac{4n^c + 1/a^2}{(4n^2 - 1/4)^{c/2}} & \frac{\omega^n}{a(2n+1/2)^{c/2}} \left(\frac{(2n-1)^2}{(2n-1)^2 - 1/4} \right)^{c/2} \\ \frac{\omega^{-n}}{a(2n-1/2)^{c/2}} & \left(\frac{(2n-1)^2}{(2n-1)^2 - 1/4} \right)^{c/2} \end{bmatrix}.
$$

Thus the analogue of (4.2), namely

$$
Z_N^{(c)} := \prod_{n=1}^{N} Y_n^{(c)}
$$

can be written, when $|\omega := b^2/a^2| = 1$,

$$
Z_N^{(c)} = \prod_{n=1}^{N} \left(I + \frac{1}{a(2n)^{c/2}} \Omega_n + O\left(\frac{1}{n^2} \right) \right).
$$

Note also, by analogy with the determinantal development prior to (4.3) we have

$$
\det \left(Z_\infty^{(c)} \right) = \left(\frac{\pi}{2} \right)^{c/2}.
$$

Theorem 8.8. *For parameter choices $a \neq 0$, $|\omega := b^2/a^2| = 1$, $\omega \neq 1$, the generalized Ramanujan fraction*

$$
\mathcal{Q}(a, b, c) := \cfrac{1^c b^2}{1 + \cfrac{2^c a^2}{1 + \cfrac{3^c b^2}{1 + \cfrac{4^c a^2}{1 + \ddots}}}}
$$

diverges for any real $c > 1$, with the even/odd parts of $\mathcal{Q}$ converging to separate limits.

Proof: We may with impunity discard outright the $O(1/n^2)$ perturbation in $Z_N^{(c)}$ by Theorem 6.1, so taking $m_j := j^{-c/2}$ in Theorem 8.1, and employing Theorem 4.1 as before (except with the determinant $\pi/2 \to (\pi/2)^{c/2}$ throughout) we obtain convergence of the even/odd parts to separate values, hence divergence. $\qquad\square$

In fact, the bound of (1.5) for the separation of convergents in terms of v_n is to be modified simply by replacing v_n with $v_n^{(c)}$.

It could be that the precise convergence region for $\mathcal{Q}(a, b, c)$ for every real $c > 1$ is identical to that (region $\mathcal{D}_1$) of the Ramanujan case ($c := 2$); we have not investigated this, except to give Theorem 8.8 above. It is interesting that even our general convergence Theorem 8.1 does not directly handle the divergence question for $\mathcal{Q}(a, b, 1)$.

We do know some exact evaluations for converging instances with $c = 1$, for example

$$\mathcal{Q}(a, a, 1) := \cfrac{1 \cdot a^2}{1 + \cfrac{2 \cdot a^2}{1 + \cfrac{3 \cdot a^2}{1 + \cfrac{4 \cdot a^2}{1 + \ddots}}}}$$

$$= \frac{4}{a e^{1/(2a^2)}(2\pi)^{1/2}\mathrm{erfc}(1/(a\sqrt{2}))} - 1,$$

where erfc is the standard complementary error function, $\mathrm{erfc}(z) := (2/\sqrt{\pi}) \int_z^\infty e^{-t^2} dt$. Perhaps surprisingly, there is another one-parameter class of exact evaluations, namely

$$\mathcal{Q}\left(\sqrt{1 + z^2}, z, 1\right) = \frac{z}{\arctan z} - 1,$$

valid at least for $\mathrm{Re}(z^2) > -1/2$ [11, 3.6.10, p. 571].

Consistent with the scope of Theorem 8.8, it appears that there are pairs a, b with $a/b \neq 1$, $|a/b| = 1$ such that $\mathcal{Q}(a, b, 1)$ actually *converges*. Using the above arctan formula, for example, we have, at least formally,

$$\mathcal{Q}\left(\frac{1}{\sqrt{2}}, \frac{i}{\sqrt{2}}, 1\right) = \frac{i}{\sqrt{2}\arctan(i/\sqrt{2})} - 1 = -0.1977218382755\ldots,$$

and numerical experiments show that the continued fraction appears to approach this value (which we say because $\mathrm{Re}(z^2) = 1/2$ in this case, which is outside the validity range given — indeed the region of validity for the arctan representation must certainly be wider than has been proven). These observations suggest that research is in order on the evidently discontinuous change in convergence behavior at real $c = 1$.

There are, of course, other directions in which to generalize the Ramanujan construct, one of the most alluring being extensions of the defining set of recurrence relations, such as the relations relevant to the construct

$$\cfrac{1^2 c^2}{1 + \cfrac{2^2 b^2}{1 + \cfrac{3^2 a^2}{1 + \cfrac{4^2 c^2}{1 + \ddots}}}}$$

that is, the n-th numerator is $n^2 \cdot (c^2, b^2, a^2)$ as $n = (1, 2, 0)$ modulo 3, so there are three essential relations (in the sense that the recurrence in our Abstract amounts to two essential relations).

Such a generalization would naturally lead into 3×3 matrix theory (so that Theorem 6.1 is intact, but such as Theorems 7.1 and 8.1 need be carefully extended). It could be that some nontrivial analogue of the "chaotic" convergents' behavior for the case $(a, b) = (i, i)$ of the Ramanujan fraction exists for the above construct—perhaps for some choice other than $(a, b, c) = (i, i, i)$—and such a finding could conceivably lead to a new class of chaotic generators.

9 The classical Pincherle and Auric theorems

The Pincherle theorem [11, Theorem 7, p. 202] says essentially that if *any* two solutions (with neither being the zero sequence $(0, 0, \dots)$) to a given recurrence have the same asymptotic behavior (in a quantifiable sense), then the associated continued fraction diverges. Moreover, the converse is true: If one can find two nonzero recurrence solutions with one dominating the other (say $t_n(t_0, t_i)/t_n(t_0', t_1') \to 0$, in which case the numerator sequence is called *minimal*), then the fraction converges. The Auric theorem [11, Theorem 10, p. 207] gives a convergence criterion (and an actual fraction value) based on a minimal recurrence solution. With a view to the Auric-Pincherle theory, one may construct an entity relevant to the (v_n) matrix iteration (4.1):

$$\frac{1}{a} \sum_{n=0}^{\infty} \frac{(-1)^n}{\bar{v}_n \bar{v}_{n-1}}, \tag{9.1}$$

where we use the notation $\bar{v}$ to signify that we are allowed to adopt initial conditions other than the standard ones in (4.4). (Note that (9.1) is equivalent to the literature presentation [11, p. 202] of this key sum.) Now (by Pincherle) if initial conditions $\bar{v}_0, \bar{v}_1$ can be found such that this sum (9.1) diverges to ∞, then $(\bar{v}_n)$ is a minimal solution of the recurrence. So the continued fraction associated with $(\bar{v}_n)$ converges if and only if (9.1) diverges to ∞. Similarly (by Auric) if (9.1) diverges to ∞ then, remarkably, the actual *value* of the continued fraction can be given in terms of the initial conditions, as $\mathcal{S}(a, b) = -a\bar{v}_0/(2\bar{v}_1)$. Note an immediate echo of one of our previous results: If the matrix Z_∞ of (4.3) exists, then (9.1) cannot diverge to ∞ (for any initial conditions on $\bar{v}$) and so the associated fraction must diverge.

Conversely, there is point of consistency: Closed-form values for *convergent* fractions in the Auric theory are entirely consistent with the developments in our Sections 2 and 3. For example, when $a = b$ with Re $(a) > 0$ it turns out that a minimal solution can be written down exactly, as

$$\bar{v}_n := \frac{\Gamma(n + 2) F_{n+1}(-a)}{\Gamma(n + 3/2) a^{n+1}}$$

(equivalently, $\bar{t}_n := F_n(-a)$ for the relevant $(\bar{t}_n)$ sequence, with F_n as in (2,3)) so that (9.1) is seen via (2.6) to diverge to complex infinity, and sure enough the fraction value $\mathcal{S}(a, a) = -a\bar{v}_0/(2\bar{v}_1)$ agrees with the hypergeometric-ratio term of (3.2).

These and some other classical ideas are mirrored in our matrix analysis; indeed, we have shown along the way that literature analyses such as Jacobsen-Masson divergence theory—which established divergence of $\mathcal{S}(i, i)$ in [4]—are also consistent with our

On the dynamics of certain recurrence relations   99

present development. We hope that our methods can add to the already voluminous theory of complex continued fractions. It would be interesting to forge a comprehensive theory—beyond, say, the particular case of the Ramanujan AGM fraction—that fuses all of these matrix methods with the classical approaches.

10 Some dynamics and their pictures

We may think of the capstone *dynamical system* $t_0 := t_1 := 1$:

$$t_n \hookleftarrow \frac{1}{n} t_{n-1} + \kappa_{n-1} \left(1 - \frac{1}{n} \right) t_{n-2}$$

where $\kappa_n = a^2, b^2$ for n even, odd respectively, of (1.2) as a *black box*. Numerically all one learns is that t_n is slowly tending to zero.

Pictorially, we learn significantly more (Fig. 1 shows the iterates swirling in towards zero, with the shading changing every few hundred iterates), and after scaling by $\sqrt{n}$, and on coloring odd and even iterates distinctly, fine structure appears—depending on whether a and/or b is a root of unity in the multiplicative group of the circle (Figs. 2 and 3 show four representative choices of iterates of $\sqrt{n}t_n$, in the Argand plane).

Indeed these pictures were part of our road to discovery. The behavior is now entirely explained by Corollary 7.6. Using (1.2) and (1.4) we have that

$$\sqrt{n+1}t_{n+1} \sim v_n \sqrt{\kappa_n}^{-n+1}.$$

When $a^2/b^2 \neq 1$, $|a/b| = 1$ we know that v_n has bifurcated convergent subsequences while $\sqrt{\kappa_n}$ alternates between a and b, so an appeal to Weyl's uniform convergence criterion explains the dynamics.

Fig. 1 The decay of t_n. Initial conditions t_o, t_1 are chosen, then a plot in the complex plane shows the amplitude $|t_n|$ decaying as $1/\sqrt{n}$

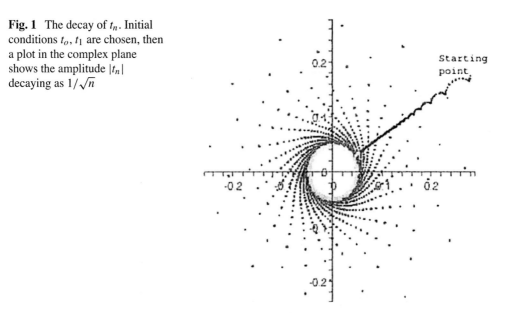

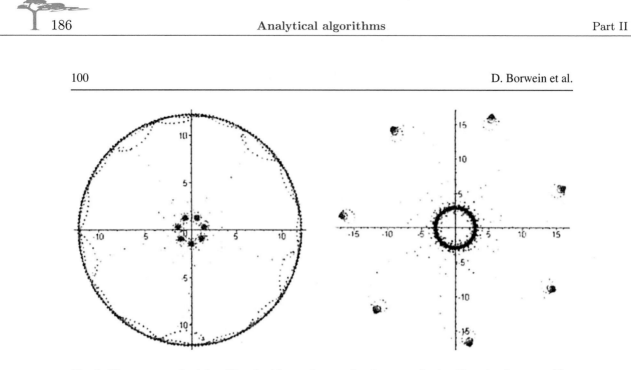

Fig. 2 The attractors for $|a| = |b| = 1$ with exactly one of a, b a root of unity. Plotted points are $\sqrt{n} t_n$, which we expect to be bounded

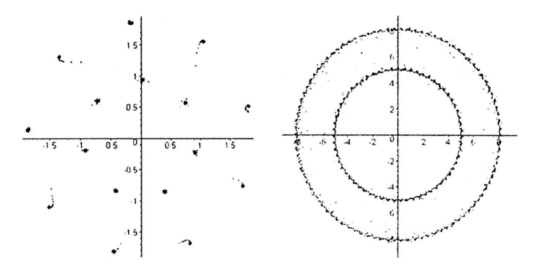

Fig. 3 The attractors for $|a| = |b| = 1$ with both or neither a, b being roots of unity. Plotted points are as in our previous figure

11 Some unresolved issues

We finish by listing some open questions.

- We again mention that we do not know a single nontrivial closed-form value of $\mathcal{R}_1(a, b)$; meaning, except for $a = 0$ or $b = 0$ or $a = b$ and some cases $\mathcal{R}_1(a, b) = \infty$ (see [4]) we have no closed forms. Could the minimal-solution scheme implicit in the Pincherle-Auric theory somehow reveal nontrivial closed forms?
- The dynamics in Figs. 2 and 3 are now explained, but what happens for more general, or even random choices of κ_n? Computational experiment suggests that (1.2) behaves similarly when the κ_n are chosen cyclically with any even period but not when chosen with an odd period. When κ_n is chosen randomly, the dynamics appear surprisingly deterministic. Why should the dynamics be so sensitive to the

On the dynamics of certain recurrence relations   101

parity (odd/even) of cycles, yet so robust when the parameters are randomly chosen? What is the nature of the convergence of t_n for cycles, odd or even, of length $c > 2$. For random κ_n, does the sequence t_n converge, almost surely or otherwise? Related results may be found in [5].

- Perhaps there is a universal, hypergeometric form for $\mathcal{Q}(a, b, 1)$? The existence of *two* one-parameter families of forms for $\mathcal{Q}(a, b, 1)$ suggests there may be more accessible forms.

- Could it be that the precise convergence domain $\mathcal{D}_0 = \mathcal{D}_1$ is also valid for the generalized Ramanujan fraction $\mathcal{Q}(a, b, c)$, any real $c > 1$? (Theorem 8.8 is consistent with this hypothesis.)

- We mentioned the other appealing generalization in which a, b, c as appearing modulo 3 or higher. Such would require 3×3 matrix theory, etc. Again computation suggests similar dynamics for yet other 3-fold variants of (1.2) such as

$$t_n = \frac{1}{n}t_{n-1} + \frac{1}{n}t_{n-2} - \frac{n-2}{n}t_{n-3}.$$

- There are at least potential cryptographic/chaos-generator applications, and likewise Monte-Carlo applications of our (t_n). In what sense is a *modular* iteration

$$t_n = (t_{n-1} - (n-1)t_{n-2}) \cdot n^{-1} \bmod N$$

chaotic, or at least useful? It is not inconceivable that such a chaotic scheme could apply to factoring (of N), much like the celebrated Pollard-rho factoring method.

Acknowledgments We thank B. Berndt, J. Buhler, J. Coope, D. Griffiths, W. Jones, L. Lorentzen, R. Luke, and S. Rayhawk for many valuable discussions.

References

1. Andrews, G., Askey, R., Roy, R.: Special Functions. Cambridge University Press (1999)
2. Borwein, J., Bailey, D., Girgensohn, G.: Experimentation in Mathematics: Computational Paths to Discovery. AK Peters (2004)
3. Borwein, J., Crandall, R., Fee, G.: On the Ramanujan AGM fraction. Part I: The real-parameter case. Experimental Mathematics, **13**, 275–286 (2004) [CECM Preprint 2003: 208/213]
4. Borwein, J., Crandall, R.: On the Ramanujan AGM fraction. Part II: The complex parameter case. Experimental Mathematics, **13**, 287–296 (2004) [CECM Preprint 2003: 214]
5. Borwein, J.M., Luke, D.R.: Dynamics of some random continued fractions, Abstract and Applied Analysis, 449–468 (2005) [CoLab Preprint 275]
6. Gill, J.: Infinite compositions of Mobius transforms. Trans. Amer. Math. Soc. **176**, 479–487 (1973)
7. Golub, G.H., van Loan, C.F.: Matrix Computations. The John Hopkins University Press (1989)
8. Jones, W., Thron, W.: Continued Fractions: Analytic Theory and Applications. Addison-Wesley (1980)
9. Knopp, K.: Theory and Application of Infinite Series. Dover (1990)
10. Luke, Y.: The special functions and their approximations, vol. I. In: Bellman, R. (ed.) Mathematics in science and engineering, vol. LIII. Academic Press, New York (1969)
11. Lorentzen, L., Waadeland, H.: Continued Fractions with Applications. NorthHolland (1992)
12. Megginson, R.E.: An Introduction to Banach Space Theory. Springer-Verlag (1998)
13. Olver, F.W.J.: Asymptotics and Special Functions. Academic Press (1974)
14. Petkovsek, M., Wilf, H., Zeilberger, D.: $A = B$, AK Peters (1996)
15. Wall, H.S.: Analytic Theory of Continued Fractions. D. Van Nostrand (1948)
16. Zygmund, A.: Trigonometric Series, vol I, Cambridge University Press (1959)

4. Effective Laguerre asymptotics

Discussion

This work has supported other lines of research in at least fourfold fashion:

1. The incomplete gamma function $\Gamma(a, z)$ occurs in the Riemann-splitting expansion $\zeta(s)$ (see "Unified algorithms for polylogarithms, L-functions, and zeta variants" in this volume). In order to use incomplete-gamma series forms *rigorously* to extreme precision, it is not enough merely to know the classical O-bounds on asymptotic terms. This author had searched fruitlessly far and wide for effective subexponential convergence bounds on continued fractions. Though there are continued-fraction subexponential-convergence bounds via such expedients as the intricate "oval theorems," effective bounds did not seem to exist. But relation (5) in this paper gives a direct way to bound error terms for incomplete-gamma itself.

2. Figure 1 can actually be realized numerically. Though many of us learned in youth the great historical machinations of complex analysis, it was never clear that some day contour integration could be an actual quadrature method giving extreme precision. Indeed, as intimated in Section 2.2, numerical quadratiue on a contour can sometimes outpace a direct defining sum for a special function!

3. Algorithm in Section 6.2 can stand as a model for other algorithms to generate effective bounds. Theorem 8 is the "effective solution" to problem of rigorous bounds. Table 1 shows actual experiments on Laguerre evaluations, including such oddities as $L_{10000000}^{-3/2}(1 - 100i)$, for which ten algorithm-generated asymptotic terms give a theoretical relative error beneath 10^{-10} and an actual, measured error of about 10^{-22}. So the bounds from this paper may not be optimal, but they at least allow rigorous arbitrary precision.

4. Section 7.2 describes means for generating Bessel series that *while manifestly convergent, agree in a certain sense with classical asymptotics*. Equation (50) for example was surprising to this author, as it is convergent and yet looks very much like the celebrated Hankel asymptotic development.

Challenges

Section 8 has various open problems. The problem of optimizing Riemann-splitting series for Riemann ζ remains an active research area. There certainly must be wider applications of the "exp-arc" theory in Section 5.

Source

D. Borwein, J. Borwein, and R. Crandall, "Effective Laguerre asymptotics," SIAM J. NUMER. ANAL. (2008).

SIAM J. NUMER. ANAL.   © 2008 Society for Industrial and Applied Mathematics
Vol. 46, No. 6, pp. 3285–3312

EFFECTIVE LAGUERRE ASYMPTOTICS*

DAVID BORWEIN[†], JONATHAN M. BORWEIN[‡], AND RICHARD E. CRANDALL[§]

Abstract. It is known that the generalized Laguerre polynomials can enjoy subexponential growth for large primary index. In particular, for certain fixed parameter pairs (a, z) one has the large-n asymptotic behavior

$$L_n^{(-a)}(-z) \sim C(a, z) n^{-a/2 - 1/4} e^{2\sqrt{nz}}.$$

We introduce a computationally motivated contour integral that allows efficient numerical Laguerre evaluations yet also leads to the complete asymptotic series over the full parameter domain of subexponential behavior. We present a fast algorithm for symbolic generation of the rather formidable expansion coefficients. Along the way we address the difficult problem of establishing effective (i.e., rigorous and explicit) error bounds on the general expansion. A primary tool for these developments is an "exp-arc" method giving a natural bridge between converging series and effective asymptotics.

Key words. Laguerre polynomials, effective asymptotic expansions, contour integrals

AMS subject classifications. 34E05, 33C65

DOI. 10.1137/07068031X

1. The challenge of "effectiveness." For primary integer indices $n = 0, 1, 2, \ldots$, we define the Laguerre polynomial thus:

$$(1) \qquad L_n^{(-a)}(-z) := \sum_{k=0}^{n} \binom{n-a}{n-k} \frac{z^k}{k!}.$$

Our use of negated parameters $-a, -z$ is intentional, for convenience in our analysis and in connection with related research, as we later explain. We shall work on the difficult problem of establishing asymptotics *with* effective error bounds for the two-parameter domain

$$\mathbb{D} := \{(a, z) \in \mathbb{C} \times \mathbb{C} : z \notin (-\infty, 0]\}.$$

That is, a is any complex number, while z is a complex number not on the negative-closed cut $(-\infty, 0]$.

Herein, we say a function $f(n)$ is of *subexponential* growth if $\log f(n) \sim C n^D$ as $n \to \infty$ for positive constants C, D, with $D < 1$. It will turn out that $\mathbb{D}$ is the precise domain of subexponential growth of $L_n^{(-a)}(-z)$ as $n \to \infty$, with (a, z) fixed. The reason for the negative-cut exclusion on z is simple: For z negative real, the Laguerre polynomial exhibits oscillatory behavior in large n and is not of subexponential growth. Note also, from the definition (1), that

$$(2) \qquad L_n^{(-a)}(0) = \binom{n-a}{n},$$

which covers the case $z = 0$; again, not subexponential growth.

---

*Received by the editors January 17, 2007; accepted for publication (in revised form) May 29, 2008; published electronically October 17, 2008.

http://www.siam.org/journals/sinum/46-6/68031.html

†Department of Mathematics, University of Western Ontario, London, ON, N6A 5B7, Canada and School of Mathematical and Physical Sciences, University of Newcastle, Callaghan, NSW 2300, Australia (dborwein@uwo.ca).

‡Faculty of Computer Science, Dalhousie University, Halifax, NS, B3H 2W5, Canada (jborwein@cs.dal.ca). This author's work was funded by NSERC and the Canada Research Chair Program.

§Center for Advanced Computation, Reed College, Portland, OR 97202 (crandall@reed.edu).

3286   D. BORWEIN, J. M. BORWEIN, AND R. E. CRANDALL

For fixed $(a, z) \in \mathbb{D}$, we shall have the large-n behavior (here and beyond we define $m := n + 1$ which will turn out to be a natural reassignment):

$$(3) \qquad L_n^{(-a)}(-z) \; \sim \; S_n(a, z) \left(1 + O\left(\frac{1}{m^{1/2}}\right)\right),$$

where the subexponential term S_n is

$$(4) \qquad S_n(a, z) \; := \; \frac{e^{-z/2}}{2\sqrt{\pi}} \frac{e^{2\sqrt{mz}}}{z^{1/4-a/2} \, m^{1/4+a/2}}.$$

In such expressions, $\Re\left(\sqrt{mz}\right)$ denotes $\sqrt{m|z|}\cos(\theta/2)$, where $\theta := \arg(z) \in (-\pi, \pi]$ (we hereby impose $\arg(-1) := \pi$), and so for $(a, z) \in \mathbb{D}$, the expression (4) involves genuinely diverging growth in n.

1.1. Research motives. There are many interdisciplinary applications of Laguerre asymptotics. The Laguerre functions appear standardly in the quantum theory of the hydrogen atom [46, Chapter 4] and in certain exactly solvable three-body problems of chemical physics [13], [14]. The so-called WKB phase of a quantum eigenstate can, in such cases and for high quantum numbers, be calculated via Laguerre asymptotics. The Hermite polynomials $H_n(x)$—closely related to the Laguerre polynomials $L_n^{(\pm 1/2)}(x^2)$—also figure in quantum analyses; e.g., one may derive Hermite asymptotics from Laguerre asymptotics. In this context there is the fascinating López–Temme representation of *any* $L_n^{(-a)}(-z)$ as a finite superposition of Hermite evaluations [23].

Our own research motive for providing effective asymptotics involves not the Laguerre–Hermite connection, rather it begins with a beautiful link to the incomplete gamma function, namely, [19]

$$(5)$$

$$\Gamma(a, z) \; = \; z^a e^{-z} \cfrac{1}{z + \cfrac{1-a}{1 + \cfrac{1}{z + \cfrac{2-a}{1 + \cdots}}}} \; = \; z^a e^{-z} \sum_{n=0}^{\infty} \frac{(1-a)_n}{(n+1)!} \frac{1}{L_n^{(-a)}(-z)\, L_{n+1}^{(-a)}(-z)},$$

where $(c)_n := c(c+1)\cdots(c+n-1)$ is the *Pochhammer symbol*. This series is valid whenever none of the Laguerre denominators has a zero. Thus an interesting sidelight is the research problem of establishing zero-free regions for Laguerre polynomials (see our Open problems section). We note that convergence of the continued-fraction can be highly problematic, especially when the full complex parameter space is allowed. One might encounter smooth convergence, or steady divergence, or even chaotic divergence, all of which possibilities are exemplified in [7, 8, 9, 22].

The problem, then, of general Laguerre asymptotics arises because there is a way to write Riemann zeta-function evaluations $\zeta(s)$ in terms of the incomplete gamma function, as was known to Riemann himself [40, p. 22]. But when s has large imaginary height, one becomes interested in the incomplete gamma's corresponding complex parameters (a, z) also of possibly large imaginary height [11], [12]. For example, in the art of prime-number counting, say, for primes $< 10^{20}$, one might need to know, for example, $\Gamma\left(3/4 + 10^{10}\, i, z\right)$ for various z also of large imaginary height, to good precision, in order to evaluate $\zeta(s)$ high up on the line $\Re(s) = 3/2$. The point is, exact computations on prime numbers must, of course, employ rigor—hence the need for effective error bounds.

Independent of Riemann-ζ considerations is the following issue: Continued-fraction theory is to this very day incomplete in a distinct sense. The vast majority of available convergence theorems are for Stieltjes, or S-fractions. Now the continued fraction in (5) is an S-fraction only when the a parameter is real [24, p. 138]. Due to the relative paucity of convergence theorems outside the S-fraction class, one encounters great difficulty in estimating the convergence rate for arbitrary $\Gamma(a, z)$; this is what led to our focus on the Laguerre asymptotics. Incidentally, we are aware that subexponential convergence results *might* be attainable via the complicated and profound work of Jacobsen and Thron [21] on oval convergence regions. In any case, our effective-Laguerre approach proves the subexponential convergence of $\Gamma(a, z)$ for any complex pairs $(a, z) \in \mathbb{D}$; moreover, this is done with effective constant factors. Separate research on these superexponential convergence issues for continued fractions is underway.

1.2. Historical results on Laguerre asymptotics. Laguerre asymptotics have long been established for certain restricted domains and usually with noneffective asymptotics.[1] For example, in 1909 Fejér established that, for z on the open cut $(-\infty, 0)$ and any real a, one has [36, Theorem 8.22.1]:

$$
(6) \qquad L_n^{(-a)}(-z) = \frac{e^{-z/2}}{\sqrt{\pi}(-z)^{1/4-a/2}\, m^{1/4+a/2}} \cos\left(2\sqrt{-mz} + a\pi/2 - \pi/4\right)
$$
$$
+ O\left(m^{-a/2-3/4}\right),
$$

where we again use index $m := n + 1$, which slightly alters the coefficients in such classical expansions, said coefficients being for powers $n^{-k/2}$, but we are now using powers $m^{-k/2}$. By 1921 Perron [35] had generalized the Fejér series to arbitrary orders, then, for $z \notin (-\infty, 0]$, established a series consistent with (3) and (4), in essentially the form [36, Theorem 8.22.3]:

$$
(7) \qquad L_n^{(-a)}(-z) = S_n(a, z) \left(\sum_{k=0}^{N-1} \frac{C_k}{m^{k/2}} + O\left(m^{-N/2}\right) \right),
$$

although this was for *real* a and so not for our general parameter domain $\mathbb{D}$. Note that $C_0 = 1$, consistent with our (3) and (4); however, one should take care that because we are using index $m := n + 1$, the coefficients C_k in the above formula differ slightly from the historical ones.

A modern literature treatment that is again consistent with the heuristic (3)–(4) is given by Winitzki in [45], where one invokes a formal generating function to yield a contour integral for $L_n^{(-a)}(-z)$. Then a stationary-phase approach yields the correct first-asymptotic term, at least for certain subregions of $\mathbb{D}$. Winitzki's treatment is both elegant and nonrigorous; there is no explicit estimate given on the $O(1/\sqrt{m})$ correction in (3).

There is an interesting anecdote that reveals the difficulty inherent in Laguerre asymptotics. Namely, W. Van Assche in a fine 1985 paper [41] used the expansion

---

[1]Some researchers use the term "realistic error bound" for big-O terms that have explicit structure. We prefer "effective bound," and when an expansion is bestowed with such a bound, we may say "effective expansion."

(7) for work on zero-distributions, only to find by 2001 that the C_1 term in that 1985 paper had been calculated incorrectly. The amended series is given in his correction note [42] as

$$L_n^{(-a)}(-z) = \frac{e^{-z/2}}{2\sqrt{\pi}} \frac{e^{2\sqrt{nz}}}{z^{1/4-a/2} n^{1/4+a/2}} \cdot$$
$$\left(1 + \left(\frac{3 - 12a^2 + 24(1-a)z + 4z^2}{48\sqrt{z}}\right)\frac{1}{\sqrt{n}} + O\left(\frac{1}{n}\right)\right),$$

or in our own notation with $m := n + 1$,

$$(8) \qquad = S_n(a,z)\left(1 + \left(\frac{3 - 12a^2 - 24(1+a)z + 4z^2}{48\sqrt{z}}\right)\frac{1}{\sqrt{m}} + O\left(\frac{1}{m}\right)\right).$$

Note the slight alteration used to obtain our $C_1/\sqrt{m}$ term. Van Assche credits T. Müller and F. Olver for aid in working out the correct $O(1/\sqrt{n})$ component.[2] This story suggests that even a low-order asymptotic development is nontrivial.

The great classical analysts certainly knew in principle how to establish effective error bounds. The excellent treatment of effectiveness for Laplace's method of steepest descent in [27] is a shining example. Also illuminating is Olver's paper [26], which explains effective bounding and shows how unwieldy rigorous bounds can be obtained. However, efficient algorithms for generating *explicit* effective big-O constants have only become practicable in recent times, when computational machinery is prevalent.

2. Contour representation. In this section we develop an efficient—both numerically and analytically—contour-integral representation for $L_n^{(-a)}(-z)$.

2.1. Development of a "keyhole" contour. A well-known integral [36, section 5.4] has contour Γ encircling $s = 1$ and avoiding the branch cut $(-\infty, 0]$:

$$(9) \qquad L_n^{(-a)}(-z) = \frac{e^{-z}}{2\pi i}\oint_\Gamma s^{-1-a}\left(1 - \frac{1}{s}\right)^{-n-1} e^{zs}\, ds.$$

This representation holds for all pairs $(a,z) \in \mathbb{C} \times \mathbb{C}$, with n a nonnegative integer.[3]

However—and this is important—we found via experimental mathematical techniques, e.g., extreme-precision evaluations, that a specific kind of contour allows very accurate, efficient, and well-behaved *numerical* Laguerre evaluations. It is not completely understood why the "keyhole" contour we are about to define does so well in numerical Laguerre evaluations; we do know that this new contour consistently provides better numerics than, say, a simple circle surrounding $s = 1$. It may be simply the internal quirks of various modern numerical integrators at work, or it may be the smooth phase behavior along our keyhole's perimeter.

For the desired contour, take $z \neq 0$, $m := n + 1$ and assume $r := \sqrt{m/z}$ has $|r| > 1/2$. Then, use a circular contour centered at $s = 1/2$ with radius $|r|$. This contour will encompass $s = 1$, so the remaining requirement is to avoid the cut $s \in (-\infty, 0]$ as can be done by cutting out a "wedge" from the negative-real arc of the circle, with apex at $s = 1/2$. We tried such schemes with high-precision integration, to

---

[2]Accordingly, we hereby name the polynomial $C_1(a,z)$ the Perron–van Assche–Müller–Olver (PAMO) coefficient.

[3]The contour representation (9) can easily be continued to noninteger n, with care taken on the $(-n-1)$th power, but our present treatment will only use nonnegative integer n.

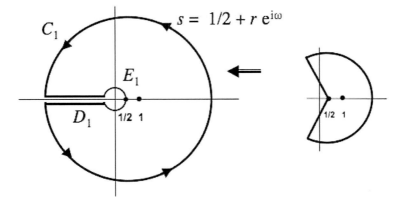

Fig. 1. *A numerically efficient "keyhole contour" for Laguerre evaluations $L_n^{(-a)}(-z)$, valid for all complex a, z, with $z \neq 0$ and $n+1 > |z|/4$. Wedges with center-1/2—as pictured at right—are experimentally accurate, leading to a "keyhole" deformation avoiding the cut $s \in (-\infty, 0]$. It turns that, for any $(a, z) \in \mathbb{D}$, the main arc C_1 gives the predominant contribution for large n, the D_1, E_1 components being subexponentially minuscule.*

settle finally on the contour of Figure 1, where the aforementioned wedge has evolved to a "keyhole" pattern consisting of cut-run D_1 and small, origin-centered circle E_1 of radius 1/2.

So, adopting constraints and nomenclature

$$z \neq 0, \quad \theta := \arg(z), \quad \omega_\pm := \pm\pi + \theta/2,$$
$$m := n+1, \quad r := \sqrt{m/z} := \sqrt{m/|z|}\,e^{-i\theta/2}, \quad R := |r| > 1/2,$$

but no other constraints, we have the following representation:

$$(10) \qquad\qquad L_n^{(-a)}(-z) = c_1 + d_1 + e_1,$$

where c_1, d_1, e_1 are the respective contributions from contour C_1, cut-discontinuity D_1, and contour E_1 from Figure 1. Exact formulae for said contributions are

$$(11) \qquad c_1 = \frac{1}{2\pi} r^{-a} e^{-z/2} \int_{\omega_-}^{\omega_+} \mathcal{H}_m(a, z, e^{-i\omega}) e^{2\sqrt{mz}\cos\omega} \, d\omega,$$

$$(12) \qquad d_1 = \frac{e^{-z}}{\pi} \sin(\pi a) \int_{1/2}^{R-1/2} T^{-1-a} \left(1 + \frac{1}{T}\right)^{-m} e^{-zT} \, dT,$$

$$(13) \qquad e_1 = -\frac{e^{-z}}{4\pi} \int_{-\pi}^{\pi} \left(2e^{-i\omega}\right)^{1+a} \left(1 - 2e^{-i\omega}\right)^{-m} e^{i\omega + \frac{z}{2}e^{i\omega}} \, d\omega,$$

and when $m > \Re(a)$, we may write this last contribution, by shrinking down the radius-1/2 contour segment to embrace the cut $(-1/2, 0]$, as

$$(14) \qquad e_1 = \frac{e^{-z}}{\pi} \sin(\pi a) \int_0^{1/2} T^{-1-a} \left(1 + \frac{1}{T}\right)^{-m} e^{-zT} \, dT.$$

For the c_1 contribution above, we have used the function $\mathcal{H}_m$ defined by

$$(15) \qquad \mathcal{H}_m(a, z, v) := v^a \left(1 + \frac{v}{2r}\right)^{-1-a} \left[F\left(\frac{v}{r}\right)\right]^m,$$

$$(16) \qquad F(t) := \left(\frac{1 + t/2}{1 - t/2}\right) e^{-t},$$

which for small t can be written $F(t) = 1 + t^3/12 + t^5/80 + \cdots = 1 + O(t^3)$.

3290   D. BORWEIN, J. M. BORWEIN, AND R. E. CRANDALL

In any event, e_1, like d_1, is subexponentially small relative to c_1 for large n and $(a, z) \in \mathbb{D}$. It is to be stressed that decomposition (10) holds for all $(a, z) \in \mathbb{C} \times \mathbb{C}, z \neq 0$, as long as $m := n + 1 > |z|/4$. We remind ourselves of (2) for $z = 0$. This means that such contour calculus applies to both oscillatory (Fejér) cases, where z is negative real, and subexponential (Perron) cases for the stated parameters.

2.2. Numerical success on the keyhole contour. The triumvirate c_1, d_1, e_1 of integrals is suitable for accurate Laguerre computations, which computations do show that *integrals d_1, e_1 tend to be subexponentially small* relative to c_1. This, of course, is our motive for so identifying the contour terms. An example, using the defining series (1) with $(a, z) := (-i, -1 - i)$ is

$$L_8^{(i)}(1+i) = -\frac{137}{288} + \frac{53}{45}i \approx -0.475694444444444444444 + 1.17777777777777777777i,$$

while the contributions from the C_1, D_1, E_1 segments of the contour of Figure 1 are

$$c_1 \approx -0.44406762576110996056 + 0.81722282272705891241i,$$

$$d_1 \approx -0.03169598827425878852 + 0.36065591639721657587i,$$

$$e_1 \approx 0.00006916959092430464 - 0.00010096134649771051i,$$

with the sum $c_1 + d_1 + e_1$ giving $L_8^{(i)}(1 + i)$ to the implied precision.

Remarkably, this "keyhole-contour" approach has the additional, unexpected feature that, for some parameter regions, the contour evaluation of $L_n^{(a)}(-z)$ is actually *faster* than direct summation of the defining series (1). A typical example is as follows: For the 13-digit evaluation

$$L_{50000}^{(25i/2)}(-30 + 3i) \approx (0.9275136583293 + 1.7406691595239i) \times 10^{1056},$$

the defining series (1) was, in our trials, slower than the integral c_1.[4]

3. Effective bounds on the contour components.

3.1. Theta-calculus. We now introduce a notation useful in effective-error analysis. When two functions enjoy

$$|f(z)| \leq |g(z)|$$

over some relevant domain of z values, we shall say that

$$f(z) = \Theta(g(z)), \quad \text{or} \quad f(z) = \Theta : g(z)$$

on said domain. Thus, the Θ-notation is an effective replacement for big-O notation. The reason for allowing the notation $\Theta : \dots$ is that a long formula g can run arbitrarily to the right of the colon.

Incidentally, there is a literature precedent for such a "theta-calculus." Namely, in computational number theory treatments such as [10], one encounters terms such as, say, $\theta x / \log^2 x$, with a stated constraint $\theta \in [-10, 10]$. This means the term is $O(x / \log^2 x)$ but with effective big-O constant bounded in magnitude by 10. In our nomenclature we would use $\Theta : 10x / \log^2 x$.

---

[4]Of course, series acceleration as in [5] would give the direct series a "leg up." Still, it is remarkable that contour integration is competitive. For the record, we compared *Mathematica*'s numerical integration to its own `LaguerreL[ ]` function.

3.2. Effective bound for e_1. First we address the integral e_1 as given in relation (13). We need some preliminary lemmas, starting with a collection of polynomial estimates to transcendental functions.

LEMMA 1. *The following inequalities hold for the respective conditions on ω:*

$$\log(5 - 4\cos\omega) \geq \frac{\log 9}{\pi^2}\omega^2, \quad \omega \in [-\pi, \pi];$$

$$\log(1 + \omega) \geq \frac{4}{5}\omega, \quad \omega \in [0, 1/2];$$

$$\arcsin\omega \leq \frac{\pi}{\sqrt{8}}\omega, \quad \omega \in [0, 1/\sqrt{2}];$$

$$\cos\omega \leq 1 - \frac{4}{\pi^2}\omega^2, \quad \omega \in [-\pi/2, \pi/2].$$

Proof. All four are straightforward calculus exercises. ☐

LEMMA 2. *Consider integrals of error-function class, specifically, for nonnegative real parameter μ,*

$$V_\mu(\alpha, \beta, \gamma) := \int_\gamma^\infty (x^2)^\mu e^{2\alpha x - \beta x^2}\, dx,$$

where each of α, β, γ is real, with $\alpha, \beta > 0$. Then we always have the bound

$$V_0(\alpha, \beta, \gamma) = \Theta : \sqrt{\frac{\pi}{\beta}}\, e^{\frac{\alpha^2}{\beta}}.$$

If in addition $\gamma > 2\alpha/\beta$, we also have

$$V_\mu(\alpha, \beta, \gamma) = \Theta : \frac{1}{2}\frac{\Gamma(\mu + 1/2)}{(\beta - 2\alpha/\gamma)^{\mu+1/2}};$$

finally, if $\gamma > 4\alpha/\beta$, we have

$$V_\mu(\alpha, \beta, \gamma) = \Theta : 2^{\mu-1/2}\frac{\Gamma(\mu + 1/2)}{\beta^{\mu+1/2}}.$$

Remark. We use the double-exponential $(x^2)^\mu$ because the first case of the theorem allows $\gamma < 0$.

Proof. Completing the exponent's square gives the first bound easily, since

$$V_0 = e^{\alpha^2/\beta}\int_\gamma^\infty e^{-\beta(x-\alpha/\beta)^2}\, dx = \Theta : \sqrt{\pi/\beta}.$$

For the second bound, it is elementary that, for $x \geq \gamma$, one has $2\alpha x - \beta x^2 \leq x^2(2\alpha/\gamma - \beta)$ so that

$$V_\mu \leq \int_0^\infty x^{2\mu}e^{-(\beta - 2\alpha/\gamma)x^2}\, dx,$$

and the resulting bound follows. The final bound follows immediately from the previous bound, because $\gamma > 4\alpha/\beta$ implies $\beta - 2\alpha/\gamma > \beta/2$. ☐

3292   D. BORWEIN, J. M. BORWEIN, AND R. E. CRANDALL

THEOREM 1. *The contour contribution e_1 defined by (13), under conditions $(a, z) \in \mathbb{D}$ and m sufficiently large in the explicit sense $m := n+1 > m_0 := |z|/4$, $m > m_1 := 5\left(|\Re(z)| + (|\Im(a)| + |\Im(z)|/2)^2\right)$, is bounded as*

$$e_1 = \Theta : e^{-z/2} 2^{\Re(a)+3} \frac{1}{\sqrt{m}}.$$

Moreover, under alternative conditions $(a, z) \in \mathbb{D}$ and $m > m_0$, $m > m_2 := \Re(a)$, $m > m_3 := -\frac{5}{4}|z|\cos\theta$ we have a bound

$$e_1 = \Theta : \frac{e^{-z}}{\pi} \sin(\pi a) \frac{2^{\Re(a)-m}}{m - \Re(a)}.$$

Proof. The proof involves careful application of Lemmas 1 and 2 to the integral forms (13), (14); for brevity we omit these details. □

3.3. Effective bound for d_1. Again we need an opening lemma.

LEMMA 3. *Consider integrals of the incomplete-Bessel class, specifically*

$$W(\alpha, \beta) := \int_0^1 e^{-\alpha x - \frac{\beta}{x}} \frac{dx}{x},$$

where each of α, β is real, with $\beta > 0$. If $\beta > \alpha$, we have a bound

$$W = \Theta : \frac{1}{2} e^{-\beta - \alpha} \sqrt{\frac{\pi}{\beta}}.$$

Proof. We write

$$W(\alpha, \beta) = \int_0^1 e^{\beta x - \alpha x} e^{-\beta x - \frac{\beta}{x}} \frac{dx}{x} \leq e^{\beta - \alpha} \int_0^1 e^{-\beta x - \frac{\beta}{x}} \frac{dx}{x}.$$

This last integral is a modified-Bessel term $K_0(2\beta)$ and has the required bound [2, Lemma 1]. □

THEOREM 2. *The contour contribution d_1 defined by (12) can be bounded, under conditions $(a, z) \in \mathbb{D}$ and $m > m_4 := 4|z|$, as*

$$d_1 = \Theta : \frac{e^{-z/2}}{\sqrt{\pi}} m^{|\Re(a)|/2 - 1/4} |z|^{-|\Re(a)|/2 - 1/4} \sin(\pi a) e^{-2\sqrt{m|z|}\cos^2\frac{\theta}{2}}.$$

Proof. Starting from (12) one applies Lemma 3; again for brevity we omit the details. □

3.4. Rigorous estimates on c_1.
Having dispensed with d_1, e_1, we next show the integration limits ω_-, ω_+ on the c_1 contribution can be changed—with a subexponentially small error penalty—to $-\pi/2, \pi/2$, respectively.

LEMMA 4. *For any v on the unit circle $\{e^{i\phi} : \phi \in (-\pi, \pi]\}$, any complex a, and any real $R > (1 + |a|)/2$ we have*

$$\left|\left(1 + \frac{v}{2R}\right)^{-1-a}\right| \leq \frac{1}{1 - \frac{1+|a|}{2R}}.$$

Proof. Via the binomial theorem,

$$\left|\left(1 + \frac{v}{2R}\right)^{-1-a}\right| \leq 1 + (1 + |a|)\frac{1}{2R} + (1 + |a|)^2\left(\frac{1}{2R}\right)^2 + \cdots = \frac{1}{1 - \frac{1+|a|}{2R}}. \quad \square$$

LEMMA 5. *Let v be on the unit circle as in Lemma 4, let $R > 1$ be real, and let m be a positive integer. For the function F appearing in (16), we have the bound*

$$\left| F\left(\frac{v}{R}\right)^m \right| \le e^{\frac{1}{6}\frac{m}{R^3}}.$$

Proof. From the definition (16) we have

$$F(v/R)^m = e^{\frac{1}{12}\frac{m}{R^3}\left(1+(3/5)/(2R)^2+(3/7)/(2R)^4+\cdots\right)}.$$

For $R := 1$, the infinite sum is no larger than 1.18 and is monotonic decreasing in R.   □

LEMMA 6. *Let v be on the unit circle as in Lemma 4 and define for nonnegative integer m*

$$K := \left(1+\frac{v}{2R}\right)^{-1-a} F\left(\frac{v}{R}\right)^m.$$

For the assignments $(a,z) \in \mathbb{C} \times \mathbb{C}, z \ne 0, R := \sqrt{m/|z|} > 1, m > m_5 := |z|(1+|a|+|z|/2)^2$, we have

$$K = \Theta : 2.$$

Proof. From Lemmas 4 and 5 we have

$$|K| \le \frac{1}{1-\frac{1+|a|}{2(1+|a|+|z|/2)}} e^{\frac{1}{6}\frac{|z|}{1+|a|+|z|/2}}.$$

The fact that the right-hand side is $\Theta(2)$ follows from the observation that the function

$$\frac{1}{1-\frac{Q}{2Q+x}} e^{\frac{1}{3}\frac{x}{2Q+x}}$$

for $Q \ge 1, x \in [0,\infty)$ is itself $\Theta(2)$. This in turn follows easily on substituting $y := x/(2Q+x)$. Now the function to be bounded is $g(y) := (2/(1+y))e^{y/3}$ on $y \in [0,1/2]$—as differentiation settles.   □

These lemmas in turn allow us to contract the range on the c_1 contour integral.

THEOREM 3. *Decompose c_1 as defined by (11) into two terms,*

$$c_1 := c_0 + c_2,$$

with c_0 involving the integral's range contracted to $[-\pi/2, \pi/2]$, namely,

$$c_0 := \frac{1}{2\pi} r^{-a} e^{-z/2} \int_{-\pi/2}^{\pi/2} \mathcal{H}_m(a,z,e^{-i\omega}) e^{2\sqrt{mz}\cos\omega} \, d\omega.$$

Then, under conditions $(a,z) \in \mathbb{D}$ and $m > m_5 := |z|(1+|a|+|z|/2)^2$, we have a bound

$$c_2 = \Theta : r^{-a} e^{-z/2} e^{\frac{3}{2}\pi|\Im(a)|}.$$

Proof. It is evident that c_2 is obtained from the definition (11) but with the integral replaced via

$$\int_{\omega_-}^{\omega_+} \to \left\{ \int_{\omega_-}^{-\pi/2} + \int_{\pi/2}^{\omega_+} \right\}.$$

3294   D. BORWEIN, J. M. BORWEIN, AND R. E. CRANDALL

Over these domains of integration, we have $e^{2\sqrt{mz}\cos\omega} = \Theta(1)$. Thus, Lemmas 4, 5, and 6 show

$$c_2 = \Theta : \frac{1}{2\pi} r^{-a} e^{-z/2} \left\{ \int_{\omega_-}^{-\pi/2} + \int_{\pi/2}^{\omega_+} \right\} 2 \left| e^{-i\omega a} \right| \, d\omega.$$

As the total support of the integrals cannot exceed π, the desired c_2 bound follows. □

3.5. Summary of the contour decomposition for $L_n^{(-a)}(-z)$. The above manipulations lead to the main result of the present section, namely, a formula that decomposes the Laguerre evaluation, as in the following.

THEOREM 4 (contour decomposition). *Let $(a, z) \in \mathbb{C} \times \mathbb{C}$ be an arbitrary parameter pair, with $z \neq 0$ (with $z = 0$ cases resolved exactly by (2)). If $m := n+1 = m_0 > |z|/4$, then the contour decomposition*

$$L_n^{(-a)}(-z) = c_0 + \mathcal{E}$$

holds, with c_0, c_2 as defined in Theorem 3 and $\mathcal{E} := c_2 + d_1 + e_1$. If an appropriate set of conditions on m across Theorems 1, 2, and 3 holds, then we can write

$$L_n^{(-a)}(-z) = c_0 + S_n(a, z)\mathcal{E}_1,$$

where $\mathcal{E}_1$ is subexponentially small, in the sense that, for any fixed positive ϵ, the large-n behavior is

$$\mathcal{E}_1 = O\left(e^{-(2-\epsilon)\sqrt{m|z|}\cos\frac{\theta}{2}} \right).$$

Moreover, an effective big-O constant is available (the proof exhibits explicit forms).

Proof. The contour calculus holds for all complex pairs (a, z), with $z \neq 0, R := \sqrt{m/|z|} > 1/2$, which assures that the point 1 is contained in the contour. It remains to analyze $\mathcal{E}_1$. Consider, then, an appropriate union of conditions from the cited theorems, say, $(a, z) \in \mathbb{D}$ and $m > \max(m_0, m_1, m_2, m_3, m_4, m_5)$, in which case we have, from Theorems 1, 2, and 3 (in that respective order of Θ terms):

$$\mathcal{E}_1 = \Theta : \frac{e^{-z/2}}{\sqrt{\pi}} \sin(\pi a) \frac{2^{1+\Re(a)-m}}{m - \Re(a)} z^{1/4-a/2} m^{1/4+a/2} e^{-2\sqrt{m|z|}\cos\frac{\theta}{2}}$$

$$+ \Theta : 2 \left(\frac{m}{|z|} \right)^{(\Re(a)+|\Re(a)|)/2} e^{\frac{3}{2}\pi|\Im(a)|} e^{-2\sqrt{m|z|}\left(\cos^2\frac{\theta}{2}+\cos\frac{\theta}{2}\right)}$$

$$+ \Theta : 2\sqrt{\pi} \, |mz|^{1/4} e^{\frac{3}{2}\pi|\Im(a)|} e^{-2\sqrt{m|z|}\cos\frac{\theta}{2}}.$$

This explicit bounding of the error term $\mathcal{E}_1$ proves the big-O statement of the theorem, while for any choice of ε, an effective big-O constant can be read off at will. □

4. Effective expansion for the $\mathcal{H}$-kernel. Theorem 4 shows the main contribution to $L_n^{(a)}(-z)$, for $(a, z) \in \mathbb{D}$, with $r := \sqrt{m/z}, |r| > 1/2$, is

(17) $$c_0 := \frac{1}{2\pi} r^{-a} e^{-z/2} \int_{-\pi/2}^{\pi/2} \mathcal{H}_m\left(a, z, e^{-i\omega}\right) e^{2\sqrt{mz}\cos\omega} \, d\omega,$$

with the integration kernel $\mathcal{H}_m$ defined, see (15) and (16), as

$$(18) \qquad \mathcal{H}_m(a,z,v) := v^a \left(1 + \frac{v}{2r}\right)^{-1-a} \left(\frac{1 + \frac{v}{2r}}{1 - \frac{v}{2r}}\right)^m e^{-mv/r},$$

where in the integral we assign $v := e^{-i\omega}$. We need to obtain the growth properties of $\mathcal{H}_m$.

4.1. Exponential form for $\mathcal{H}_m$.

LEMMA 7. *For $|v| = 1$ and $m > |z|/4$, the $\mathcal{H}$-kernel can be cast in the exponential form*

$$(19) \qquad \mathcal{H}_m := v^a \exp\left\{\sum_{k \geq 1} \frac{a_k}{k} \frac{1}{m^{k/2}}\right\},$$

where

$$(20) \qquad a_k := (1+a)(-1)^k \left(\frac{v\sqrt{z}}{2}\right)^k + \left(1 - (-1)^k\right) \frac{k}{k+2} \left(\frac{v\sqrt{z}}{2}\right)^{k+2}.$$

Moreover, we have the general coefficient bound

$$(21) \qquad |a_k| \leq \left(\frac{\sqrt{z}}{2}\right)^k (1 + |a| + |z|/2).$$

Proof. (18) can be recast, with $\rho := v/(2r)$, as

$$(22) \quad \mathcal{H}_m := v^a \exp\left\{-(1+a)\log(1+\rho) - 2m\rho + m\left(\log(1+\rho) - \log(1-\rho)\right)\right\}.$$

Since $|r| > 1/2$ and $|v| = 1$, the logarithmic series converge absolutely and we have

(23)

$$\mathcal{H}_m := v^a \exp\left\{\sum_{k \geq 1} \left[\frac{(1+a)(-1)^k}{k}\left(\frac{v\sqrt{z}}{2}\right)^k g^k + \frac{2}{2k+1}\left(\frac{v\sqrt{z}}{2}\right)^{2k+1} g^{2k-2}\right]\right\},$$

where $g := 1/\sqrt{m}$, and the precise form (20) for the a_k follows immediately. The given bound on $|a_k|$ is also immediate from (20). □

4.2. Exponentiation of series.

Though Lemma 7 is progress, we still need to exponentiate a series, in the sense that we want to know, for the following expansion, given the sequence (a_k),

$$\exp\left\{\sum_{k \geq 1} \frac{a_k}{k} x^k\right\} =: \sum_{h \geq 0} A_h x^h,$$

how the A_h depend on the a_k. The combinatorial answer is

$$A_h = \sum_{j=0}^{h} \frac{1}{j!} G_h(j;\vec{a}), \qquad G_h(j;\vec{a}) := \sum_{h_1 + \cdots + h_j = h} \frac{a_{h_1} \cdots a_{h_j}}{h_1 \cdots h_j},$$

3296   D. BORWEIN, J. M. BORWEIN, AND R. E. CRANDALL

with the understanding that $G_h(0, \vec{a}) := \delta_{0h}$ and that such combinatorial sums involve positive integer indices h_i. One useful result is the following.

LEMMA 8. *If all coefficients $a_k = 1$, then, for $j > 0$,*

$$G_h(j; \vec{1}) = \sum_{h_1 + \cdots + h_j = h} \frac{1}{h_1 \cdots h_j} = \Theta : \frac{1}{h} (2H_{h-j+1})^{j-1} = \Theta : \frac{1}{h} (2\gamma + 2 \log h)^{j-1}.$$

Here, $H_p := 1 + 1/2 + \cdots + 1/p$ is the pth harmonic number, $H_0 := 0$, and γ is the Euler constant.

Remark. It turns out that G here enjoys a closed form of sorts, namely,

$$G_h(j; \vec{1}) = \frac{j!}{h!} (-1)^{h-j} \mathcal{S}_h^{(j)},$$

where $\mathcal{S}$ denotes the Stirling number of the first kind, normalized via $x(x-1)\cdots(x - h + 1) =: \sum_{j=0}^{h} \mathcal{S}_h^{(j)} x^j$. So one byproduct of our lemma is a rigorous bound on the growth of Stirling numbers; see [1, 24.1.3,III] and [37] for research on Stirling asymptotics.

Proof. The first Θ-estimate arises by induction. For notational convenience we omit the vector $\vec{1}$ and just use the symbol $G_h(j)$. Note that $G_N(1) = 1/N$ and $G_N(2) = \frac{2}{N} H_{N-1}$. Generally we have

$$G_N(J) = \sum_{j=1}^{N-J+1} \frac{1}{j} G_{N-j}(J-1).$$

Now, assume by induction that $G_h(j) = \Theta : \frac{1}{h} (2H_{h-j+1})^{j-1}$ holds for all $j < J$. Then

$$G(N, J) \leq \sum_{j=1}^{N-j+1} \frac{2^{J-2} H_{N-J-j+2}^{J-2}}{j(N-J)} \leq \frac{2^{J-2}}{N} H_{N-J+1}^{J-2} \sum_{j=1}^{N-J+1} \left(\frac{1}{j} + \frac{1}{N-j} \right).$$

Now this last parenthetical term is $H_{N-J+1} + H_{N-1} - H_J$ which, because $H_a - H_b \leq H_{a-b}$ for any positive integer indices $a > b$, is bounded above by $2H_{N-J+1}$, which proves the first Θ-bound of the theorem. For the second Θ-bound it suffices, since H_j is increasing, to show that $H_{n-1} > \gamma + \log(n)$, which is an elementary calculus problem. $\square$

Though we do not use Lemma 8 directly in what follows, it is useful in proving convergence for various sums $\sum A_h x^h$ and may well matter in future research along our lines.

LEMMA 9. *Let $y \geq 1$ and $x \in (-1, 1)$ be real. Then in the expansion*

$$\exp \left\{ y \sum_{k \geq 1} \frac{x^k}{k} \right\} =: \sum_{h \geq 0} Y_h x^h$$

the coefficients Y_h enjoy the bound $Y_h = \Theta : y^h$.

Proof. First, $Y_0 = 1 \leq 1$. The left-hand side is $\exp(-y \log(1 - x)) = (1 - x)^{-y}$ whose binomial expansion has hth coefficient ($h \geq 1$) equal to

$$\frac{y(y+1)\cdots(y+h-1)}{h!} \leq y^h \frac{1}{1} \frac{1+1/y}{2} \cdots \frac{1+(h-1)/y}{h} \leq y^h. \quad \square$$

4.3. Effective expansions of exponentiated series. We are now in a position to derive an effective expansion for an exponentiated series, starting with the following.

LEMMA 10. *Assume complex vector $\vec{b}$ of defining coefficients b_k bounded as $|b_k| \leq cd^k$ for positive real c, d with $c \geq 1$. Assume also $|x| < 1/(2cd)$. Then, for any order $N \geq 0$, we have an effective expansion*

$$\exp\left\{\sum_{k \geq 1} \frac{b_k}{k} x^k\right\} = \sum_{h=0}^{N-1} B_h x^h + \Theta : 2c^N d^N x^N,$$

where

$$B_h = \sum_{j=0}^{h} \frac{G_h(j; \vec{b})}{j!}$$

are the usual coefficients of the full formal exponentiation.

Proof. Denoting $f(x) := \exp\left\{\sum_{k \geq 1} \frac{b_k}{k} x^k\right\}$, we have

$$f(x) = \sum_{h=0}^{N-1} B_h x^h + T_N,$$

with remainder $T_N = \sum_{h \geq N} B_h x^h$, having B_h coefficients given by a G-sum as in Lemma 10. Now,

$$|B_h| \leq \sum_{j=0}^{h} \frac{|G_h(j; \vec{b})|}{j!} \leq \sum_{j=0}^{h} \frac{|G_h(j; \vec{f})|}{j!},$$

where $\vec{f} = (cd^k : k \geq 0)$. By Lemma 9 we know that $|B_h| \leq (cd)^h$. Therefore

$$|T_N| \leq \sum_{h \geq N} (cd)^h x^h = \frac{c^N d^N x^N}{1 - cdx} = \Theta : 2(cdx)^N. \qquad \square$$

Finally we arrive at a general expansion—with effective remainder—for the $\mathcal{H}$-kernel.

THEOREM 5 (effective expansion for $\mathcal{H}_m$). *For general complex $(a, z) \in \mathbb{C} \times \mathbb{C}$, assume that $m > m_5 := |z|(1 + |a| + |z|/2)^2$ and $|v| = 1$. Then, for any expansion order $N \geq 0$, we have*

$$\mathcal{H}_m(a, z, v) = v^a \left(\sum_{h=0}^{N-1} \frac{A_h}{m^{h/2}} + \Theta : 2\left(\frac{m_5}{4m}\right)^{N/2} \right),$$

where, on the basis of the defining coefficients a_k given in (20),

$$A_h := \sum_{j=0}^{h} \frac{G_h(j; \vec{a})}{j!}, \qquad G_h(j; \vec{a}) := \sum_{h_1 + \cdots + h_j = h} \frac{a_{h_1} \cdots a_{h_j}}{h_1 \cdots h_j}.$$

Proof. The result follows immediately from Lemma 10, on assigning $\vec{b} = \vec{a}$, with $x := 1/\sqrt{m}$, $c := 1 + |a| + |z|/2$, $d := (1/2)\sqrt{|z|}$. $\quad \square$

Note that Theorem 5 in the instance $N = 0$ implies our previous Lemma 6. It is interesting and suggestive that the threshold $m_5 := |z|(1 + |a| + |z|/2)^2$ appears in these results rather naturally.

3298   D. BORWEIN, J. M. BORWEIN, AND R. E. CRANDALL

4.4. Effective integral form for c_0. To obtain a useful form for c_0, the dominant component of $L_n^{(-a)}(-z)$, we use Theorem 5 to obtain the following.

THEOREM 6. *For $(a, z) \in \mathbb{D}$ and $m > m_5$, the dominant component of Theorems 3 and 4, namely,*

$$c_0 := \frac{1}{2\pi} r^{-a} e^{-z/2} \int_{-\pi/2}^{\pi/2} \mathcal{H}_m(a, z, e^{-i\omega}) e^{2\sqrt{mz}\cos\omega} \, d\omega,$$

can be given an effective form for any order $N \geq 0$, as

$$c_0 = \frac{1}{2\pi} r^{-a} e^{-z/2} \sum_{h=0}^{N-1} \frac{1}{m^{h/2}} \int_{-\pi/2}^{\pi/2} e^{-i\omega a} A_h e^{2\sqrt{mz}\cos\omega} \, d\omega$$
$$+ \; S_n(a, z) \, \mathcal{E}_{2,N},$$

where the error term is bounded as

$$\mathcal{E}_{2,N} = \Theta : \frac{\pi}{\sqrt{2}} \left(\frac{m_5}{4m}\right)^{N/2} \exp\left(\frac{\pi^2 \Im(a)^2 \sec\frac{\theta}{2}}{32\sqrt{m|z|}}\right) \sec^{1/2}\frac{\theta}{2}$$

and the A_h are to be calculated as the first N coefficients of

$$\sum_{h=0}^{\infty} A_h x^h := \exp\left\{\sum_{k\geq 1} \frac{a_k}{k} x^k\right\},$$

via (20), with $v := e^{-i\omega}$.

Proof. Inserting the effective $\mathcal{H}$-kernel expansion from Theorem 5 directly into the c_0 integral gives the indicated sum over $h \in [0, N-1]$ plus an error term

$$\Theta : \frac{1}{2\pi} r^{-a} e^{-z/2} 2 \left(\frac{m_5}{4m}\right)^{N/2} \int_{-\pi/2}^{\pi/2} e^{\omega \Im(a)} e^{-2\sqrt{m|z|}\cos(\theta/2)\cos\omega} \, d\omega.$$

Using Lemma 1 on $\cos\omega$ and the V_0-part of Lemma 2, we obtain the $\mathcal{E}_{2,N}$ bound of the theorem. $\square$

With Theorem 6 we have come far enough to see that a Laguerre evaluation can be obtained—up to a subexponentially small relative error—via the A_h terms in said theorem. To this end, an inspection of the defining relations reveals that, in general, we can decompose an A_h coefficient in terms of powers of $v := e^{-i\omega}$, namely, we define $\alpha_{h,\mu}$ terms via

(24) $$A_h =: \sum_{u=0}^{h} \alpha_{h,u}(a, z) v^{h+2u}.$$

For example,

$$\alpha_{00} = 1, \quad \alpha_{10} = -\frac{1+a}{2} z^{1/2}, \quad \alpha_{11} = \frac{z^{3/2}}{12}, \quad \alpha_{31} = \frac{1}{480}\left(5a^2 + 15a + 16\right) z^{5/2}.$$

The point being, we now have special formulae for the dominant contribution c_0, namely,

(25)

$$c_0 = \frac{1}{2\pi} r^{-a} e^{-z/2} \sum_{h=0}^{N-1} \frac{1}{m^{h/2}} \sum_{u=0}^{h} \alpha_{h,u}(a, z) \, \mathcal{I}(2\sqrt{mz}, a + h + 2u) \; + \; S_n(a, z) \, \mathcal{E}_{2,N},$$

where the integral

$$(26) \qquad \mathcal{I}(p,q) := \int_{-\pi/2}^{\pi/2} e^{-iq\omega} e^{p\cos\omega} \, d\omega$$

thus emerges as a fundamental entity for the research at hand.

5. $\mathcal{I}$-integrals and an "exp-arc" method. Having reduced the problem of subexponential Laguerre growth to a study of the $\mathcal{I}$-integrals (26), we next develop a method that is effective for both their numerical and theoretical estimation. This method amounts to the avoidance of stationary-phase techniques, employing instead various forms of *exponential-arcsine* (exp-arc) series, as we see shortly.

Let us first define, for any complex pair (p,q) and $\alpha, \beta \in (-\pi, \pi)$,

$$(27) \qquad I(p,q,\alpha,\beta) := \int_{\alpha}^{\beta} e^{-iq\omega} e^{p\cos\omega} \, d\omega$$

so that our special case (26) is simply $\mathcal{I}(p,q) := I(p,q,-\pi/2,\pi/2)$.

Importantly, one may write the Bessel functions J_n of integer order n in terms of $\mathcal{I}$-integrals:

$$(28) \qquad J_n(z) = \frac{1}{2\pi} \left(e^{-i\pi n/2} \mathcal{I}(iz,n) + e^{i\pi n/2} \mathcal{I}(-iz,n) \right),$$

and the modified Bessel function, again of integer order n:

$$(29) \qquad I_n(z) = \frac{1}{2\pi} \left(\mathcal{I}(z,n) + (-1)^n \mathcal{I}(-z,n) \right),$$

about which representations we shall have more to say in a later section.[5]

5.1. Essentials of the exp-arc method. Now we investigate what we call exp-arc series. First, for any complex τ and $x \in [-1,1]$, one has a remarkable, absolutely convergent expansion (see [6]):

$$(30) \qquad e^{\tau \arcsin x} = \sum_{k=0}^{\infty} r_k(\tau) \frac{x^k}{k!},$$

where the coefficients depend on the parity of the index

$$r_{2m+1}(\tau) := \tau \prod_{j=1}^{m} \left(\tau^2 + (2j-1)^2 \right), \qquad r_{2m}(\tau) := \prod_{j=1}^{m} \left(\tau^2 + (2j-2)^2 \right).$$

By differentiating with respect to x, we obtain

$$\frac{e^{\tau \arcsin x}}{\sqrt{1-x^2}} = \frac{1}{\tau} \sum_{k=0}^{\infty} r_{k+1}(\tau) \frac{x^k}{k!},$$

valid for $x \in (-1,1)$. In particular, we have the important expansion (using a function G, in passing)

$$(31) \qquad G(\tau,x) := \frac{\cosh(\tau \arcsin x)}{\sqrt{1-x^2}} = \sum_{k=0}^{\infty} g_k(\tau) \frac{x^{2k}}{(2k)!},$$

---

[5]The $\mathcal{I}$-integral can also be written in compact terms of an *Anger function* **J** and *Weber function* **E**; see [1].

3300   D. BORWEIN, J. M. BORWEIN, AND R. E. CRANDALL

where

$$g_k(\tau) := \prod_{j=1}^{k} \left((2j-1)^2 + \tau^2 \right).$$

These exp-arc expansions may be applied to the $I(p, q, \alpha, \beta)$ integrals as follows.[6] From (27) and its subsequent manipulations we have

$$I(p, q, \alpha, \beta) = e^p \int_{\alpha}^{\beta} e^{-iq\omega} e^{-2p \sin^2(\omega/2)} \, d\omega = 2e^p \int_{\sin \frac{\alpha}{2}}^{\sin \frac{\beta}{2}} \frac{e^{-2iq \arcsin x}}{\sqrt{1-x^2}} e^{-2px^2} \, dx$$

$$(32) \qquad = i \frac{e^p}{q} \sum_{k=0}^{\infty} \frac{r_{k+1}(-2iq)}{k!} \int_{\sin \frac{\alpha}{2}}^{\sin \frac{\beta}{2}} x^k e^{-2px^2} \, dx.$$

Even though we eventually consider asymptotic expansions, the convergence of such an I-series is the rule.

LEMMA 11. *For $\alpha, \beta \in (-\pi, \pi)$ and any complex pair (p, q), the series (32) converges absolutely.*

Proof. Define $\delta := \max(|\sin(\alpha/2)|, |\sin(\beta)/2|)$ so that the integral in (32) is bounded in magnitude by $2 \exp(2\Re(p))\delta^{k+1}/(k+1)$. But this means the kth summand in (32) is, up to a k-independent multiplier, bounded in magnitude by $r_{k+1}(2|q|)\delta^{k+1}/(k+1)!$. This summand is of the same form as that in the defining series (30), so absolute convergence is assured. □

Likewise, from (31) and (32) we have an absolutely convergent expansion for the special-case $\mathcal{I}$ integral:

$$(33) \qquad \mathcal{I}(p, q) = 4e^p \int_0^{1/\sqrt{2}} G(-2iq, x) e^{-2px^2} = 4e^p \sum_{k=0}^{\infty} \frac{g_k(-2iq)}{(2k)!} B_k(p),$$

where B_k is an error–function–class integral

$$(34) \quad B_k(p) := \int_0^{1/\sqrt{2}} x^{2k} e^{-2px^2} \, dx = \frac{1}{2} \frac{1}{(2p)^{k+1/2}} \left\{ \Gamma(k+1/2) - \Gamma(k+1/2, p) \right\}.$$

A computationally key recurrence relation accrues: with $B_0(p) = \sqrt{\frac{\pi}{8p}} \, \text{erf} \left(\sqrt{p} \right)$ and $k > 0$, we have

$$(35) \qquad B_k(p) := \frac{2k-1}{4p} B_{k-1} - \frac{2^{-k-3/2}}{p} e^{-p}.$$

5.2. Effective expansion for the cosh-arc G-series. Our asymptotic analysis of representation (33) begins with a lemma that reveals how the G function (31) has an attractive self-similarity property. Namely, the function appears naturally in modified form within its own error terms.

LEMMA 12. *For any complex τ, the G function (31) can be given an effective expansion to any integer order $N \geq 0$, as*

$$(36) \, G(\tau, x) := \frac{\cosh(\tau \arcsin x)}{\sqrt{1-x^2}} = \sum_{k=0}^{N-1} g_k(\tau) \frac{x^{2k}}{(2k)!} + g_N(\tau) \frac{x^{2N}}{(2N)!} \left(1 + T_N(\tau, x)\right),$$

---

[6]An equivalent approach to expanding I uses a series for $e^{iq\omega} \sec(\omega/2)$ in terms of $x := \sin(\omega/2)$, as in [20, p. 276].

EFFECTIVE LAGUERRE ASYMPTOTICS   3301

with the error term T_N conditionally bounded over real $x \in [0, 1/\sqrt{2}]$ in the form

$$T_N = \Theta : 1 \qquad\qquad\qquad\quad \text{if } N \geq 2x^2|\tau|^2 - 1;$$
$$= \Theta : (\sqrt{2}\ x^2 + x|\tau|)e^{|\tau|\arcsin x} \quad \text{otherwise.}$$

Proof. The error term is, by the definition of the $g_k(\tau)$, given by the absolutely convergent sum

$$T_N = h_N + h_N h_{N+1} + h_N h_{N+1} h_{N+2} + \cdots,$$

where

$$h_k := \frac{(2k+1)^2 + \tau^2}{(2k+1)(2k+2)} x^2.$$

When $|\tau|^2 x^2 \leq (N+1)/2$ and since $x^2 \leq 1/2$, it is immediate that, for $k \geq N$, we have a bound:

$$|h_k| \leq \frac{(4k^2 + 4k + 1)/2 + (N+1)/2}{4k^2 + 5k + 2} \leq 1/2,$$

whence $T_N = \Theta : 1/2 + 1/4 + 1/8 + \ldots$, settling the first conditional bound of the theorem. In any case, i.e., any complex τ and any $x \in [0, 1/\sqrt{2}]$, we have for $j \geq 0$:

$$h_N h_{N+1} \cdots h_{N+j} = x^{2j} \prod_{k=0}^{j} \frac{(2N+2k+1)^2 \left((2k+1)^2 + \tau^2 \frac{(2k+1)^2}{(2N+2k+1)^2}\right)}{(2N+2k+1)(2N+2k+2)(2k+1)^2}$$
$$= \Theta : \frac{g_{j+1}(|\tau|)}{(2j+1)!!^2}.$$

However, it is elementary that $(2j+1)!!^2 \geq (2j+1)!$ by simple factor-tallying, so

$$|T_N| \leq \frac{\left(1^2 + |\tau|^2\right) \cdot 2}{2!} x^2 + \frac{\left(1^2 + |\tau|^2\right)\left(3^2 + |\tau|^2\right) \cdot 4}{4!} x^4 + \ldots = x\frac{\partial}{\partial x}G(|\tau|, x),$$

where we have noticed that the right-hand series here is itself a differentiated "cosh-arc" series. Thus

$$T_N = \Theta : x\frac{\partial}{\partial x}\frac{\cosh(|\tau|\arcsin x)}{\sqrt{1 - x^2}} = \Theta : \frac{x^2}{2\left(1 - x^2\right)^{3/2}}\left(e^u + e^{-u}\right) + \frac{x|\tau|}{2\left(1 - x^2\right)}\left(e^u - e^{-u}\right),$$

where $u := |\tau|\arcsin x$. Now by excluding $2x^2|\tau|^2 \leq N+1$ for the second conditional bound of the lemma, we have $|\tau| \geq 1$, whence the e^{-u} terms can be ignored over $x \in [0, 1/\sqrt{2}]$, and the second conditional bound follows. □

5.3. Effective expansion of the $\mathcal{I}$-integral. We first invoke a classical lemma that bounds the incomplete gamma function for certain parameters.

LEMMA 13. *For integer $M \geq 0$, $\Re(z) \geq 0, z \neq 0$ and $|z| \geq 2M - 1$, we have*

$$\Gamma(M + 1/2, z) = \Theta : 2z^{M-1/2}e^{-z}.$$

Proof. Proofs of results such as these on incomplete-gamma bounds appear in various texts on special functions, e.g., [27, section 2.2, p. 110]. □

The results of the present section may now be applied to a general, effective expansion of the $\mathcal{I}$-integral whenever $\Re(p)$ is sufficiently positive.

THEOREM 7 (effective $\mathcal{I}$ expansion). *For the integral*

$$\mathcal{I}(p,q) := I(p,q,-\pi/2,\pi/2) = \int_{-\pi/2}^{\pi/2} e^{-iq\omega} e^{p\cos\omega}\, d\omega,$$

assume an integer expansion order $N \geq 1$. Assume $\phi := \arg(p) \in (-\pi/2, \pi/2)$ and conditions

$$\Re(p) \geq 2N+1, \quad \Re(p) \geq 2\pi|q|^2.$$

Then we have an effective expansion

$$(37)\qquad \mathcal{I}(p,q) = \sqrt{\frac{2\pi}{p}}\, e^p \left\{ \sum_{k=0}^{N-1} \frac{g_k(-2iq)}{k!\, 8^k} \frac{1}{p^k} + \Theta : \sqrt{\frac{8}{\pi p}}\, e^{-p} \cosh\left(\frac{\pi}{2}|q|\right) \right.$$

$$\left. + \frac{g_N(-2iq)}{N!\, 8^N} \frac{1}{p^N} \left(1 + \Theta : u_N \sec^{N+1/2}\phi \right) \right\},$$

where the g_k are as in the cosh-arc expansion (31), and we may take

$$(38)\qquad u_N := 1 + 2^N + \frac{2^{3N+1}}{\pi\binom{2N}{N}}\,;$$

however, on the extra condition $N \geq 4|q|^2 - 1$, taking $u_N := 1$ suffices.

Proof. The insertion of the series of Lemma 12 into representation (33) results in

$$\mathcal{I}(p,q) = 4e^p \left\{ \sum_{k=0}^{N} \frac{g_k(-2iq)}{(2k)!} B_k(p) + \frac{g_N(-2iq)}{(2N)!} \int_0^{1/\sqrt{2}} x^{2N} T_N(-2iq,x) e^{-2px^2}\, dx \right\},$$

where the $B_k(p)$ are given by (34). The sum over $k \in [0, N]$ here is thus

$$\frac{1}{2}\sum_{k=0}^{N} \frac{g_k(-2iq)}{(2k)!} \frac{\Gamma(k+1/2)}{(2p)^{k+1/2}} + \Theta : e^{-p} \sum_{k=0}^{N} \frac{|g_k(-2iq)|}{(2k)!} \frac{1}{2^{k+1/2}p},$$

where the Θ-term here follows from Lemma 13 on our condition $\Re(p) \geq 2N+1$. But this very Θ-term is bounded above by

$$\frac{e^{-p}}{p\sqrt{2}} \sum_{k=0}^{\infty} \frac{g_k(2|q|)}{(2k)!} \frac{1}{2^k} = \frac{e^{-p}}{p} \cosh\left(2|q|\arcsin\left(1/\sqrt{2}\right)\right),$$

so we have settled the summation and the $\cosh(\pi|q|/2)$ term in (37). Now consider the integral term

$$I_0 := \int_0^{1/\sqrt{2}} x^{2N} T_N(-2iq,x) e^{-2px^2}\, dx.$$

Define $\gamma := |q|^{-1}\sqrt{(N+1)/8}$. If $\gamma \geq 1/\sqrt{2}$, then by Lemma 12, we know that $T_N = \Theta(1)$, and our theorem follows in the $u_N := 1$ case. Otherwise, $\gamma < 1/\sqrt{2}$, and

EFFECTIVE LAGUERRE ASYMPTOTICS   3303

we bound I_0 using two integrals

$$|I_0| \leq \int_0^{1/\sqrt{2}} x^{2N} e^{-2\Re(p)x^2}\, dx + \int_\gamma^{1/\sqrt{2}}$$
$$\left(\sqrt{2}\, x^{2N+2} + 2|q|\, x^{2N+1} \right) e^{2|q| \arcsin x - 2\Re(p)x^2}\, dx.$$

From Lemma 1 we know that the exponent here can be taken to be $\pi|q|x/\sqrt{2} - 2\Re(p)x^2$. For the assignments $\alpha := \pi|q|/\sqrt{8}$, $\beta := 2\Re(p)$, we have $\beta > 4\alpha/\gamma$ so that by Lemma 2 there is a bound

$$I_0 \leq \frac{1}{2}\frac{\Gamma(N+1/2)}{(2\Re(p))^{N+1/2}} + \sqrt{2}\, V_{N+1}(\alpha,\beta,\gamma) + 2|q|\, V_{N+1/2}(\alpha,\beta,\gamma)$$
$$\leq \frac{1}{2}\frac{\Gamma(N+1/2)}{(2\Re(p))^{N+1/2}} + 2^N \frac{\Gamma(N+3/2)}{(2\Re(p))^{N+3/2}} + 2^{N+1}|q|\frac{\Gamma(N+1)}{(2\Re(p))^{N+1}}.$$

Using $|q|^2 \leq \Re(p)/(2\pi)$, $\Re(p) \geq 2N+1$ with $\Re(p) = |p| \cos\phi$ yields the N-dependent u_N form of the theorem. $\quad\square$

6. Effective asymptotics for $L_n^{(-a)}(-z)$. At last we may provide explicit terms for Laguerre expansions in the subexponential-growth regime, which regime turns out to be precisely characterized by the parameter-pair requirement: $(a,z) \in \mathbb{D}$.

First, for convenience, we recapitulate the thresholds for sufficiently large $m := n+1$ from our previous theorems:

$$m_0 := |z|/4, \quad m_1 := 5\left(|\Re(z)| + (|\Im(a)| + |\Im(z)|/2)^2\right),$$

$$m_2 := \Re(a), \quad m_3 := -(5/4)|z|\cos\theta, \quad m_4 = 4|z|, \quad m_5 := |z|(1 + |a| + |z|/2)^2,$$

$$m_6 := (2N+1)^2\, \frac{\sec^2(\theta/2)}{|z|}, \quad m_7 := 4\pi^2\, \frac{\sec^2(\theta/2)}{|z|}(|a| + 3N - 3)^4.$$

Here, $\theta := \arg(z)$ as before, while m_6, m_7 involve an asymptotic expansion order $N \geq 1$. We are aware that the m_i bounds here are interdependent, and some are masked by others (e.g., m_0 is masked by m_4). The important quantity in our central result (Theorem 8, below) is simply $\max_{i\in[0,7]} m_i$.

Before delving into the central result, let us remind ourselves of previous nomenclature:

$$(39) \qquad g_k(\tau) := \prod_{j=1}^{k} \left((2j-1)^2 + \tau^2\right),$$

$$(40) \qquad a_k := (1+a)(-1)^k \left(\frac{v\sqrt{z}}{2}\right)^k + \left(1 - (-1)^k\right)\frac{k}{k+2}\left(\frac{v\sqrt{z}}{2}\right)^{k+2},$$

$$(41) \qquad \sum_{h=0}^{\infty} A_h x^h := \exp\left\{\sum_{k\geq 1} \frac{a_k}{k} x^k\right\},$$

$$(42) \qquad A_h := \sum_{u=0}^{h} \alpha_{h,u}(a,z) v^{h+2u}.$$

Note that the $\alpha_{h,u}$ coefficients are thus implicitly defined in terms of the original a_k functions.

3304   D. BORWEIN, J. M. BORWEIN, AND R. E. CRANDALL

6.1. The general subexponential expansion. Using Theorem 7 with $p :=$ $2\sqrt{mz}$ and inserting this into formula (25), we arrive at our desired effective Laguerre expansion.

THEOREM 8 (effective Laguerre expansion). *Assume $(a, z) \in \mathbb{D}$. For asymptotic expansion order $N \geq 1$ and $m := n + 1$ sufficiently large in the sense $m >$ $\max_{i \in [0,7]} m_i$, we have the expansion*

$$(43) \quad L_n^{(-a)}(-z) = \frac{1}{2\sqrt{\pi}} \frac{e^{-z/2} e^{2\sqrt{mz}}}{z^{1/4-a/2} \, m^{1/4+a/2}} \left\{ \sum_{j=0}^{N-1} \frac{C_j}{m^{j/2}} + \frac{\overline{C}_N}{m^{N/2}} + \mathcal{E}_1 + \mathcal{E}_{3,N} \right\},$$

where the expansion coefficients C_j are given in finite form:

$$(44) \quad C_j := \sum_{k=0}^{j} \frac{1}{16^k} \frac{1}{k!} \frac{1}{z^{k/2}} \sum_{u=0}^{j-k} \alpha_{j-k,u}(a, z) \, g_k(-2i(a + j - k + 2u)),$$

while the error term $\overline{C}_N$ is bounded as

$$\overline{C}_N = \Theta : \sum_{v=1}^{N} \frac{1}{|z|^{v/2} 16^v v!} \left(1 + u_v \sec^{v+1/2} \frac{\theta}{2} \right)$$

$$\sum_{u=0}^{N-v} |\alpha_{N-v,u}(a, z) \, g_v(-2i(a + N - v + 2u))|$$

$$+ \, \Theta : 4(m_5/4)^{N/2} \sec^{1/2} \frac{\theta}{2},$$

with u_v taking the v-dependent form of (38) in Theorem 7 (with $q := a$ there). Finally, the term $\mathcal{E}_1$ is subexponentially small (from Theorem 4), as is

$$\mathcal{E}_{3,N} = \Theta : \sqrt{\frac{4}{\pi\sqrt{mz}}} e^{-2\sqrt{mz}} \sum_{h=0}^{N-1} \frac{1}{m^{h/2}} \sum_{u=0}^{h} |\alpha_{h,u}(a, z)| \cosh\left(\frac{\pi}{2} |a + h + 2u| \right).$$

Proof. For brevity we leave out the details—all of which are straightforward, if tedious, applications of the previous theorems and formulae.   □

We now have a resolution of the domain of subexponential growth as follows.

COROLLARY 1. *For $(a, z) \in \mathbb{D}$, the Laguerre polynomial grows subexponentially in the sense that, for order $N \geq 1$ and any $\varepsilon > 0$,*

$$L_n^{(-a)}(-z) = S_n(a, z) \left\{ \sum_{j=0}^{N-1} \frac{C_j}{m^{j/2}} + O\left(\frac{1}{m^{N/2}} \right) + O\left(e^{-(2-\varepsilon)\sqrt{m|z|} \cos(\theta/2)} \right) \right\},$$

with all coefficients and the implied big-O constant effectively bounded via our previous theorems. Moreover, for $(a, z) \notin \mathbb{D}$, the large-n growth is not subexponential. Thus, the precise domain of subexponential growth is characterized by $(a, z) \in \mathbb{D}$.

Proof. The given subexponential formula is a paraphrase of Theorem 8. Now assume that $z \in (-\infty, 0]$. We already know that $z = 0$ does not yield such growth (see (2)). Now, for z negative real, note that the integrals in (12, 14) are both decaying in large m. Finally, the integral in (11) has phase factor $|\exp(2\sqrt{mz} \cos \omega)| \leq 1$, and Lemma 6 show that c_1 also cannot grow subexponentially in m.   □

EFFECTIVE LAGUERRE ASYMPTOTICS   3305

This corollary echoes, of course, the classical Perron result (7), and we again admit that historical efforts derived the C_j coefficients in principle. The new aspects are (a) we have asymptotic coefficients and effective bounds for general $(a, z) \in \mathbb{D}$ parameters, and (b) we can develop in a natural way an algorithm for symbolic generation of said coefficients.

6.2. Algorithm for explicit asymptotic coefficients. Theorem 8 indicates that, to obtain actual C_j coefficients, we need the cosh-arc numbers $g_k(\tau)$ and the $\alpha_{h,u}(a, z)$ coefficients. Observe that the chain of relations starting with (39) is a prescription for the generation of the C_k. All of this can proceed via symbolic processing, noting that v is simply a placeholder throughout.

Remarkably, there is a fast algorithm that bypasses much of the symbolic tedium. First, we have an explicit recursion for A_h, with $A_0 := 1$, as

$$(45) \qquad A_k = \frac{1}{k} \sum_{j=0}^{k-1} A_j\, a_{k-j}$$

as follows from differentiating (41) logarithmically and then comparing terms. Second, when we use (42) together with (45), we obtain a recursion *devoid* of the symbolic placeholder v, as

$$(46) \qquad \alpha_{k,u} = \frac{1}{k} \sum_{j=0}^{k-1} \left(\alpha_{j,u} b_{k-j} + \alpha_{j,u-1} d_{k-j} \right),$$

where these new recursion coefficients are

$$b_h := (-1)^h (1+a) \left(\frac{\sqrt{z}}{2} \right)^h, \quad d_h := \left(1 - (-1)^h \right) \frac{h}{h+2} \left(\frac{\sqrt{z}}{2} \right)^{h+2}.$$

In practice we define $\alpha_{0,0} := 1$ and force any $\alpha_{j,u}$, with $u > j$ or $u < 0$ to vanish. In this sense, the collection of $\alpha_{k,u}$ make up a *lower-triangular matrix*, e.g., the entries for $k \leq 3$ appear thus:

$$\begin{pmatrix} 1 & 0 & 0 & 0 \\ -\dfrac{(a+1)}{2} z^{1/2} & \dfrac{1}{12} z^{3/2} & 0 & 0 \\ \dfrac{a^2 + 3a + 2}{8} z & -\dfrac{(a+1)}{24} z^2 & \dfrac{1}{288} z^3 & 0 \\ -\dfrac{a^3 + 6a^2 + 11a + 6}{48} z^{3/2} & \dfrac{5a^2 + 15a + 16}{480} z^{5/2} & -\dfrac{a+1}{576} z^{7/2} & \dfrac{1}{10368} z^{9/2} \end{pmatrix},$$

where $\alpha_{3,3}$ is the lower-right element here.

These observations lead to a fast algorithm for the computation of the asymptotic coefficients.

ALGORITHM 1 (fast computation of Laguerre asymptotic coefficients).
For given $(a, z) \in \mathbb{D}$ and desired expansion order N, this algorithm returns the asymptotic coefficients $(C_k : k \in [0, N])$ of relation (44), Theorem 8.
1. Set $\alpha_{0,0} := 1$ and, for desired order N, calculate the lower-triangular matrix elements
$(\alpha_{k,u} : 0 \leq u \leq k \leq N)$ via a recursion such as
$\alpha(k, u)\ \{$

$\mathrm{if}(k == 0)$ return $\delta_{0,u}$;

return $\frac{1}{k}\sum_{j=0}^{k-1}\left(\alpha_{j,u}b_{k-j}+\alpha_{j,u-1}d_{k-j}\right)$;

}

or use an unrolled, equivalent loop (i.e., one may generate the left-hand column of the α-matrix, then fill in one row at a time, lexicographically).

2. Use the recursion $g_k(\tau)=\left((2k+1)^2+\tau^2\right)g_{k-1}(\tau)$ and the lower-triangular matrix of α values to generate the C coefficients via (44). □

We observe that the expensive sums in this algorithm are all *acyclic convolutions*. Thus, for numerical input (a,z), the algorithm complexity turns out to be $O\left(N^{2+\epsilon}\right)$ arithmetic operations, with the "2" part of the complexity power arising from the area of the lower-triangular sector.[7]

We employed the algorithm to generate exact asymptotic coefficients as follows:

$$C_0 = 1, \qquad C_1 = \frac{1}{48\sqrt{z}}\left(-12a^2-24za+4z^2-24z+3\right),$$

$$C_2 = \frac{1}{4608z}\left(144a^4+576za^3+480z^2a^2+1728za^2-360a^2-192z^3a\right.$$
$$\left.+1152z^2a+1584za+16z^4-192z^3+312z^2+432z+81\right),$$

and so on. Note that the C_1 form here agrees with the PAMO coefficient in (8). We were able to generate the full, symbolic $C_{64}(a,z)$ in about one minute of CPU on a typical desktop computer. To aid future researchers, we report that the numerator of C_{64} has degree 128 in both a,z, while the denominator is an integer divisible by every prime not exceeding 64, namely,

$$2^{319}\cdot3^{94}\cdot5^{25}\cdot7^{13}\cdot11^7\cdot13^5\cdot17^4\cdot19^3\cdot23^3\cdot29^2\cdot31^2\cdot37\cdot41\cdot43\cdot47\cdot53\cdot59\cdot61.$$

While formidable, the bivariate $C_k(a,z)$ coefficients evidently have intricate structure and patterning. Future research into said structure would be of interest.

6.3. Generating and verifying effective bounds. For the first nontrivial effective bound, we can employ the rigorous bound $\overline{C}_1$ in Theorem 8, with $N=1$, to get an effective version of the original asymptotic (3) as

$$L_n^{(-a)}(-z)=S_n(a,z)\left(1+\frac{\overline{C}_1}{\sqrt{m}}+\mathcal{E}_1+\mathcal{E}_{3,1}\right),$$

with

$$\overline{C}_1=\Theta:\frac{|1-4a^2|}{16|z|^{1/2}}\left(1+6\sec^{3/2}\frac{\theta}{2}\right)+2|z|^{1/2}(1+|a|+|z|/2)\sec^{1/2}\frac{\theta}{2},$$

and we remind ourselves that $\mathcal{E}_{3,1}$ (Theorem 8) and $\mathcal{E}_1$ (Theorem 4) are both subexponentially small.

At last we have an effective numerator, then, for the $1/\sqrt{m}$ asymptotic term. Though this effective numerator is almost surely nonoptimal, we are evidently on the right track, because the exact C_1 asymptotic coefficient above (see (8)) has very much the same form as does the Θ-expression for $\overline{C}_1$ here (i.e., the same degrees

---

[7]For example, floating-point FFT-based convolutions of length L require $O(L\log L)$ operations, of complexity less than $O\left(L^{1+\epsilon}\right)$.

TABLE 1

Examples of rigorous accuracy versus asymptotic order. For the indices n and parameters $(a, z) \in \mathbb{D}$, numbers $\mathcal{L}_{n,N}$ are computed via the asymptotic series through term $C_{N-1}/(n+1)^{(N-1)/2}$ as in Theorem 8. The relative error (RE) and rigorous upper bound (RRE) on the relative errors in taking these terms is reported. All decimal values are reported as correct to ± 1 in the last displayed digit of $\Re, \Im$ parts. Such rigorous bounds can be used to establish indisputable accuracies for various other functions.

n	a, z	N	Approximation $\mathcal{L}_{n,N}$ to $L_n^{(-a)}(-z)$	RE	RRE
1300	$0, -1 + 4i$	1	$(7 - i) \cdot 10^{37}$	0.0038	0.43
10^7	$0, 1$	1	$1.59 \cdot 10^{2744}$	0.00011	0.00099
		2	$1.595190 \cdot 10^{2744}$	$1.4 \cdot 10^{-9}$	$2.8 \cdot 10^{-7}$
		3	$1.5951907764 \cdot 10^{2744}$	$5.4 \cdot 10^{-13}$	$9.2 \cdot 10^{-11}$
10^6	$\frac{1}{2} + i, \frac{1}{2} + 4i$	1	$(-1.5 - 2.0i) \cdot 10^{1303}$	0.0016	0.021
		2	$(-1.565 - 2.041i) \cdot 10^{1303}$	$1.6 \cdot 10^{-6}$	$9.8 \cdot 10^{-5}$
		3	$(-1.562555 - 2.041423i) \cdot 10^{1303}$	$1.4 \cdot 10^{-9}$	$4.2 \cdot 10^{-7}$
		4	$(-1.56255567 - 2.041423269i) \cdot 10^{1303}$	$1.3 \cdot 10^{-12}$	$1.4 \cdot 10^{-9}$
		5	$(-1.56255567881 - 2.04142326969i) \cdot 10^{1303}$	$1.3 \cdot 10^{-15}$	$5.9 \cdot 10^{-12}$
10^7	$-\frac{3}{2}, 1 - 100i$	1	$(2 + i) \cdot 10^{19520}$	0.027	0.41
		2	$(1.8 + 0.9i) \cdot 10^{19520}$	0.00035	0.034
		3	$(-1.85 + 0.95i) \cdot 10^{19520}$	$3.1 \cdot 10^{-6}$	0.0028
		4	$(-1.851 + 0.957i) \cdot 10^{19520}$	$2.1 \cdot 10^{-8}$	0.00023
		5	$(-1.8510 + 0.9575i) \cdot 10^{19520}$	$1.1 \cdot 10^{-10}$	$2.0 \cdot 10^{-5}$
		10	$(-1.8509993203 + 0.9575616603i) \cdot 10^{19520}$	$7.1 \cdot 10^{-23}$	$7.6 \cdot 10^{-11}$

of appearance for a, z, and similar coefficients). And, it is easy to see that the Θ-expression here is an upper bound on $|C_1|$ itself, as must of course be true; it is necessary that $\overline{C}_N \geq |C_N|$ for every asymptotic order N.

Table 1 shows the numerical instances of our asymptotic series from Theorem 8. It displays various indices n and parameters a, z, together with corresponding asymptotic orders N from the theorem, using the following nomenclature. Denote, for asymptotic order N,

$$T_{n,N} := 1 + \frac{C_1}{m^{1/2}} + \cdots + \frac{C_{N-1}}{m^{(N-1)/2}},$$

so, for S_n being the assignment (4), we have an approximation $\mathcal{L}_{n,N}$ defined by

$$\mathcal{L}_{n,N} := S_n T_{n,M} \approx L_n^{(-a)}(-z).$$

Thus, $\mathcal{L}_{n,1}$ is simply S_n, as $T_{n,1} = 1$ always. How good are these $\mathcal{L}_{n,N}$ approximations? The important right-most columns of Table 1 report relative error (call it RE), with

$$\text{RE} := \left| \frac{\mathcal{L}_{n,N} - L_n^{(-a)}(-z)}{L_n^{(-a)}(-z)} \right|,$$

and a *rigorous* upper bound on relative error (call it RRE), expressed in terms of Theorem 8's various error bounds, with

$$\text{RRE} := \left| \frac{\epsilon}{T_{n,N}} \right|,$$

where $\epsilon > 0$ can be taken to be any number not exceeding $\left|\overline{C}_N/m^{N/2} + \mathcal{E}_1 + \mathcal{E}_{3,N}\right|$. Note that the three error components in this last expression are all magnitude-bounded via Theorem 8; this is how the RRE table entry was constructed.

A convenient way to view RRE is to realize that $L_n^{(-a)}(-z)$ is rigorously known to lie in the interval $\mathcal{L}_{n,N} \cdot (1 - \mathrm{RRE}, 1 + \mathrm{RRE})$. The decimal representations in Table 1 are reflexive; i.e., there is ambiguity only in the respective final digits. (When a reported value is $(a + ib) \cdot 10^c$, then both a and b are correct to within ± 1 in their respective last digits.)

One might ask about the "first-missing" asymptotic term $C_N/m^{N/2}$—a term of interest in many asymptotic theories. This term is typically close to, but often greater than, the RE. As is common to many asymptotic developments, rules about the behavior of these first-missing terms are hard to establish when parameters (in our case a, z) roam over the complex plane. In any case, the first-missing term is often an order of magnitude or more greater than the RRE. Yet this discrepancy between first-missing and RRE terms is not as harmful as might appear, as we discuss next.

6.4. Example application of effective bounds: Rigor for incomplete gamma. We have mentioned research motives in section 1.1. A worked example of *rigorous* incomplete gamma evaluations is the following. Let us use our very first entry from Table 1, namely,

$$L_{1300}^{(0)}(1 - 4i) \approx (7 - i) \cdot 10^{37},$$

with a rather large RRE = 0.43. Though this is an order of magnitude larger than the relative error, the penalty paid is only a few digits of precision for the relevant incomplete gamma. Indeed, from (5),

$$\Gamma(0, -1 + 4i) = e^{1-4i} \sum_{n=0}^{1299} \frac{1}{n + 1} \frac{1}{L_n^{(0)}(1 - 4i)L_{n+1}^{(0)}(1 - 4i)} + \delta,$$

where the error δ is bounded, using RRE = 0.43 and (4), by

$$(47) \qquad |\delta| < \frac{\left|e^{1-4i}\right|}{(1 - \mathrm{RE})^2} \sum_{n \geq 1300} \frac{1}{n + 1} \frac{1}{|S_n S_{n+1}|} < 10^{-76},$$

the last bound using (4) and careful estimates on the summation over $n \geq 1300$. Thence the sum for $n \in [1, 1299]$ gives a rigorous, 76–digit-accurate value for $\Gamma(0, -1 + 4i)$, ($\approx 0.497 + 0.415\,i$). Correspondingly, the actual absolute error in using the sum over $n \in [0, 1299]$ is about 10^{-78}, so the apparently poor RRE bound brings only a 2-digit penalty.

An additional application of Theorem 8 would be to use effective bounds to rule out zeros of L_n in the crossed (a, z)-plane, that is, to establish an m_0 such that the term $1 + \overline{C}_1/\sqrt{m}$ must be positive for all $m > m_0$ (see our Open problems section).

7. Brief remarks on oscillatory regimes. The exp-arc method has led to rigorous asymptotics for subexponential growth but not for the Fejér form (6) for $z \neq 0$ on the cut $(-\infty, 0]$. Such oscillatory behavior can be dealt with, but other techniques come into play. For one thing, contour integrals must be handled differently.

7.1. When z is on the negative real cut. Even on $z \in (-\infty, 0)$ the contour prescription of Figure 1 is valid, and the Laguerre polynomial is exactly the sum

$c_1 + d_1 + e_1$, with $R := \sqrt{m/|z|} > 1/2$ being the only requirement for contour validity. However—and this is important—the dominant contribution (17) has to change to involve an expanded integration interval; in fact, now we must use the contour integral c_1 itself as the main contribution. For reasons of brevity, we simply state the complete asymptotic result stemming from contour integral c_1 as

$$(48) \qquad L_n^{(-a)}(-z) \sim \frac{e^{-z/2}}{\sqrt{\pi}(-z)^{1/4-a/2}\, m^{1/4+a/2}}$$

$$\times \left\{ \left(\sum_{k=0}^{\infty} \frac{A_k}{m^k} \right) \cos\left(2\sqrt{-mz} + a\pi/2 - \pi/4 \right) \right.$$

$$\left. + \left(\sum_{k=0}^{\infty} \frac{B_k}{m^{k+1/2}} \right) \sin\left(2\sqrt{-mz} + a\pi/2 - \pi/4 \right) \right\},$$

where these oscillatory-series coefficients are directly related to the coefficients in Theorem 8 by

$$A_k := C_{2k}(a, z), \quad B_k := iC_{2k+1}(a, z).$$

It transpires that, for a real, every A_k, B_k is real, whence the asymptotic has all real terms. The Fejér–Perron–Szegö expansion in [36, Theorem 8.22.2] for the oscillatory Laguerre mode is stated there in a fashion structurally different from our asymptotic (48); notwithstanding this Szegö's own $A_{\text{odd}}, B_{\text{even}}$—also not defined quite like ours —vanish.[8]

7.2. Brief remarks on Bessel functions I_n, J_n. There is vast literature on Bessel asymptotics [44], [39], resulting in the *Hankel asymptotic series* [1, section 9.2.5] and effective bounds on error terms (for certain parameter domains). In many cases these restricted error bounds are nevertheless optimal [44]. It is instructive to explore, at least briefly, the application of our exp-arc method to Bessel expansions. We recall (29) which leads to

$$(49)$$

$$I_n(z) = \frac{2}{\pi} \sum_{k \geq 0} \frac{g_k(-2in)}{(2k)!} \left\{ e^z B_k(z) + (-1)^n e^{-z} B_k(-z) \right\} \sim \frac{e^z}{\sqrt{2\pi z}} \sum_{k=0}^{\infty} \frac{g_k(-2in)}{k!\, 8^k} \frac{1}{z^k},$$

(again we omit proofs) where the first sum is convergent, exact, and the second sum agrees with the classical Hankel asymptotic [44], [1]. For $\arg(z) \in (-\pi/2, \pi/2)$, at least, one may derive effective bounds on I_n error terms using the exp-arc techniques.

Computationalists have known for decades that one way to evaluate Bessel functions uniformly in the argument z is to use the standard ascending series for small $|z|$, but an asymptotic series for large $|z|$. However, via the exp-arc method one can establish a converging series whose evaluation only involves *a single error-function evaluation*, followed by recursion and elementary algebra. In fact, the relations (34), (35) can be used to calculate $J_n(z)$ from relation (28) via the sum

$$(50) \qquad J_n(z) = \frac{2}{\pi} \sum_{k=0}^{\infty} g_k(-2in)\left(b_k \cos\chi - c_k \sin\chi \right),$$

---

[8]Our oscillatory asymptotic (48) has been verified to several terms by Temme; also he verifies our claim that the classical Szegö coefficients do vanish for the parities indicated [38].

with

$$\chi := z - \pi n/2 - \pi/4, \quad b_k := B_k(iz)e^{i\pi/4}$$
$$+ B_k(-iz)e^{-i\pi/4}, \quad ic_k := B_k(iz)e^{i\pi/4} - B_k(-iz)e^{-i\pi/4}.$$

Note that if z is real, then each b_k, c_k is real, whence our series here has all real terms. Note that our recursion (35) likewise ignites a recursion amongst the b_k, c_k.

Note that (50) is actually the Hankel asymptotic if we replace B_k by its first term in (34), namely, $(1/2)(2iz)^{-k-1/2}\Gamma(k+1/2)$; however, we already know that the sum (50) is always convergent. It is remarkable that we are using the same structure as the classical asymptotic, yet convergence for all complex z is guaranteed. Moreover, the $B_k(iz)$ are independent of the order n and so can be reused if multiple $J_n(z)$ are desired for fixed z.

Those acquainted with the intricacies of Bessel theory may observe that our convergent expansion (49) is at least reminiscent of the convergent *Hadamard expansion* found in [44, p. 204] for the modified Bessel function I_ν. Though both expansions are absolutely convergent, there are some important differences between this Hadamard expansion and our exp-arc forms (49, 50). For example, we have given our convergent sum only for integer ν. Moreover, the exp-arc expansion is geometrically convergent, while the Hadamard expansion is genuinely slower.

The research area of convergent expansions related to classical, asymptotic ones has been pioneered in large part by Paris, whose works cover real and complex domains, saddle points, and the like [29], [30], [31], [32], [33], [34]. We should point out that Paris was able to develop within the last decade some similar, linearly convergent Bessel series by modifying the "tails" of Hadamard-class series. Finally, we note that the problem of generalizing such unconditionally convergent series as our (50) to noninteger indices—and with comparison to the recent work of Paris—is analyzed in a new work [3].

8. Open problems.
- How might one proceed with the exp-arc theory to obtain effective error bounds for oscillatory Laguerre modes and-or oscillatory Bessel modes? We know that previous researchers have described how to give effective bounds in these cases (e.g., to our asymptotic (48), as in [36, section 8.72]), but once again we stress, How can this be done explicitly and for full parameter ranges?
- Where are the zeros in the complex z-plane—for fixed a—of $L_n^{(-a)}(-z)$? Are "most of" the zeros along some a-dependent ray, in some sense? There is a considerable literature on this zero–free-region topic, especially for polynomials in real variables. For example, with $a := 0$ the Laguerre zeros are all real and negative; see [24, Chapter X] and references therein. There is also an interesting connection between Laguerre zeros and eigenvalues of certain (large) matrices [15]. Certainly the theorems of the Saff–Varga type are relevant in this context [24, Theorem 11].
- How can Laguerre asymptotics be gleaned from standard recurrence relations amongst the $L_n^{(-a)}(-z)$? One may ask the same question for the Laguerre differential equation as starting point. A promising research avenue for a discrete-iterative approach to asymptotics is [43]; see also [4] for the asymptotic analysis of certain complex continued fractions. As for differential-equation approaches, there is the classical work of Erdélyi and Olver, plus modern work on the combinations of differential, discrete, and saddle-point theory [16], [18].

- The efficient "keyhole" contour of Figure 1 was discovered experimentally. What other analytical problems might be approached in this fashion? For that matter, how might one properly use the celebrated Watson loop-integral lemma with error term [27, Theorem 5.1] on our keyhole contour to obtain similar effective asymptotics?
- Here is a scenario suggested by Temme [38]. When n is small (in some sense not made rigorous here), and, say, $|z| \ll |a|$, one should expect an asymptotic behavior

$$L_n^{(-a)}(-z) \sim \binom{n-a}{n}\left(1 + \frac{z}{1-a}\right)^n.$$

The problem is, How does one make this rigorous, with effective error bounds?

Acknowledgments. We are indebted to J. Buhler, R. Mayer, O-Yeat Chan, R. Paris, and N. Temme for the generous lending of expertise, ideas, and suggestions. We are equally grateful to two referees for their insightful remarks.

REFERENCES

[1] M. AMBRAMOWITZ AND I. STEGUN, *Handbook of Mathematical Functions*, NBS (now NIST), Washington, D.C., 1965.

[2] D. H. BAILEY, J. M. BORWEIN, AND R. E. CRANDALL, *Integrals of the Ising class*, J. Phys. A, 39 (2006), pp. 12271–12302.

[3] D. BORWEIN, J. BORWEIN, AND O-Y. CHAN, *The evaluation of Bessel functions via exp-arc integrals*, J. Math. Anal. Appl., 341 (2008), pp. 478–500.

[4] D. BORWEIN, J. BORWEIN, R. CRANDALL, AND R. MAYER, *On the dynamics of certain recurrence relations*, Ramanujan J., 13 (2007), pp. 65–104.

[5] J. M. BORWEIN AND P. B. BORWEIN, *Pi and the AGM: A Study in Analytic Number Theory and Computational Complexity*, John Wiley, New York, 1987. Paperback edition 1998.

[6] J. BORWEIN AND M. CHAMBERLAND, *Integer powers of Arcsin*, Int. J. Math. Math. Sci., 2007, Article ID 19381, http://www.hindawi.com/GetArticle/aspx?doi=10.1155/2007/19381.

[7] J. BORWEIN, R. CRANDALL, AND G. FEE, *On the Ramanujan AGM fraction. Part I: The real-parameter case*, Experiment. Math., 13 (2004), pp. 275–286.

[8] J. BORWEIN AND R. CRANDALL, *On the Ramanujan AGM fraction. Part II: The complex-parameter Case*, Experiment. Math., 13 (2004), pp. 287–296.

[9] C. BOSBACH AND W. GAWRONSKI, *Strong asymptotics for Laguerre polynomials with varying weights*, J. Comput. Appl. Math., 99 (1998), pp. 77–89.

[10] K. CHAO AND R. PLYMEN, *A new bound for the smallest x with $\pi(x) > \mathrm{li}(x)$*, arXiv:math/0509312v5.

[11] R. CRANDALL, *Unified algorithms for polylogarithm, L-series, and zeta variants*, http://www.reed.edu/~crandall, 2008, preprint.

[12] R. CRANDALL, *Alternatives to the Riemann–Siegel formula*, http://www.reed.edu/~crandall, 2008, preprint.

[13] R. CRANDALL, R. BETTEGA, AND R. WHITNELL, *A class of exactly soluble three-body problems*, J. Chem. Phys., 83 (1985), pp. 698–702.

[14] R. CRANDALL, R. WHITNELL, AND R. BETTEGA, *Exactly soluble two-electron atomic model*, Amer. J. Phys., 52 (1984), pp. 438–442.

[15] H. DETTE, *Strong approximation of eigenvalues of large dimensional Wishart matrices by roots of generalized Laguerre polynomials*, J. Approx. Theory, 118 (2002), pp. 290–304.

[16] A. ERDÉLYI, *Asymptotic forms for Laguerre polynomials*, J. Indian Math. Soc., Golden Jubilee Commemoration Volume, 24 (1960), pp. 235–250.

[17] W. FOSTER AND I. KRASIKOV, *Explicit bounds for Hermite polynomials in the oscillatory region*, LMS J. Comput. Math., 3 (2000), pp. 307–314.

[18] C. L. FRENZEN AND R. WONG, *Uniform asymptotic expansions of Laguerre polynomials*, SIAM J. Math. Anal., 19 (1988), pp. 1232–1248.

[19] P. HENRICI, *Applied and Computational Complex Analysis: Special Functions, Integral Transforms, Asymptotics, Continued Fractions, Vol. 2 (Classics Library)*, John Wiley, New York, 1974 (reprint 1988).

3312   D. BORWEIN, J. M. BORWEIN, AND R. E. CRANDALL

[20] E. HOBSON, *A Treatise on Plane Trigonometry*, Cambridge University Press, London, 1925.
[21] L. JACOBSEN AND W. THRON, *Oval convergence regions for continued fractions $K(a_n/1)$*, preprint, 1986.
[22] A. KUIJLAARS AND K. MCLAUGHLIN, *Riemann–Hilbert analysis for Laguerre polynomials with large negative parameter*, Comput. Methods Func. Theory, 1 (2001), pp. 205–233.
[23] J. LÓPEZ AND N. TEMME, *Approximations of orthogonal polynomials in terms of Hermite polynomials*, Methods Appl. Anal., 6 (1999), pp. 131–146.
[24] L. LORENTZEN AND H. WAADELAND, *Continued Fractions with Applications*, North-Holland, Amsterdam, 1992.
[25] E. LOVE, *Inequalities for Laguerre functions*, J. Inequal. Appl., 1 (1997), pp. 293–299.
[26] F. OLVER, *Why steepest descents?* SIAM Rev., 12 (1970), pp. 228–247.
[27] F. OLVER, *Asymptotics and Special Functions*, Academic Press, New York, 1974.
[28] R. B. PARIS, *private communication*, 2007.
[29] R. B. PARIS, *On the use of Hadamard expansions in hyperasymptotic evaluation of Laplace-type integrals. I: Real variable*, J. Comput. Appl. Math., 167 (2004), pp. 293–319.
[30] R. B. PARIS, *On the use of Hadamard expansions in hyperasymptotic evaluation of Laplace-type integrals. II: Complex variable*, J. Comput. Appl. Math., 167 (2004), pp. 321–343.
[31] R. B. PARIS, *On the use of Hadamard expansions in hyperasymptotic evaluation of Laplace-type integrals. III: Clusters of saddles*, J. Comput. Appl. Math., to appear.
[32] R. B. PARIS, *On the use of Hadamard expansions in hyperasymptotic evaluation of Laplace-type integrals. IV: Poles*, Comput. Appl. Math., to appear.
[33] R. B. PARIS, *Exactification of the method of steepest descents: The Bessel functions of large order and argument*, Proc. R. Soc. Lond. Ser. A Math. Phys. Eng. Sci., 460 (2004), pp. 2737–2759.
[34] R. B. PARIS AND D. KAMINSKI, *Hadamard expansions for integrals with saddles coalescing with an endpoint*, Appl. Math. Lett., 18 (2005), pp. 1389–1395.
[35] O. PERRON, *Über das Verhalten einer ausgearteten hypergeometrischen Reihe bei unbegrenztem Wachstum eines Parameters*, Journal für die reine und angewandte Math, 151 (1921), pp. 63–78.
[36] G. SZEGÖ, *Orthogonal Polynomials*, Amer. Math. Soc. Colloq. Publ., 4th ed., 23, American Mathematical Society, Providence, RI, 1975.
[37] N. TEMME, *Asymptotic estimates of Stirling numbers*, Stud. Appl. Math., 89 (1993), pp. 223–243.
[38] N. TEMME, *private communication*, 2006.
[39] S. THANGAVELU, *Lectures on Hermite and Laguerre Expansions*, Math. Notes 42, Princeton University Press, Princeton, N.J., 1993.
[40] E. TITCHMARSH, *The Theory of the Riemann Zeta Function*, Clarendon Press, Oxford, 1951 & 1967.
[41] W. VAN ASSCHE, *Weighted zero distribution for polynomials orthogonal on an infinite interval*, SIAM J. Math. Anal., 16 (1985), pp. 1317–1334.
[42] W. VAN ASSCHE, *Erratum to "Weighted zero distribution for polynomials orthogonal on an infinite interval"*, SIAM J. Math. Anal., 32 (2001), pp. 1169–1170.
[43] Z. WANG AND R. WONG, *Linear difference equations with transition points*, Liu Bie Ju Center for Math. Sci., preprint, 2004.
[44] G. WATSON, *Treatise on the Theory of Bessel Functions*, 2nd ed., Cambridge Math. Lib. Ed., Cambridge University Press, Cambridge, 1966.
[45] S. WINITZKI, *Computing the incomplete gamma function to arbitrary precision*, in Lecture Notes in Comput. Sci. 2667, Springer–Verlag, New York, 2003, pp. 790–798.
[46] S. WONG, *Computational Methods in Physics and Engineering*, Prentice–Hall, New Jersey, 1992.

5. Theory of ROOF walks

Discussion

Imagine a random-walker who steadily depletes its store of, say "jumping power." The resulting ROOF (running-out-of-fuel) walk gives rise to fascinating statistical phenomena. For example, it is possible to use a *finite* amount of fuel and yet meander arbitrarily far in either direction—essentially, to "diverge to infinity." And this, as argued in regard to Figure 1, can happen even if there be steady friction.

One peculiar development is that one can obtain probability densities $f(x)$ for a theoretically relevant[1] ROOF walk such that f is never zero, yet in vast regions obeys

$$f(x) \; < \; e^{-Ae^{Bx}e^{Ce^x}}.$$

Moreover, this triply-exponential density depends on the distribution of prime numbers!

This paper is previously unpublished.

Challenges

Numerous open problems appear in Section 8. An especially interesting problem is whether a particle-physics model can be developed, whereby the ROOF effect is due to, say, photon emission, and therefore a natural means for depletion of walking energy.

Source

Crandall, R.E., "Theory of ROOF walks," (preprint 2008).

---

[1] Relevant in the sense of connecting with the "sinc surprises" of R. Baillie and J. Borwein.

Theory of ROOF walks

Richard E. Crandall*

May 23, 2008

Abstract: We define and analyze what we call "running-out-of-fuel" (ROOF) walks, whereby each random step uses progressively less fuel (variance), in such a way that the total fuel expenditure is finite. In spite of this fuel constraint, a ROOF walk might still meander arbitrarily far. Herein we analyze the probability-density functions f_n for the walker's position after n steps, establishing in this way connections with other fields of analysis. One interdisciplinary aspect is a probabilistic view on the "sinc surprises" of Baillie–Borwein–Borwein [5] that arise in their study of certain sums and integrals. Aspects of the asymptotic density function $f := f_\infty$ remain mysterious. Yet, for a certain canonical ROOF walk we are able to prove that the density $f(x)$ decays at least superexponentially, in the sense that for certain positive constants A, B, C and sufficiently large x,

$$0 \; < \; f(x) \; < \; Ae^{-Be^{Cx}}.$$

We also discuss a "primes" walk whose density function f is even more exotic—in fact triply superexponentially decaying—with the relevant analysis combining statistics and number theory.

*Center for Advanced Computation, Reed College, Portland OR, `crandall@reed.edu`

1   Running out of fuel

We define a running-out-of-fuel (ROOF) walk as follows.[1] Let a coordinate after n random jumps be denoted

$$x_n = \Delta_1 + \Delta_2 + \cdots + \Delta_n,$$

where the random jumps Δ_j are pairwise statistically independent, with expectations satisfying

$$\langle \Delta_j \rangle \;=\; 0, \tag{1}$$
$$\langle |\Delta_1| \rangle \;>\; \langle |\Delta_2| \rangle \;>\; \langle |\Delta_3| \rangle \;>\; \ldots, \tag{2}$$
$$\sum_j \langle \Delta_j^2 \rangle \;<\; \infty. \tag{3}$$

The idea is, we speak of a finite amount of "fuel," being the last sum (3) above, which is, at least heuristically, $\langle x_\infty^2 \rangle$. This fuel store can—as we shall see—be identified in certain physics models as a random walker's total expended energy. The inequality chain (2) tells us that the random jumps (to left or right) have essentially decreasing magnitude. This is all reasonable, since we desire a scenario where the random walker continually expends fuel, but can execute infinitely many jumps in doing so.

It is well known—in fact since the days of Rademacher—that x_n approaches a definite limit, with probability one, merely on the assumption of finite total variance (3) above. Throughout the present work, when we talk of this "almost sure" coordinate $x := x_\infty$ and its density function, we shall assume certain existence proofs; for a more thorough discussion of relevant existence/convergence theorems and their converses, see the delightful monographs [15] [17].

In what follows, we denote by $p_j(\Delta)$ the density function of the j-th jump, and by $f_n(x)$ the density function of x_n. For convenience, we shall identify $f(x) := f_\infty(x)$, which ultimate density function does exist for any of our example ROOF walks. It is also the case that, for all of our ROOF examples below, at least, $f_n(x)$ is, for every $n = 1, 2, 3, \ldots$, monotone decreasing away from $x = 0$, as follows from the fact of monotonicity being preserved under convolution (see [17]).

2   Classes of ROOF walks

Herein we cite some examples of ROOF walks. A relatively trivial, classical ROOF walk has jump density function

$$p_n(\Delta) = \frac{1}{2} \left(\delta(\Delta - 1/2^n) + \delta(\Delta + 1/2^n) \right),$$

---

[1]The ROOF moniker is the present author's, intended for physicists. Previous names in the literature are "fatigued random walker" [15] and named special cases such as "random harmonic series" [17]. The physical interpretations in the present treatment arose more than a decade ago on the author's idea of a particle model with finite-absorption/random-emission, and subsequent discussions of same with O. Bonfim and S. Wolfram.

where the Dirac delta-function notation means that the n-th coordinate is

$$x_n = \pm\frac{1}{2} \pm \frac{1}{2^2} \pm \frac{1}{2^3} \pm \cdots \pm \frac{1}{2^n},$$

with independently random sign choices. It is well known [15] [17] that the ultimate density for this "binary" ROOF walk is

$$f(x) := f_\infty(x) = \frac{1}{2}\mathbf{1}_{[-1,1]}(x),$$

i.e., an equidistribution pedestal on $x \in [-1,1]$. This result is intuitively clear, on the realization that balanced -binary representation of reals is unique.

The above example—in giving a pedestal for ultimate density—has all bounded walks, for any choice of signs. Because of wide connections with other mathematical domains, we next define:

Class-D ROOF walks: These are "diverging" ROOF walks in the following sense: Along with the defining relations (1)-(3), we also posit

$$\sum \langle |\Delta_j| \rangle = \infty.$$

Note that this divergence condition is not satisfied by the "binary" walk above (where the relevant sum over $\langle |\Delta_j| \rangle$ is simply 1). But for a ROOF walk of class D, we see that *the walker can go arbitrarily far away from the origin.* It is generally true, in fact, that a class-D walk has an ultimate density $f := f_\infty$ that is everywhere positive.[2] We next turn to specific subclasses of class-D walks.

D_0 walks: A relatively trivial—albeit interesting—subclass of class-D ROOF walks involves the n-th-step density function

$$p_n(\Delta) = \frac{1}{\sqrt{2\pi v_n}}e^{-\Delta^2/(2v_n)},$$

i.e., ever-thinner, centered Gaussian jumps with real, positive variances $v_1 > v_2 > \ldots$, satisfying $\sum v_j < \infty$. In this instance we reside in class D if

$$\sum \sqrt{v_i} = \infty.$$

We know that any two Gaussian densities convolve into a Gaussian density of summed variance, so the *final* density function for the ultimate coordinate x_∞ involves the assumed-finite sum $\sum v_n$, as

$$f_\infty(x) = \frac{1}{\sqrt{2\pi \sum v_n}}e^{-x^2/(2\sum v_n)}.$$

---

[2]We leave to others the issue of proving (or disproving) positivity of $f := f_\infty$ for *every* class-D walk; yet, proofs for our main examples are found in [17].

It is interesting where this ROOF scenario is basically the only kind for which the classical central-limit theorem can be (conceptually, accidentally if you will) applied: Indeed, for none of our other examples does the final density function have Gaussian character. So walks in subclass D_0 are completely solvable in an obvious sense—and yet, we pay a price for said solvability. Namely, the concept of "meandering arbitrarily far" as one runs out of fuel is trivial since on any one step the walker can go arbitrarily far under its instantaneous Gaussian.

D_1 walks: This subclass we define to involve discrete jumps, in that the density function for jump Δ_j is two superimposed Dirac-delta spikes:

$$p_j(\Delta) = \frac{1}{2}\left(\delta(\Delta - b_j) + \delta(\Delta + b_j)\right),$$

where the real, positive b_j satisfy $b_1 > b_2 > \ldots$, with $\sum b_j^2 < \infty$, and class-D membership is then assured by

$$\sum b_j = \infty.$$

Note that expectations for class D_1 are given simply by

$$\langle |\Delta_j| \rangle = b_j, \quad \langle \Delta_j^2 \rangle = b_j^2.$$

Now we can give the notion of "arbitrary meandering while running out of fuel" some teeth: The density function $f(x) := f_\infty(x)$ is defined and everywhere positive (see, as before, [17]). We should admit that the phenomena attendant on sublass D_1 remain to some extent shrouded in mystery. For example, we *do not* know the exact density f for any explicit set of b_j, and we do not even know an exact $f(0)$, say, for any D_1 walk.

D_2 walks: This subclass we define to involve ever-shrinking jumps again, but with each density function being a pedestal:

$$p_j(\Delta) = \frac{1}{2c_j}\mathbf{1}_{[-c_j, c_j]}(\Delta),$$

where the real, positive c_j satisfy $c_1 > c_2 > \ldots$, with $\sum c_j^2 < \infty$, and class-D membership is then assured by

$$\sum c_j = \infty.$$

The expectations for class D_2 are:

$$\langle |\Delta_j| \rangle = \frac{1}{2}c_j, \quad \langle \Delta_j^2 \rangle = \frac{1}{3}c_j^2.$$

One of the many delightful surprises that ROOF walks offer is this: It can happen that a class-D_2 walk *has the very same final density function* $f := f_\infty$ as a corresponding walk from class-D_1. This remarkable fact—known to other researchers in one guise or another—will be shown from the probabilistic perspective in what follows.

4

3   Physics models for ROOF walkers

The present author proposes the following physics model for any ROOF walker of class D_1. We refer to the Figure 1 pictorial to aid our intuition. Ponder a *Gedanken*—"thought experiment"— involving a sled on a frictive track. The law of motion we contemplate is actually reasonable, and used in many physics applications, namely

$$m\frac{d^2x}{dt^2} = -\gamma\frac{dx}{dt}.$$

Here m is sled mass and γ is the coefficient of sliding friction. There is on-board fuel, yet to simplify the analysis we shall neglect the initial fuel mass compared to m and simply assume m is constant for all time.

    The equation of motion is not enough, of course; we also need imagine a "demon" on the sled, and the demon has an algorithm for precisely how to expend on-board fuel. To this end, consider first the space-time trajectory under initial condition $x(0) = 0$, $dx/dt(0) = v$. The exact solution to the frictive motion is

$$x(t) = \frac{mv}{\gamma}\left(1 - e^{-\gamma t/m}\right).$$

Note that this motion, once begun with initial velocity v—say by a quick burst of fuel-burn— will never stop as $t \to \infty$; rather, the sled will asymptotically approach a jump distance of mv/γ. So let us assume a fixed, finite jump-time τ, so that in time τ the sled moves $x(\tau)$, at which time τ we use more fuel-burn to *brake* the sled from its velocity $dx/dt(\tau) = ve^{-\gamma\tau/m}$ to a standstill. Thus, the braking action works opposite to the sign of initial velocity v.

    Now, to execute a ROOF walk, the control demon knows an infinite set of initial speeds $(v_1, v_2, \dots)$ and randomly selects direction $\pm$ for the initial velociti $v = v_n$ on the n-th jump. Thus, the jump distances that occur every time interval τ are

$$\Delta_n = \pm\frac{mv_n}{\gamma}\left(1 - e^{-\gamma\tau/m}\right).$$

Moreover, each jump operation involves energy outlay

$$E_n = \frac{1}{2}mv_n^2 + \frac{1}{2}mv_n^2\,e^{-2\gamma\tau/m},$$

with the two components here being, respectively, the initial thrust energy to achieve speed $|v_n|$ and the braking energy to remove the speed after time τ.[3]

---

[3]It could be argued that this energy budget neither takes into account thrusting into the frictive medium, or braking into said medium. However, if the thrusting and braking is done with infinitesimally brief energy pulses (and so, virtually infinite power pulses yet with the finite energy), then the energy needed is effectively that of giving/taking the kinetic energy of the sled.

Now for generality within the class D_1: For the defining sets of incremental distances $b_1, b_2, \ldots$ for a class-D_1 walk, choose physical constants and units such that

$$m = \gamma := 1$$

and the time quantum

$$\tau := \log\left(2 + \sqrt{3}\right),$$

in which case the control demon can assign velocities

$$v_n := \frac{1 + \sqrt{3}}{2} b_n$$

and so expend successive fuel energies

$$E_n := b_n^2$$

to effect the ROOF walk. This model—which, by thus assigning physical units appropriately—covers all ROOF walks of class D_1, and this is what Figure 1 shows. It is an interesting scaling property of class D_1 that the time quantum $\log(2 + \sqrt{3})$ can be fixed, across the entire class.

We presume that class-D_0 ROOF walks can be modeled similarly; e.g., for class D_0 one might imagine the demon has Maxwell-distributed gas on-board, and somehow allows Gaussian-distributed velocities for emanating gas particles.[4]

There are other interesting *Gedankens* that generate ROOF walks. For the "convex" friction law

$$m\frac{d^2x}{dt^2} = -\gamma\frac{\frac{dx}{dt}}{\sqrt{\left|\frac{dx}{dt}\right|}}$$

one can show that a sled starting with an initial velocity v_n actually comes to a halt in *finite* time. Thus, the sled demon can employ initial velocity bursts only, needing no braking mechanism. Taking $\gamma = m = 1$, an exact trajectory, given initial speed $v_n > 0$, is

$$x(t) = \frac{2}{3}v_n^{3/2} - \left(\left(\frac{2}{3}\right)^{1/3}v_n^{1/2} - \frac{1}{3}\left(\frac{3}{2}\right)^{2/3}t\right)^3,$$

from which we may casually read off the total distance-to-halt, assuming random choice of direction (left-right), as

$$\Delta_n = \pm\frac{2}{3}v_n^{3/2}.$$

If we choose velocities $v_n := (3/(2n))^{2/3}$, we have divergence of $\sum\langle|\Delta_j|\rangle = 1 + 1/2 + 1/3 + \ldots$. So this frictive walk is in class D_1: The variance is $\sum\langle\Delta_j^2\rangle = \pi^2/6$, yet the sled can go "anywhere"

---

[4]Of course, such a scenario requires gas molecules of zero mass—or some other appropriate "law of the infinitesimal"—to avoid just a single emanating molecule dissipating by itself all the finite energy!

on finite fuel. It is of interest that for this convex-friction model, the total energy burn is not the same as the total (finite) variance; rather

$$E = \sum_n \frac{1}{2}mv_n^2 = \frac{3^{4/3}}{2^{7/3}}\zeta(4/3),$$

which is itself finite.

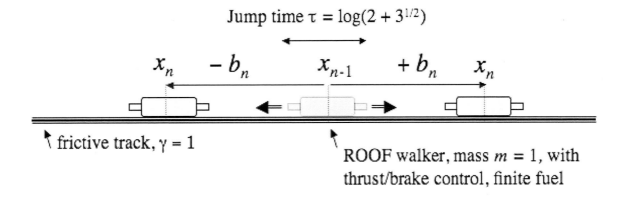

Figure 1: A physics *Gedanken* for class-D_1 ROOF walks. On the n-th jump, a fuel (i.e. energy) parcel b_n^2 is used to move the sled a distance $\pm b_n$, under the friction law $m\, d^2x/dt^2 = -\gamma\, dx/dt$. Remarkably, by burning thus a *finite* total fuel $\sum b_n^2$, and in spite of the friction, the sled still reaches arbitrarily remote positions with nonzero probability, being as $\sum b_n$ diverges. For $b_n := \frac{1}{2n}$, the ultimate position density will be the everywhere-positive $f(x)$ in Figure 2 (see Theorem (1) for equivalence of D_1, D_2 classes).

4   Density-function calculus

From our original prescription for the n-th position,

$$x_n = \Delta_1 + \Delta_2 + \cdots + \Delta_n,$$

we may write the n-th density function, for any integer $m \in [2, n]$, as the convolution

$$f_n(x) = \int_{-\infty}^{\infty} f_{m-1}(x-y)g_{m,n}(y)\, dy, \tag{4}$$

where $g_{m,n}$ here is the density function for the random sum

$$y = \Delta_m + \Delta_{m+1} + \cdots + \Delta_n.$$

For example, for $m = n$ we have the more transparent relation

$$f_n(x) = \int_{-\infty}^{\infty} f_{n-1}(x - y)p_n(y) \, dy,$$

where $p_n(y) = g_{n,n}(y)$ is, as before, the transitional density for the single jump Δ_n. We shall soon be exploiting the freedom of choice in (4) for the intermediate index m. Note that all densities are normalized in this theory; i.e., for all positive integers n and $m \in [2, n]$

$$\int_{x \in R} p_n(x) \, dx = \int_{x \in R} f_n(x) \, dx = \int_{x \in R} g_{m,n}(x) \, dx = 1.$$

In a classical vein, a complementary view on density convolution involves characteristic functions; namely, for the given transition densities $p_n(\Delta)$ we consider the Fourier transform

$$\hat{p}_n(\omega) := \int_{-\infty}^{\infty} p_n(\Delta)e^{-i\omega\Delta} \, d\Delta.$$

Then the familiar classical result is that convolution in configuration space turns into mutliplication in spectral space, that is

$$f_n(x) = \frac{1}{2\pi} \int_{-\infty}^{\infty} \prod_{j=1}^{n} \hat{p}_j(\omega)e^{i\omega x} \, d\omega. \tag{5}$$

Now we can cite specific integral forms for class-D_1 and class-D_2 density functions, namely:

Density function for class-D_1 walks (discrete-Dirac-delta jumps by $\pm b_n$):

$$
\begin{aligned}
\hat{p}_n(\omega) &= \cos(b_n\omega), \\
f_n(x) &= \frac{1}{2\pi} \int_{-\infty}^{\infty} e^{i\omega x} \, d\omega \prod_{j=1}^{n} \cos(b_j\omega).
\end{aligned}
\tag{6}
$$

Density function for class-D_2 walks (pedestal jumps of widths $2c_n$):

$$
\begin{aligned}
\hat{p}_n(\omega) &= \mathrm{sinc}(c_n\omega), \\
f_n(x) &= \frac{1}{2\pi} \int_{-\infty}^{\infty} e^{i\omega x} \, d\omega \prod_{j=1}^{n} \mathrm{sinc}(c_j\omega).
\end{aligned}
\tag{7}
$$

We should state right off one of the remarkable results found in previous literature. We shall say that two processes are *statistically equivalent* if their respective density functions ϕ, f satisfy a scaling law $\phi(x) = cf(cx)$ for some positive constant c.

Theorem 1 *The D_1 walk with $b_n := \frac{1}{2n}$ is statistically equivalent to the D_2 walk with $c_n := \frac{1}{2n-1}$, in the sense that both walks have the same asymptotic density function $f := f_\infty$.*

Proof: This follows immediately from (6) and (7), on the observation that for all real ω

$$\prod_{j=1}^{\infty} \cos\left(\frac{\omega}{2j}\right) = \prod_{j=1}^{\infty} \text{sinc}\left(\frac{\omega}{2j-1}\right).$$

This is established in [6] [8]; moreover, [17] has a probabilistic argument for the statistical equivalence of the two walks, which argument exploits the fact that an odd-reciprocal sequence is naturally buried in the harmonic series.   **QED**

But there are a great many other properties to be gleaned from Fourier representations, as we shall see.

5   Focus on class-D_2 ROOF walks

Referring to Figure 2, we envision the canonical class-D_2 walk, that is, the figure assumes $c_n := \frac{1}{2n-1}$; yet, we now proceed for general c_n for the class.

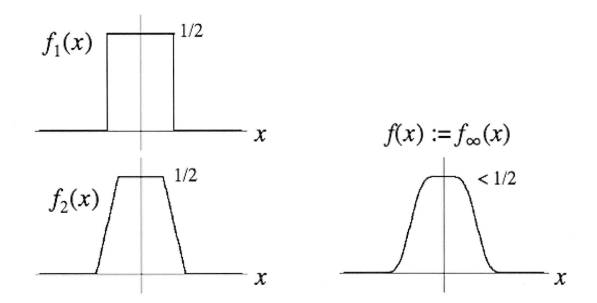

Figure 2: The class-D_2 ROOF walk with canonical assignments $c_n := 1/(2n-1)$. The upper-left plot is for the first random jump: A pedestal density over $[-1, 1]$ with height $1/2$. The lower-left plot is the density after 2 jumps: The top-height is still $1/2$, but straight lines connect $f_2(c_1 - c_2 = 2/3) = 1/2$, $f_2(c_1 + c_2 = 4/3) = 0$. At lower-right is a plot of $f := f_\infty$, which plot remarkably has $f(0) < 1/2$, yet barely so. It is also (barely) true that $f(1) < 1/4$, and that for large x, the decay of (the everywhere-positive) f is at least superexponential; for example, we prove $f(2\pi) < 10^{-13680}$.

For any c_n appropriate to class D_2, a useful representation also following from the convolution theory is

$$f_n(x) = \frac{1}{2^n} \int_{[-1,1]^n} \delta(c_1 z_1 + c_2 z_2 + \cdots + c_n z_n - x) \, \mathcal{D}\vec{z}. \tag{8}$$

Here, $\mathcal{D}\vec{z}$ is the volume element $dz_1 dz_2 \cdots dz_n$, and the integration is over the n-dimensional centered unit cube. This geometrically powerful representation can be derived easily from iteration of (4), or from the subsequent Fourier theory using $\delta(\omega) = 1/(2\pi) \int e^{i\omega z} dz$. We shall later use this delta-representation (8) in connection with the Baillie–Borwein–Borwein theory of sinc integrals. We note that relations such as (8) have been used previously in the study of multidimensional sinc integrals [9].

Net we aim to prove three salient bounds relevant to class-D_2 walks. The asymptotic (final) density function $f := f_\infty$ of the walk has properties:

- $f(0) < \frac{1}{2c_1}$,

- $f(c_1) < \frac{1}{4c_1}$,

  - For the canonical case $c_n := 1/(2n - 1)$, the decay of $f(x)$ is at least superexponential, meaning that for sufficiently large x, and positive constants A, B, C, we have

    $$0 < f(x) < Ae^{-Be^{Cx}}.$$

We note right off that various of our literature references have already established the two bounds for $f(0), f(c_1)$ respectively; our intent is to prove these via the present nomenclature and conceptual framework. The third, superexponential-decay result is new as far as the present author is aware.

To begin our bounding analysis , we employ the convolution relation (4) in the specific ($m = 2$) form

$$f_n(x) = \int_{-\infty}^{\infty} f_1(x - y) g_{2,n}(y) \, dy. \tag{9}$$

In what follows, we denote probabilities by $\mathcal{P}$, and recall that for a class-D_2 walk, the first density function is $f_1(x) = p_1(x) = \frac{1}{2c_1} \mathbf{1}_{[-c_1.c_1]}(x)$. Also, let us denote generally the tails $y_m = \Delta_m + \cdots + \Delta_n$; e.g., the $g_{2,n}$ function in (9) above is the density for random variable y_2.

Theorem 2 *For a class-D_2 ROOF walk, we have, for all positive integers n and also $n = \infty$,*

$$f_n(0) = \frac{1}{2c_1} \quad \text{if } c_2 + \ldots c_n \le c_1,$$

$$< \frac{1}{2c_1} \quad \text{otherwise.}$$

10

Similarly,

$$f_n(c_1) = \frac{1}{4c_1} \qquad \text{if } c_2 + \ldots c_n \le 2c_1,$$

$$< \frac{1}{4c_1} \quad \text{otherwise.}$$

We also have exact sum rules

$$\sum_{k \in Z} f_n(2kc_1) = \sum_{k \in Z} f_n((2k+1)c_1) = \frac{1}{2c_1},$$

and so we have an overall sum rule

$$\sum_{k \in Z} f_n(k) = \frac{1}{c_1}.$$

Finally, for integer $m > 1$ and $x > c_1 + \ldots c_{m-1}$,

$$f_n(x) \le \frac{1}{2c_1} \mathcal{P}(x < y_m + c_1 + \ldots c_{m-1}.)$$

Remark: Note that the sum rules themselves imply the first two inequalities for $f(0), f(c_1)$—i.e., for $n = \infty$ ultimate densities. The sum rules also yield interesting theoretical and numerical side-connections (see Corollary 2 and Appendix).

Proof: We have immediately from (9) and the stated form for f_1 the relation

$$f_n(x) = \frac{1}{2c_1} \mathcal{P}(y_2 \in [x - c_1, x + c_1]) \tag{10}$$

From this we infer

$$f_n(0) = \frac{1}{2c_1} \mathcal{P}(y_2 \in [-c_1, c_1]) = \frac{1}{2c_1}(1 - 2\mathcal{P}(y_2 > c_1)), \tag{11}$$

and this settles both branches of the $f_n(0)$ claim of the theorem. Similarly,

$$f_n(c_1) = \frac{1}{2c_1} \mathcal{P}(y_2 \in [0, 2c_1]) = \frac{1}{4c_1}(1 - 2\mathcal{P}(y_2 > 2c_1)), \tag{12}$$

thus settling the claims for $f_n(c_1)$.

Now to prove the sum rules. Using (10) again, we have

$$f_n(0) + 2f_n(2c_1) + \cdots = \frac{1}{2c_1} \left(\mathcal{P}(|y_2| \in [0, c_1]) + \mathcal{P}(|y_2| \in [c_1, 3c_1]) + \mathcal{P}(|y_2| \in [3c_1, 5c_1]) + \ldots \right)$$

$$= \frac{1}{2c_1} \cdot 1.$$

11

A similar argument reveals that

$$f_n(c_1) + f_n(3c_1) + \cdots = \frac{1}{4c_1},$$

and this establishes the sum rules. (Not only do the sum rules hold for all positive integers n, they have to hold for the ultimate density f, i.e. for $n = \infty$, because it is known (see introductory section) the sequence $f_n \to f$ pointwise.)

Finally, going back to (4) for general index m and given that $x > c_1 + \ldots c_{m-1}$, we have

$$f_n(x) \leq \left(\sup_z f_n(z)\right) \mathcal{P}(y_m > x - c_1 - c_2 - \cdots - c_m).$$

However, the sup is bounded above by $1/(2c_1)$, as follows trivially from (10), and we obtain the theorem's claimed bound on $f_n(x)$.

QED

In order to render the general bound on $f_n(x)$ in Theorem 2 more useful, we next cite a result concerning expectations of exponential arguments. The following theorem is a generalization of the expectation theory found in [17]:

Theorem 3 *For a class-D_2 ROOF walk, take $n \geq m$, set $v := \sum_{k=m}^n c_k^2$, and denote by $g_{m,n}(y)$ the density for the random variable $y = \Delta_m + \cdots + \Delta_n$. Then for any real t, under the density $g_{m,n}$ we have a bounded expectation*

$$\langle e^{ty} \rangle \;\leq\; e^{\frac{1}{6}vt^2}.$$

Moreover, for positive real z we have, under the stated density, a probability bound

$$\mathcal{P}(y > u) \;\leq\; e^{-\frac{3u^2}{2v}}.$$

Proof:

$$\langle e^{ty} \rangle = \prod_{j \in [m,n]} \frac{1}{2c_j} \int_{-c_j}^{c_j} e^{ty}\, dy = \prod_{j \in [m,n]} \frac{\sinh(c_j t)}{c_j t} \leq e^{\frac{1}{6}vt^2}.$$

This last bound follows immediately from

$$\frac{\sinh z}{z} = 1 + \frac{z^2}{3!} + \frac{z^4}{5!} + \cdots \leq 1 + \frac{z^2/6}{1!} + \frac{z^4/6^2}{2!} + \cdots \leq e^{z^2/6}.$$

Next we invoke a beautiful theorem of Markov (see [17]), saying that for $u > 0$,

$$\mathcal{P}(y > u) \leq \inf_{t>0} \langle e^{ty} \rangle e^{-tu}.$$

This can readily be seen in our present context by writing

$$\mathcal{P}(y > u) = \int_u^\infty g_{m,n}(y)\,dy \le \int_{-\infty}^\infty g_{m,n}(y)e^{t(y-u)}\,dy.$$

But all of this means we can take $t := 3u/v$ in the inf argument and thereby derive the resulting bound for $\mathcal{P}(y > u)$.

QED

Theorem 4 *For the class-D_2 ROOF walk, denote for $m > 1$ the quantities $a_m := \sum_{k=1}^{m-1} c_k$ and $V_m := \sum_{k=m}^\infty c_k^2$. Then the density $f := f_\infty$ satisfies*

$$0 < \frac{1}{2c_1} - f(0) < \frac{1}{c_1}e^{-\frac{3c_1^2}{2V_2}},$$

$$0 < \frac{1}{4c_1} - f(c_1) < \frac{1}{2c_1}e^{-\frac{6c_1^2}{V_2}}.$$

Moreover, for $x > c_1$,

$$0 < f(x) < \frac{1}{2c_1}\inf_{a_m < x} e^{-\frac{3}{2}\frac{(x-a_m)^2}{V_m}}.$$

Proof: Recall that $f(x)$ is everywhere positive [17], so that the first two inequality chains, for $f(0), f(c_1)$ respectively, follow directly from relations (11), (12) and the final statement of Theorem 3. Then for $x > 1$, we can apply the last statements of Theorems 2 and 3 to get the desired inf inequality.

QED

Corollary 1 *For the canonical Class-D_2 ROOF walk (assignments $c_n := 1/(2n-1)$) the asymptotic density $f := f_\infty$ satisfies*

$$0 < \frac{1}{2} - f(0) < e^{-\frac{12}{\pi^2-8}} < 10^{-3},$$

$$0 < \frac{1}{4} - f(1) < e^{-\frac{48}{\pi^2-8}} < 10^{-11},$$

$$0 < f(x > 1) < \frac{1}{2}e^{-\frac{12(x-1)^2}{\pi^2-8}}.$$

Moreover, for large x the density decays at least superexponentially, in the sense that for $x \ge 3$

$$0 < f(x) < \frac{1}{2}e^{12}e^{-\frac{6}{e^4}e^{2x}}.$$

Remark: The bounds are nonoptimal; in truth, $1/2 - f(0) \approx 10^{-6}$ and $1/4 - f(1) \approx 10^{-43}$. The superexponential-decay bound says $f(3) < 10^{-14}$ while the numerical value is evidently $f(3) \approx 5 \cdot 10^{-43}$. A special case we discuss later is $f(2\pi) < 10^{-13680}$, and for this we do not know an accurate mantissa, even to one significant decimal. Interestingly: It could be that for *general* class-D_2 ROOF walks $f(x)$ decays at least doubly exponentially, i.e. perhaps we always have $f(x) < A \exp(-B \exp(Cx))$ within class D_2; we do not know if this is true, although this corollary for our canonical case is suggestive. (Later we mention a ROOF walk for which f decays *triply* superexponentially, and so still bounded above by a doubly superexponential decay.)

Proof: For this canonical ROOF walk, $V_{2,\infty} = 1/3^2 + 1/5^2 + \cdots = \pi^2/8 - 1$ and so the first three inequality chains of the corollary result from Theorem 4. As for the last, superexponential inequality, observe that for $m \geq 2$,

$$a_m < 1 + \frac{1}{2}\log m,$$

$$V_m < \frac{1}{4(m-1)}.$$

These can be proved using elementary integral bounds on the function $1/(2z-1)$. Now for $x \geq 3$ set

$$m := \lfloor e^{2(x-2)} \rfloor,$$

so that $m \geq 2$, thus

$$f(x) < \frac{1}{2}e^{-(3/2)(x-1-(1/2)\cdot 2(x-2))^2 \cdot 4(m-1)} \leq \frac{1}{2}e^{-(3/2)\cdot(4(m-1))}$$

$$\leq \frac{1}{2}e^{12}e^{-6e^{2x-2}},$$

which is the desired superexponential bound.

QED

6   Numerical evaluation of density functions

For class D_2 ROOF walks, one might contemplate using the Fourier decomposition (7), performing numerical integration so as to approximate $f_n(x)$ numerically. And, with enough good fortune, perhaps some high-precision values of $f(x)$ for the asymptotic density function f can be obtained—via tight approximations for the infinite sinc-product. Actually there are two related approaches, but ones that differ interestingly in the details.

One approach is used in the work [3] [4] to evaluate integrals of the type that appear in our characteristic relation (6). Those authors were working in a scenario statistically equivalent to a ROOF model having parameters $b_n := 1/(2n)$. (We have hereby scaled the jumps b_n with respect to [3] for our own purposes, yet statistical equivalence prevails.) The two referenced works employ the expansion (valid for $|\omega| < \pi/2$)

$$\log \cos \omega = \sum_{j=1}^{\infty} \frac{(-1)^j \left(2^{2j} - 1\right) B_{2j}}{(2j)! j} \omega^{2j}$$

with a partitioned integral

$$f_{\infty}(x) = \frac{1}{2\pi} \int_{-\infty}^{\infty} e^{i\omega x} \, d\omega \; e^{\sum_{k=m(\omega)}^{\infty} \log \cos(\omega/(2k))} \prod_{j=1}^{m(\omega)-1} \cos(\omega/(2j)). \tag{13}$$

Here, $m(\omega)$ is an integer-valued function that bounds $|\omega|/\pi$ from above—so that the log-cos series is always valid during numerical integration. They were able to obtain in this way what is equivalent to an extreme-precision value for $f(0) = f_{\infty}(0)$ for this canonical class-D_1 walk (see our Appendix for listed values).

A second approach is to exploit the equivalence Theorem 1 by considering instead the canonical class D_2 walk with $c_n := 1/(2n - 1)$, for which

$$f_{\infty}(x) = \frac{1}{2\pi} \int_{-\infty}^{\infty} e^{i\omega x} \, d\omega \; e^{\sum_{k=m(\omega)}^{\infty} \log(\operatorname{sinc}(\omega/(2k-1)))} \prod_{j=1}^{m(\omega)-1} \operatorname{sinc}(\omega/(2j - 1)). \tag{14}$$

In this alternative formulation we would use

$$\log \operatorname{sinc} \omega = \sum_{j=1}^{\infty} \frac{(-1)^j 2^{2j-1} B_{2j}}{(2j)! j} \omega^{2j},$$

valid this time for the range $|\omega| < \pi$.

Note that we are saying the f_{∞} in (14) is the *same* function as the f_{∞} in (13).

In [17], where the two-ROOF-walk equivalence we are exploiting here is argued probabilistically, appears the germ of a fine idea, as follows: That the characteristic kernel for either of the equivalent asymptotic walks can be partitioned as a product of trigonometric (cos or sinc) terms—corresponding to the random variable $\Delta_1 + \cdots + \Delta_{m-1}$—times a product of characteristics for the "tail variable" $y_m := \Delta_m + \Delta_{m+1} + \ldots$. However, it is argued in [17] that *the distribution of y_m for large m is essentially Gaussian.* But this would mean a reasonable approximation might be the relatively simple construction

$$f_{\infty}(x) \approx \frac{1}{2\pi} \int_{-\infty}^{\infty} e^{i\omega x} e^{-v_m \omega^2/2} \, d\omega \prod_{j=1}^{m(\omega)-1} \operatorname{sinc}(\omega/(2j - 1)),$$

wherein one witnesses the term $e^{-v_m \omega^2/2}$—the characteristic function for a Gaussian of variance $v_m = \langle y_m^2 \rangle$. A beautiful aspect of this argument is that it is actually addressing the *first* term of the log-sinc series. Indeed, take for example the log-sinc series through a few terms:[5]

$$\log \operatorname{sinc} \omega = -\frac{\omega^2}{6} - \frac{\omega^4}{180} - \frac{\omega^6}{2835} - \frac{\omega^8}{37800} - \frac{\omega^{10}}{467775} - \cdots$$

and note that the tail y_m claimed to be near-Gaussian—has density even closer to

$$g_m(y) = \frac{1}{2\pi} \int_{-\infty}^{\infty} e^{i\omega y} e^{-a_2 \omega^2 - a_4 \omega^4 - \cdots} \, d\omega,$$

where

$$a_2 := \frac{1}{6} \sum_{k \geq m} \frac{1}{(2k-1)^2},$$

$$a_4 := \frac{1}{180} \sum_{k \geq m} \frac{1}{(2k-1)^4},$$

and so on. The $g_m(y)$ integral here is problematic. All the present author knows along such lines is that by keeping only a_2, a_4, the result $g_m(0)$ is a Bessel-$K_{1/4}$ evaluation.

It is interesting that even though the log-cos and log-sinc series have finite radii of convergence, we can in these instances ignore that and simply choose optimal cutoff m for a desired final precision on the desired $f_\infty(x)$. This freedom is due to the happy fact of all series elements in both of these log-series being negative, even outside the convergence domain. The downside to this way of thinking is that we do not yet know an analytical relation giving the optimal parameters such as cutoff m and series cutoff degree for a desired ultimate precision for $f_\infty(x)$. Still, it appears possible in principle to obtain arbitrary precision—albeit at steeply ramping cost—via empirical tuning of the relevnt parameters, as we have done for our Appendix.

7   Experimental and theoretical surprises

7.1   The "sinc surprises" of Baillie–Borwein–Borwein

A recent work by R. Baillie, D. Borwein and J. Borwein [5] on certain sums and integrals has the word "surprising" in its very title, for indeed, those researchers considered various product functions—in our probabilistic language these are characteristic functions of class-D_2 ROOF walks—of a form we denote

$$\Pi^{(n)}(\omega) := \prod_{j=1}^{n} \operatorname{sinc}(c_j \omega)$$

---

[5]The notion that all terms here are ω-powers divided by integers is specious: It may well be that the term shown with ω^{10} is the last such "pure" term.

and managed to build identities that sometimes fail with—shall we say—surprisingly minuscule errors.[6]

We remind ourselves that there are ROOF walks for which the c_j for j up through any finite n can be "anything monotonic," i.e. subject only to the constraint $c_j > c_k$ for $j > k$, and still represent the initial pedestal-half-widths of a finite-total-variance walk, as in the chain (2). That is, only when we contemplate infinite products, $n \to \infty$, denoting these Π_∞ when they exist, are we concerned about convergence issues for the c_j and the ensuing products $\Pi^{(n)}$. For finite n, then, any pairwise distinct real constants c_j in the definition of $\Pi^{(n)}$ are effectively allowed, since they can always be sorted into decreasing order.[7] In what follows, therefore, we shall assume that in the construction of a sinc-product $\Pi^{(n)}$ the multiplicands adopt the class-D_2 ROOF walk's natural sort, with the real, positive c_j strictly decreasing.

Now, two interesting constructs residing at the core of the previous research [5] are what we shall call a sinc-sum and a sinc-integral, respectively

$$S_n := \sum_{k \in Z} \Pi^{(n)}(k), \qquad (15)$$

$$I_n := \int_{\omega \in R} \Pi^{(n)}(\omega) \, d\omega. \qquad (16)$$

We reserve the freedom to formally allow $n = \infty$, when, of course, Π_∞ converges as an infinite product. This notion is tantamount, in our probabilistic view, to discussing the density function limit $f_n \to f_\infty =: f$.

Of great interest are sim-integral error terms addressed in [5] and which we equivalently denote

$$\rho_n := S_n - I_n.$$

In regard to such error terms, the aforementioned authors of [5] made surprising discoveries of the following type: It can happen that the sum-integral error here is $\rho_n = 0$ for some very large initial set of n indices, but suddenly some ρ_{n_0} does *not* vanish. Similarly, there can be long stretches of n values for which not only $\rho_n = 0$ but both S_n, I_n maintain a steady value, said value being invariant with increasing n until we reach some sudden index n_1 where both S_{n_1}, I_{n_1} depart(s) from the steady value (and so the error ρ_{n_1} at that sudden juncture may or may not vanish).

---

[6]The philosopher Wittgenstein said "There can never be surprises in logic," yet when mathematicians speak of surprise we all know what is really meant: something "unexpected." The present author maintains that believing in "unexpectedness" does not violate Wittgenstein's thesis, in that the sheer unpredictability of where analysis might lead pervades all of mathematics. After all, the eminent Carl Ludwig Siegel did say "One cannot guess the real difficulties of a problem before having solved it"—in such a sense, surprise perhaps really does have a place in mathematics, logic notwithstanding.

[7]Evidently nothing prevents us from using.more generally, nonincreasing c_j-sequences; i.e., the pairwise-distinctness condition is not fundamentally required. However, much of our analysis goes through more efficiently when no c_j pairs coincide.

In the ensuing subsections we cover some observations and results in regard to these "sinc surprises" and related analyses.

7.2   Exact finite forms for sinc-sums S_n

It turns out to be possible, for arbitrary c_j, to evaluate any S_n (for finite n) in finite form. We shall derive the general formula and indicate later how this sometimes leads to immediate results for the integrals I_n. Note that this manner of sum evaluation is foreshadowed in [5]; here we attempt to generalize such ideas to cover arbitrary sinc-sums (for finite n).

First there is a general polylogarithmic identity, starting from

$$\Pi^{(n)}(\omega) = \frac{1}{(2i\omega)^n \prod_{j=1}^n c_j} \prod_{j=1}^n \left(e^{ic_j\omega} - e^{-ic_j\omega} \right)$$

with a sinc-sum emerging as a combinatorial entity

$$S_n = 1 + 2 \frac{1}{(2i)^n T(\vec{c})} \sum_{\vec{\beta} \in \{-1,1\}^n} T(\vec{\beta}) \, \mathrm{Li}_n \left(e^{i\vec{\beta}\cdot\vec{c}} \right), \tag{17}$$

where for any vector $\vec{r}$, we denote $T(\vec{r}) := \prod_{j=1}^n r_j$, while the vector index $\vec{\beta}$ runs over all 2^n balanced-binary n-vectors, $\vec{c} := (c_1, \ldots, c_n)$, and $\mathrm{Li}_n(z) := \sum_{k \geq 1} z^k / k^n$ is the standard polylogarithm. Happily, $\Re(\mathrm{Li}_n), \Im(\mathrm{Li}_n)$ here can be evaluated in finite form as n is even, odd respectively. In fact, for real z one has polynomial evaluations (see [5], [11] and references therein)

$$\Re \left(\mathrm{Li}_n \left(e^{iz} \right) \right) = \sum_{k=1}^\infty \frac{\cos kz}{k^n} = Q_n B_n \left(\left\{ \frac{z}{2\pi} \right\} \right); \quad n \text{ even}, \tag{18}$$

$$\Im \left(\mathrm{Li}_n \left(e^{iz} \right) \right) = \sum_{k=1}^\infty \frac{\sin kz}{k^n} = Q_n B_n \left(\left\{ \frac{z}{2\pi} \right\} \right); \quad n \text{ odd}, \tag{19}$$

where B_n is the standard Bernoulli polynomial, { } indicates (mod 1) fractional part, and

$$Q_n := (-1)^{\lfloor n/2 \rfloor - 1} 2^{n-1} \frac{\pi^n}{n!}.$$

In spite of the fact of (17) having 2^n summands if naively summed, this manner of polylogarithmic evaluation does indeed give a finite form for S_n. We do not yet know a method that will significantly accelerate this summation, so down to polynomial complexity in n rather than 2^n, although something should be possible because there is so much redundancy in the combinatorics. (Of course, the manner in which the alternating signs of the terms in $\vec{\beta} \cdot \vec{c}$ enter tells us that the basic complexity of a ROOF walk is somehow embedded in S_n itself.)

Using the prescription (17), (18), (19) for evaluation of S_n one can establish the following: For the class-D_2 ROOF-walk assignment $c_j := 1/(2j-1)$ we have sinc-sum evaluations

$$S_1, S_2, \ldots, S_7 = \pi,$$

whereas

$$S_8 = \pi \frac{467807924713440738696537864469}{467807924720320453655260875000}, \tag{20}$$

a striking result from earlier research; see [5], and note that the corresponding sinc-integrals I_n will "track" the above S_n values—in fact the matching goes on up to surprisingly large n, as we shall see. It will turn out for example that the origin density of the canonical ROOF walk in question, after 8 jumps, has $f_8(0) = S_8/(2\pi)$, i.e. a value just a touch below $1/2$.

More closed forms based on the polylogarithm decomposition appear in our Appendix, where also are found stultifying factorizations for some of the nontrivial multiples of π. For example, the prime factorization of the "surprising" S_8 above is

$$S_8 = \pi \frac{4322433877 \cdot 108227896140339439297}{2^3 \cdot 3^{12} \cdot 5^6 \cdot 7^7 \cdot 11^6 \cdot 13^6}.$$

7.3 Probabilistic approach for sinc-integrals

In regard to the connection between ROOF walks and sinc-integrals, there is an immediate interdisciplinary observation of great utility:

Theorem 5 *The sinc-integrals defined by (16) can be identified as*

$$I_n = 2\pi f_n(0),$$

where f_n is the density function for n steps of the class-D_2 ROOF walk having pedestal widths $2c_j$.

Proof: This is immediate from the characteristic relation (7).   **QED**

This theorem connects the probabilistic notions with the theory of sinc-integrals, as exemplified in the following interpretation of the Mares integrals, named after B. Mares [3] [8]:

Corollary 2 *Starting with the function*

$$C(\omega) := \prod_{j=1}^{\infty} \cos\left(\frac{\omega}{j}\right),$$

the Mares integrals we hereby define

$$M_q := \int_0^{\infty} C(\omega) \cos(2q\omega) \, d\omega$$

19

have the probabilistic interpretation

$$M_q = \frac{\pi}{2} f(q),$$

where f is the ultimate density of the canonical ($c_j := 1/(2j-1)$) class-D_2 ROOF walk. Thus every $M_q, q = 0, 1, 2, 3, \ldots$ is positive, with particular bounds

$$M_0 := \int_0^\infty C(\omega) \, d\omega < \frac{\pi}{4},$$

$$M_2 := \int_0^\infty C(\omega) \cos(2\omega) \, d\omega < \frac{\pi}{8}.$$

In addition, we have sum rules

$$M_0 + 2M_2 + 2M_4 + \cdots = \frac{\pi}{4},$$

$$M_1 + M_3 + M_5 + \cdots = \frac{\pi}{8},$$

and the overall sum rule

$$\sum_{q \in Z} M_q = \frac{\pi}{2}.$$

Proof: Reminding ourselves of the equivalence Theorem 1, we see from the very proof of said theorem and the characteristic relation (6) that $M_q = \frac{\pi}{2} f(q)$. But by bounding Theorem 2, we have $f(0) < 1/2$ and the result follows. For the second integral, the same argument goes through involving Theorem 2 again, with $M_1 = \frac{\pi}{2} f(1) < \frac{\pi}{8}$. Finally, the sum rules for the Mares integrals M_q follow immediately from the probabilistic sum rules in Theorem 2. **QED**

Note that if desired, one may establish rigorous bounds for the Mares integrals, along the lines of our Corollary 1. As for numerical estimates, our Appendix automatically has high-precision evaluations for the Mares integrals, being as they are the given multiples of f values in Corollary 2.

The next observation is a powerful one appearing in various references such as [3] [5]. Our present intent is to give this previously established result a probabilistic connection:

Theorem 6 *For rational, monotonic decreasing c_j, the sinc-integral defined in (16) is a rational multiple of π—alternatively, the origin probability density $f_n(0)$ ($= \frac{1}{2\pi} I_n$) is rational—that is*

$$I_n := \pi \frac{1}{2^{n-1} c_1} \mathrm{Vol}(\mathcal{W}),$$

where $\mathcal{W}$ is the $(n-1)$-dimensional polyhedron defined by the constraints

$$|c_2 z_2 + \ldots c_n z_n| \leq c_1, \quad |z_k| \leq 1.$$

Proof: The theorem follows immediately from Theorem 5 and the convolution (8).   **QED**

Evidently, then, $I_n = \pi/c_1$ if the polyhedron under its defining constraints does not—to use language from [8]—"bite into" the centered hypercube; however, when the constraint is "active" we have a sudden reduction $I_n < \pi/c_1$ and this explains some of the surprises in the realm of sinc-integrals [3]. Note that this geometrical picture involving intersecting regions is consistent with the constraints involved in our Theorem 2.

There is also a finite form available for sinc-integrals at a certain threshold, namely [8, Ex. 27, p. 123], this form being reminiscent of the finite form for sinc-sums, our prescription (17), (18), (19). We cite this result in our probabilistic context:

Theorem 7 *For a class-D_2 ROOF walk, consider an index N such that*

$$c_2 + \cdots + c_N = c_1 + \delta$$

with $0 < \delta \leq c_N$. Then the probability density at the origin after N steps is

$$f_N(0) = \frac{1}{2c_1}\left(1 - \frac{\delta^{N-1}}{2^{N-2}(N-1)!\prod_{j=2}^{N}c_j}\right).$$

Proof: We assume the validity of the published exercise, which starts with Bernoulli identities of the type we used above for sinc-sums. Alternatively, one can prove the theorem directly, via our relation (8) for class-D_2 walks, on the notion that the "bite" of the polyhedron is relatively trivial under the given δ-constraints.   **QED**

An example of Theorem 7 in action is as follows. Note that for the canonical class-D_2 ROOF walk, it happens that

$$\frac{1}{3} + \frac{1}{5} + \cdots + \frac{1}{15} \approx 1.0218,$$

so that we have

$$f_8(0) = \frac{1}{2}\left(1 - \frac{6879714958723010531}{46780792472032045365526087500}\right)$$

$$= \frac{1}{2}\frac{46780792471344073869653786469}{46780792472032045365526087500},$$

which sure enough involves the same rational as did S_8 in (20).

We next begin an analysis intended to rigorously compare sinc-sums and sinc-integrals.

7.4   Theory of sum-integral error terms ρ_n

An analysis of the sinc-sum-integral error terms starts with Poisson transformation. As in [10] we can Poisson-transform the sum over $k \in Z$, as

$$S_n := \sum_{k \in Z} \Pi^{(n)}(k) = \sum_{\mu \in Z}\int_{\omega \in R} \Pi^{(n)}(\omega)e^{2\pi i\mu\omega}\,d\omega$$

$$= I_n + \sum_{\mu \neq 0} \int_{\omega \in R} \Pi^{(n)}(\omega) e^{2\pi i \mu \omega} \, d\omega,$$

and so the desired error term is

$$\rho_n = 2 \sum_{\mu=1}^{\infty} \int_{-\infty}^{\infty} \Pi^{(n)}(\omega) e^{2\pi i \mu \omega} \, d\omega. \tag{21}$$

This transformation algebra amounts to a proof of the following

Theorem 8 *For a sinc-product* $\Pi^{(n)}(\omega) := \prod_{j=1}^{n} \mathrm{sinc}(c_j \omega)$ *where the* c_j *are associated with a class-*D_2 *ROOF walk, the sinc-sum-integral error is given by*

$$\rho_n := S_n - I_n = 4\pi \sum_{\mu \geq 1} f_n(2\pi\mu),$$

where f_n *is the density function for* n *steps of the ROOF walk.*

This exact series representation for the error ρ_n leads, then, to a result known essentially to previous authors; again we endeavor to cast a result in the language of our present framework:

Corollary 3 *For the ROOF-walk* c_j *of Theorem 8, the error* $\rho_n := S_n - I_n$ *is never negative, and is positive if and only if* $c_1 + c_2 + \cdots + c_n > 2\pi$.

Proof: In Theorem 8, the constraint $\cdots > 2\pi$ determine whether there be any contributions from $f(2\pi\mu)$ for positive integer μ. **QED**

When the converse constraint $c_1 + \cdots + c_n < 2\pi$ succeeds, ρ_n vanishes and I_n can be calculated in finite form as an instance of S_n from (17); or conversely, S_n can be evaluated via an instance of I_n per Theorem 6. In any case, Corollary 3 determines the precise threshold at which S_n, I_n differ.

As an example, we finally have derived that for the canonical assignment $c_j := 1/(2j-1)$, the errors $\rho_n; \ n = 1, \ldots 8$ all valish, which means I_8—remarkably enough—has the same peculiar value as S_8, in (20). It will turn out below that the index n must be taken *much* farther to yield a nonzero ρ_n.

But in a different direction for the moment, we find that Theorem 8 has an interesting implication regarding sum-integral identities that are *always* true:

Corollary 4 *For a class-*D_2 *ROOF walk with all rational* c_j, *there is a sinc-sum/sinc-integral relation that is true for all finite* $n = 1, 2, \ldots,$ *namely*

$$S_n := \sum_{k \in Z} \Pi^{(n)}(k) = \int_{\omega \in R} \Pi^{(n)}(\omega) \, \frac{\sin(\pi(K + 1/2)\omega)}{\sin(\pi\omega/2)} \, d\omega,$$

where we define the integer

$$K := \left\lfloor \frac{c_1 + \cdots + c_n}{2\pi} \right\rfloor.$$

Remark: Note once again: If the c_j sum is less than 2π, then $K = 0$ and we recover the equality $S_n = I_n$, hence $\rho_n = 0$. Also, here and elsewhere we use rational c_j simply to avoid the sometimes delicate phenomenon that a sum of c_j values actually equals a multiple of π. Still, appropriate modifications of the analysis should be able to handle irrational c_j.

Proof: This corollary follows from Theorem 8 and a trigonometric identity valid for any nonnegative integer K:

$$\sum_{\mu=-K}^{K} e^{2\pi i \mu \omega} = \frac{\sin(\pi(K + 1/2)\omega)}{\sin(\pi\omega/2)}.$$

QED

One of the attractive results from [5] is that for the choice $c_k := 1/(2k-1)$, which assignment we have associated with the canonical class-D_2 ROOF walk, the error $\rho_n := S_n - I_n$ has the surprising property

$$\rho_1, \rho_2, \ldots, \rho_{40249} = 0,$$

whereas

$$\rho_{40250} > 0.$$

(Note that in their work, the index n is offset by 1 with respect to our probability models.) This phenomenon can be interpreted, via Corollary 3, as that of the sum

$$\frac{1}{1} + \frac{1}{3} + \cdots \frac{1}{2N - 1} = 2\pi + \delta$$

having

$$0 < \delta < 1/(2N - 1)$$

precisely when $N := 40250$. That is, the sum through c_{n-1} does not quite reach the threshold 2π. To be clear with respect to our previous discussions about S_n and I_n, we remind ourselves that even though ρ_n thus vanishes for thousands of low-lying n, we know that both S_n, I_n fall below the value π at $n = 8$, because we have already given enough finite forms for some low-lying S_n (and see Appendix for specifics).

To quantify the error ρ_n in such threshold cases, we establish

Theorem 9 *Assume rational c_j elements for a class-D_2 ROOF walk, with $c_1 < 2\pi$. Then there always exists a unique threshold index N such that*

$$c_1 + c_2 + \cdots + c_N = 2\pi + \delta,$$

with $0 < \delta < c_N$. Moreover, the exact sum-integral error at this threshold is

$$\rho_N := S_N - I_N = \frac{\pi \delta^{N-1}}{2^{N-2}(N-1)! \prod_{j=1}^{N} c_j}.$$

Remark: Minuscule as this "threshold error" ρ_N can be experimentally, it is nevertheless a rational polynomial in π.

Proof: The sum of the c_j diverges for the ROOF walk, and monotonicity of the c_j settles the existence of the claimed δ and threshold N. From Theorem 8 and convolution (8), we therefore have

$$\rho_N = \frac{\pi}{2^{N-2}} \int_{[-1,1]^N} \delta(c_1 z_1 + c_2 z_2 + \cdots + c_N z_N - 2\pi) \, \mathcal{D}\vec{z},$$

that is, only the density terms $f_N(2\pi)$, $f_N(-2\pi)$ can possibly contribute to the error ρ_N. However, also because of the constraint $\delta < c_N$, the delta-function argument $c_1 z_1 + c_2 z_2 + \cdots + c_N z_N - 2\pi$ cannot vanish if any z_k be negative. Therefore we can change variables $z_k \to 1 - u_k$ and write

$$\rho_N = \frac{\pi}{2^{N-2}} \int_{[0,1]^n} \delta(c_1 u_1 + c_2 u_2 + \cdots + c_N u_N - \delta) \, \mathcal{D}\vec{u},$$

where the new domain is, importantly, $[0,1]^N$. Integrating over just u_1 gives

$$\rho_N = \frac{\pi}{2^{N-2} c_1} \int_{u_j \geq 0; \; c_2 u_2 + \cdots + c_N u_N < \delta} du_2 \cdots du_N.$$

This polyhedral integral is easily doable, yielding the desired formula of the theorem.   **QED**

Now, in the canonical case ($c_j := 1/(2j-1)$) we obtain the surprising scenario of Baillie et a;. [5], namely, with the aforementioned threshold $N = 40250$ we calculate

$$\sum_{j=1}^{40250} \frac{1}{2j-1} = 2\pi + (\delta = 0.000000234727\ldots),$$

with $\delta < 1/(2 \cdot 40251 - 1)$. We already know that all errors $\rho_1, \ldots, \rho_{40249}$ vanish. But Theorem 9 gives the sum-integral error at threshold, as

$$\rho_{40250} = S_{40250} - I_{40250} = \frac{\pi \delta^{40249}}{2^{40248}} \frac{80499\,!!}{40249\,!}$$

$$\approx 8.42 \cdot 10^{-226577}.$$

Here, !! means odd-factorial, as in $5!! := 5 \cdot 3 \cdot 1$. This phenomenon tells us that sheer experimental computation even in the region of 100000 decimal digits would not catch this kind of "surprise!"

What can be said about an arbitrary error ρ_n in the canonical case? A summary result follows immediately from the effective superexponential bound of Corollary 1, together with Theorem 8, as

Theorem 10 (Summary of the Baillie et al. error terms) *For the canonical case ($c_j :=$ $1/(2j-1)$), the error $\rho_n := S_n - I_n$ satisfies:*

$$\rho_1, \ldots, \rho_{40249} = 0,$$

$$0 < \rho_{40250} < 10^{-226576},$$

and for all other n (including $n = \infty$)

$$0 < \rho_n < 10^{-13679}.$$

Remark: In spite of the rigor of these bounds, we do not know over what regions ρ_n might be monotonic in n; neither do we know ρ_∞ nor S_∞ nor I_∞.

7.5   A "primes" ROOF walk

For certain exotic ROOF walks, such as the class-D_2 walk having $c_j := 1/p_j$ where p_j is the j-th prime, the "surprise" effect is yet more striking. Note first that $\sum 1/p_j, \sum 1/p_j^2$ diverges, converges respectively, as is classically known, so class-D_2 membership is assured. We may also assign $b_j = 1/p_j$ to get a class-D_1 "primes" walk. For class D_2 we can obtain a threshold error that is remarkably small, as follows. With a view to Theorem 9, we require a threshold index N—determining the prime p_N—such that

$$\frac{1}{2} + \frac{1}{3} + \frac{1}{5} + \cdots + \frac{1}{p_N} = 2\pi + \delta, \tag{22}$$

with $0 < \delta < 1/p_N$. An immediate question is, if we know N, p_N, what is the threshold error ρ_N from Theorem 9? Happily, there are known rigorous bounds on the n-th prime number; said bounds having been developed in the classic work of Rosser and Schoenfeld [16], and more recently by Dusart [13]. We know that for $n \geq 6$,

$$n \log n \; < \; p_n \; < \; n \log n + n \log \log n, \tag{23}$$

and also some bounds on the prime-counting function: For real $x \geq 599$,

$$\frac{x}{\log x}\left(1 + \frac{0.992}{\log x}\right) \; \leq \; \pi(x) \; \leq \; \frac{x}{\log x}\left(1 + \frac{1.2762}{\log x}\right). \tag{24}$$

The lower bound of (23) leads to

Theorem 11 *For the "primes" ROOF walk threshold N described by (22), the Baillie–Borwein–Borwein error satisfies*

$$0 < \rho_N < \frac{1}{N^N}.$$

Proof: Clearly by direct computation, $N \geq 28$. Since $\delta < 1/p_N$ we have, from Theorem 9 and the bounds (23),

$$\rho_N < \frac{\pi \cdot 2 \cdot 3 \cdot 5 \cdot 7 \cdot 11}{(N \log N)^{N-1}} \frac{1}{2^{N-2}(N-1)!} \prod_{j=6}^{N} j(\log j + \log \log j)$$

$$< \frac{\pi \cdot 2 \cdot 3 \cdot 5 \cdot 7 \cdot 11}{(N \log N)^{N-1}} \frac{1}{2^{N-2}(N-1)!} \frac{N!}{5!} (\log N + \log \log N)^{N-5}.$$

The right-hand side is easily seen to be $< 1/N^N$ for $N \geq 28$.   **QED**

The problem now is to estimate N. Baillie et al. [5] estimated, on the basis of the Mertens theorem

$$\sum_{p \leq x} \frac{1}{p} = \log \log x + B + o(1),$$

with B being the Mertens constant, that the threshold N and the prime p_N are *very* large—in fact, each turns out to well exceed a googol (10^{100}), as we shall eventually prove. Incidentally, B is resolvable to high precision, for example

$$B = 0.2614972128476427837554268386086958590515666482611992061920 6421392 \ldots$$

accrues easily from an algorithm of E. Bach that, curiously enough, does not employ primes directly, rather using Riemann-zeta evaluations at integers [12, Ex. 1.90].

Let us first approach this threshold problem on the assumption of the Riemann hypothesis (RH), under which it is known that for $x \geq 13.5$,

$$\left| \sum_{p \leq x} \frac{1}{p} - \log \log x - B \right| < \frac{3x + 4}{8\pi \sqrt{x}},$$

and that the prime count $\pi(x)$ is, for $x \geq 2.01$, well approximated by the logarithmic integral li, in the sense

$$|\pi(x) - \mathrm{li}(x)| < \sqrt{x} \log x.$$

These facts are found in [1, Equ. 7.1] and [12, Ex. 1.37], and all date back to the celebrated work of Schoenfeld and Rosser. From these (RH-conditional) bounds, one can derive

$$2\pi + \delta - \log \log p_N - B \in [-\epsilon, \epsilon],$$

and so

$$e^{e^{2\pi - B + \delta - \epsilon}} < p_N < e^{e^{2\pi - B + \delta + \epsilon}},$$

where ϵ is certainly less then 10^{-50}. It follows—again, under the RH—that at threshold, the prime and index are bounded in the forms

$$p_N \approx 10^{179.05 \pm 0.01},$$

26

$$N \approx 10^{176.44 \pm 0.01}.$$

These offsets ± 0.01 are quite conservative, yet it follows from Theorem 11 that the Baillie–Borwein–Borwein error has, again on the RH,

$$\rho_N < 10^{-10^{178}}.$$

Let us endeavor, then, to make such arguments unconditional (independent of the RH), to rigorously analyze this fascinating "primes" ROOF walk. We start with

Theorem 12 (Unconditional, no RH) *There exist nonnegative, monotone-nonincreasing sequences* $(c_1, c_2, \dots), (d_1, d_2, \dots)$, *both converging to zero, such that for a fixed m the reciprocal-prime sum is constrained for all $n > m$ by*

$$\sum_{j=1}^{n} \frac{1}{p_j} - \log\log n \ \in \ [B - c_m, \ B + d_m].$$

In particular, for $n > 3 \cdot 10^8$, the constraint interval can be taken to be $[B - 0.047, B + 0.147]$, while for $n > 4.41 \cdot 10^{16}$ we may use $[B - 0.025, B + 0.093]$.

Remark: Even though the Mertens theorem standardly says $\sum_{p \le x} 1/p - \log\log x - B = o(1)$, i.e. the $\log\log$ argument is the upper limit on primes and not their upper index, the notion that both $c_m, d_m \to 0$ is equivalent to said theorem, since $\log\log(x := p_n) - \log\log n = o(1)$. Also, the present author conjectures that the c_m can all be taken to be 0; equivalenetly, for $n > 1$ the difference $\sum_{j \le n} 1/p_j - \log\log n$ is *always* $\ge B$.

Proof: It is elementary that for $h(x)$ positive, continuous, and monotone decreasing on $[k, n]$,

$$\int_{k+1}^{n+1} h(x)\, dx \ \le \ \sum_{j=k+1}^{n} h(j) \ \le \ \int_{k}^{n} h(x)\, dx.$$

(This can be shown quickly by replacing $h(j) \to h(\lfloor x \rfloor)$ in the summation, then converting to an integral over $h(\lfloor x \rfloor)$.) Take an integer $k \in [6, n-1]$. The integral bounds can be applied along with $p_n > n \log n$ from (23), taking $h(x) := 1/(x \log x)$, to obtain

$$\sum_{j=1}^{n} \frac{1}{p_j} \le \log\log n + B + \inf_{k \in [6, n-1]} \left(\sum_{j=1}^{k} \frac{1}{p_j} - \log\log(k) - B \right).$$

But we also have, using the other bound $p_n > n(\log n + \log\log n)$ from (23),

$$\sum_{j=1}^{n} \frac{1}{p_j} \ge \frac{1}{2} + \frac{1}{3} + \cdots + \frac{1}{p_{k-1}} + \int_{k}^{n+1} \frac{dx}{x(\log x + \log\log x)}.$$

27

Now we observe

$$\int_a^b \frac{dx}{x(\log x + \log\log x)} = \int_{\log a}^{\log b} \frac{du(1 + 1/u)}{u + \log u} - \int_{\log a}^{\log b} \frac{du}{u(u + \log u)}.$$

It follows that

$$\sum_{j=1}^n \frac{1}{p_j} \geq \log\log n + B + \sup_{k \in [6, n-1]} \left(\sum_{j=1}^{k-1} \frac{1}{p_j} - \log(\log k + \log\log k) - \frac{1}{\log k} - B \right).$$

Now, c_m, d_m for $m \geq 6$ can simply be defined respectively as the $\inf_{k \in [6,m]}(\ )$, $\sup_{k \in [6,m]}(\ )$ above, so the existence of properly monotonic c_m, d_m is established; the classical Mertens theorem says $c_m, d_m \to 0$. Taking $k = m = 3 \cdot 10^8$ in the argments of the sup, inf terms, we can settle by direct summation of $1/p$ the bounding interval $[-0.047, 0.147]$. Taking $k = m$ corresponding to the Bach–Sorenson prime $p_m = 1801241230056600523$ (see text), we have from (24) the constraint $m = \pi(p_m) \in [4.39 \cdot 10^{16}, 4.41 \cdot 10^{16}]$ and this, used in the sup, inf terms, is enough to achieve the final explicit constraint interval of the theorem.

QED

These bounds on reciprocal-primes sums leads to a rigorous resolution of the threshold problem, in the form

Corollary 5 (Unconditional) *The threshold index N from (22) satisfies*

$$10^{163} < N < 10^{184},$$

and hence, the Baillie–Borwein–Borwein error satisfies

$$\rho_N < 10^{-10^{165}}.$$

Moreover, the "primes" ROOF walk has ultimate probability density decaying at least triply superexponentially; indeed, for $x \geq 2$,

$$0 < f(x) < e^{-e^{2(x-1.36)}e^{e^{x-1.36}}}.$$

In particular,

$$0 < f(2\pi) < 10^{-10^{63}}.$$

Thus, for any n exceeding the threshold N, we have

$$0 < \rho_n < 10^{-10^{62}}.$$

Remark: These results mean, unconditionally, that the threshold N exceeds a googol, and the threshold error ρ_N is less than 1/googolplex.

Proof: The bound on N is an easy computation from Theorem 12 with its specific interval $[B - 0.025, B + 0.093]$; we have

$$e^{e^{2\pi - B + 0.0251}} > N > e^{e^{2\pi - B - 0.093}}.$$

Then the bound on ρ_N follows from Theorem 11. The bound on $f(x)$ is a consequence of Corollary 1, with the assignment

$$m := \left\lfloor e^{e^{x - 1 - B - 0.093}} \right\rfloor,$$

the only extra bound needed being

$$\sum_{j=m}^{\infty} \frac{1}{p_j^2} < \frac{1}{(m-1)\log^2 m},$$

as follows from the lower bound on p_j from (23). The last bound on $\rho_{(n > N)}$ follows from Theorem 8.

QED

Incidentally, the situation in regard to *exact* knowledge of reciprocal-prime sums is not infinitely hopeless. Recently, E. Bach and J. Sorenson [1] devised a clever scheme for assessing the threshold index m for such as

$$\frac{1}{2} + \frac{1}{3} + \frac{1}{5} + \cdots + \frac{1}{p_m} = 4 + \delta,$$

with $\delta < 1/p_m$, where one notes here a change of our previous threshold $2\pi \rightarrow$ threshold 4. They found the precise threshold prime as

$$p_m = 1801241230056600523,$$

for which those authors derived

$$\sum_{j=1}^{m} \frac{1}{p_j} \approx 4.00000000000000000021,$$

as we used in the computations for the proof of our Theorem 12 above. The authors of [1] did not need to determine the exact m for their purposes, although they could have. This is why we used for our Theorem 12 the unconditional bounds (24) on $\pi(x)$ to sufficiently closely estimate the index m.

Can the Baillie–Borwein–Borwein threshold be assessed by sieving, then? The authors of [1] do say:

> "*We note that [the threshold prime for reciprocal-prime sum > 5] is about $4.2 \cdot 10^{49}$, so its precise value may remain unknown for all eternity.*"

Those authors are, of course, referring to the shortcomings of prevailing sieve methods and machinery in 50-digit regions. This having been said, the notion of locating the explicit threshold for $\sum 1/p > 2\pi$, as is fundamental to the theory of the "primes" ROOF walk, looms even more stultifying.

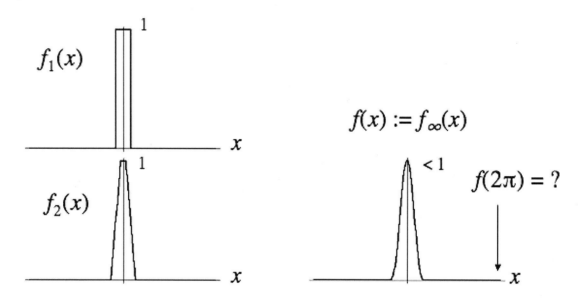

Figure 3: Probability density for a "primes" ROOF walk. On its n-th step the walker jumps uniformly an increment $[-1/p_n, 1/p_n]$ where p_n is the n-th prime number. The upper-left plot is the density f_1, the lower-left is f_2. (The x-axis is compressed; the initial density f_1 is positive on $x \in [-1/2, 1/2]$.) The lower-right plot is for $f := f_\infty$, the everywhere-positive, ultimate density for this walk. The decay of f is *triply superexponential*; for example, we prove that $f(2\pi) < 1/\exp(\exp(\exp(4.99)))$ (Corollary 5).

8   Open problems

- What is an exact evaluation of the discrepancy $1/2 - f(0)$ for the canonical (or for that matter, any!) class-D_2 walk? (In the canonical case, an equivalent question is, what is the Mares integral M_0 in closed form?) Is the situation—that we do not know any closed forms for a single $f(x)$–for class-D_1 walks—just as dificult?

- What can be said about the class-D_2 walk with $c_n = 1/n$? In the present paper this particular "fuel assignment" was not touched upon at all. There could—or could not—be a trivial scaling to analyze this walk in previous terms.

- It is possible to calculate higher moments of ROOF walks. For example, the canonical class-D_1 walk ($b_n := 1/(2n)$ is the jump length) has, as we know, $\langle x_\infty^2 \rangle = \frac{1}{4}\zeta(2)$. It can be derived [15] that $\langle x_\infty^4 \rangle = \frac{3}{4}\zeta(2)^2 - \frac{1}{8}\zeta(4)$, and so on. The question is, how can knowledge of the structure of $f(x)$ accrue from, say, the knowledge of many such moments?

- We advise that ROOF walks outside class D, e.g. walks with *converging* sum $\sum \langle |\Delta_j| \rangle$, are difficult to analyze and many mysteries abound. Though our example walk with $\Delta_n = \pm 1/2^n$ was easy, intuitive, a walk with $\Delta_n = \pm 1/3^n$ is already intricate, conjuring up such entities as the Cantor set of fractional dimension and evidently intractable characteristic. The references [17] [15] discuss some open problems for such ROOF classes.

- When is a class-D_1 walk equivalent to a class-D_2 walk, in the sense of the two walks' ultimate density functions coinciding at all x? Equivalence *does* hold if for every positive integer n we have
$$\sum_{j \geq 1} c_j^{2n} = (4^n - 1) \sum_{j \geq 1} b_j^{2n},$$
as is certainly true for our canonical cases $c_j := 1/(2j-1), b_j := 1/(2j)$.

- Presumably there is a ROOF walk of class D having *quadruply superexponential* decay—so how does one develop a theory of ROOF walks that have arbitrary, cascaded superexponential bounds?

- One aspect of density-function dynamics we have not addressed at all is diffusion theory. Heuristically, for the canonical class-D_1 ROOF walk, having $b_n := 1/(2n)$, we might approximate
$$f_n(x) \approx \frac{1}{2}f_{n-1}(x - 1/(2n)) + \frac{1}{2}f_{n-1}(x + 1/(2n)).$$
This leads to the heuristic differential equation (setting $n := t$, a "time"):
$$\frac{\partial f}{\partial t} = \frac{1}{8t^2}\frac{\partial^2 f}{\partial x^2}.$$

An exact solution is

$$f(x,t) = \frac{1}{\sqrt{a - 1/t}} e^{-2x^2/(a-1/t)},$$

for constant a. But this suggests a limiting Gaussian distribution as $t \to \infty$, which we know to be incorrect (the true $f(x, \infty)$ decays superexponentially in x, after all). Perhaps this exact solution should be used as a propagation kernel, connecting successive finite t values.

• What is the "natural" quantum potential $V(x)$ such that the canonical class-D_2 density $f(x)$ is the Schrödinger ground state of V? We demand that the Schrödinger equation holds, i.e.

$$-\frac{d^2 f}{dx^2} + Vf = E_0 f,$$

where the energy scale is set by $V(0) := 0$. This means that the potential V is given explicitly by

$$V(x) = E_0 + \frac{f''(x)}{f(x)},$$

with eigenvalue $E_0 = -f''(0)/f(0)$. This appears to be a difficult problem. What we can say is that this "natural" energy E_0 is given by

$$E_0 = \frac{4}{M_0} \int_0^\infty \omega^2 C(\omega) \, d\omega,$$

where C is the Mares kernel as in Corollary 2. D. Bailey [2] has calculated this fundamental quantum energy as

$$E_0 \approx 0.013990359117147734511346575283227759543878661010906634135035.$$

(Actually, Bailey can obtain hundreds of digits beyond this.) Also of interest: The graph of $V(x)$ apparently itself has zero-crossings.

• How would one develop a physics theory of a fundamental ROOF particle—say in a full, 3-dimensional setting—that, upon photon-energy absorption, dissipates the absorbed energy in the style of a ROOF walk, thus ending up at a probabilistic position?

9 Appendix

For the numerical and symbolic results herein, we adopt the canonical prescription

$$\pi^{(n)}(\omega) := \prod_{j=1}^{n} \text{sinc}\left(\frac{\omega}{2j-1}\right),$$

and define the sinc-sums and sinc-integrals, respectively:

$$S_n := \sum_{k \in Z} \pi^{(n)}(k), \quad I_n := \int_{\omega \in R} \pi^{(n)}(\omega) \, d\omega.$$

9.1 Exact sinc-sums

We recall that for the canonical assignments $c_j := 1/(2j-1)$ the sinc sums are

$$S_1, S_2, ..., S_7 = \pi.$$

Use of relations (17), (18), (19) give the following, for $n \in [8, 16]$ in the form

$$\frac{S_n}{\pi} = \frac{\text{num}}{\text{den}},$$

where num, den are integers—and the prime factorization of num is given when known. We observe that separately interesting is the factorization of the numerator of $1 - \text{num}/\text{den}$; we do not report these herein, although we do caution researchers that the first few such factorizations are trivial—and that triviality is misleading. (We do give one instance of factorization of $1 - \text{num}/\text{den}$, for the last example $n = 16$ below.)

$$S_8/\pi = \frac{467807924713440738696537864469}{467807924720320453655260875000},$$
$$\text{num} = 4322433877 \cdot 108227896140339439297.$$

$$S_9/\pi = \frac{17708695183056190642497315530628422295569865119}{17708695394150597647449176493763755467520000000},$$
$$\text{num is prime.}$$

$$S_{10}/\pi = \frac{8096799621940897567828686854312535486311061114550605367511653}{8096800377970649960875919032857634716820075076062381575000000},$$
$$\text{num} = 109 \cdot 307 \cdot 24196275354692937177864168945738682982132687190480 8456131,$$

$$S_{11}/\pi = \frac{2051563935160591194337436768610392837217226815379395891838337765936509}{2051564503724359411435325207087513361930253427318374450656960000000000},$$
$$\text{num} = 5167 \cdot 397051274464987651313612689880083769540783204060266284466486891027.$$

33

$$S_{12}/\pi = 3719316770169049234444819453328348890204104923676043830296516790118732385138484006728 7863$$
$$/$$
$$37193188390019359679267753038304609065247968318560293442237453760993482625662395625000000,$$
$$\text{num} = 457 \cdot 704477 \cdot 423614372509 \cdot 4481160013802705926237 \cdot$$
$$6085818709117459611112775500677879032519627 5699.$$

$$S_{13}/\pi = 543110896461169846307682746504491201561304453085191937514671790535975762663132049434 50734539904399598201 24079$$
$$/$$
$$5431113911376064346013898379192680475298012990888015359173701456675706968166358939600 7450000000000000000000000,$$
$$\text{num is composite.}$$

$$S_{14}/\pi = 936685793682547700229015382776792629410143680302554911141990675902213231444012222480 80717527867670545886934135527429468069$$
$$/$$
$$93668702988747470522353716916858309344196014509267648553881576518613841590965134255669 3677990179313694763183593750000 0000,$$
$$\text{num is composite.}$$

$$S_{15}/\pi = 110658275165671269070539632401115425920752643832418158711753605989327076464755451468 091430698696898117300223589930956816496969864394787533954587510478 87537$$
$$/$$
$$1106584704540421452539632837696465219054283911745133113615555073406768770441455630777 50327648920069167541590183 0608$$

$$316356778613513375000000000000000000000000,$$

$$\text{num} = 313 \cdot 1031 \cdot 2903 \cdot 5569 \cdot 482011381136627 \cdot$$

$$49022141298582194203 \cdot 15265526308002685361 \cdot$$

$$2483545910815547560028581364947 \cdot$$

$$2367680487134139644000178219983941204010506257 3952696269811.$$

$$S_{16}/\pi = 102885742414551619757366905943828719620462539 9240253519$$

$$781651784876399870694224539789974108272189041614 8925398988$$

$$870835066166007744052418638591660237591063020045 3822268852$$

$$643263934963$$

$$/$$

$$102885971988537707660519059741757265548279300360 88790172$$

$$468690003333625836111565713233727527852994542095 91370945$$

$$435907838981606429475270406888624733780606156179 428100$$

$$58593750000000000,$$

num is composite.

For S_{16}/π above, the factorization of the numerator of $1 - \text{num}/\text{den}$ is

$$449 \cdot 20310271 \cdot 47617241426617180429 \cdot P,$$

where P is prime. It does appear that factoring $1 - \text{num}/\text{den}$ is generally easier than factoring num/den (see Theorem 7 for hints in such a direction), but this ease may well vanish for large n.

9.2   D. Bailey's calculations and probability densities

D. Bailey's computed values [2] [8, p. 101] [3, p. 220] for the Mares integrals M_q (see our Corollary 2) are as follows:

M0 = 0.78538055729863287349258301146733252476165283080347360800414
69176993352354790521476036620853656602506509045709197834959317463160036034276655
46557930704783174112197353022911012758523047668096248212698667484074483746778911
92997656890603879924119300411734744121340251835043120675312327518451159416662
632805150039797969274686767754964900877151205000474117515374260860864341412734623
40203432904856702365298120786854984823645457071868728389420781473030254405185
15375759834490183474894245260192233482609683185 ...

M1 = 0.39269908169872415480783042290993786052464543418723159592681
22851620932471399385461790165127474553667775073955731238984035820271548568876093
65093419344489869213452474548876333219399274808697766142856120560186166274834092
18692277622673687237679213072113312270198904062690726232070655634843283152846783
57323338980594711041571395979629698980697925450710450859796765509555398866516988
21686714081774640884165053767665111199338842259445367394476127437532497167621503
6520551311534938104429251335903210314221422...,

M2 = 8.80304940771806153891717627159814381975952015142361979461518
88094330462500510266733104849633802432456562152428086721551505581766844516377733
90776901434071417861692191677212042572173056208731364371349877184619296084267231
01072728338583600869437127114540073951502201596913466726744862119921797863346537
43859794160145405366702241344870916181373869997979606515612997763140147748848631
66195830614143156552424897822869503887336907416418939863181063500235661487794848
0011522239215564331511436166575... * 10^(-6),

M3 = 7.40734656631695055788876383803645837578649487840420309266720
26111524103638992730470342133693516960057650667157027046783303828886598191879568
07691708532234416249934152056834075885791366798024864755188667695175100833688763
91916669044470124470325768076002674823483234395231078406950464210485421372958836
01371009512677702949186091745208557009830587755124628010952859448214460795560571
39288109987968151997628148468330927609747923843840476146613600675890625622...*
10^(-43),

M4 = 1.42548695386600110540119621157777475570100223045903273417681
71385742382197942574762915224148607435095431482216254593190009582987372569676751
05280447773922905712332170849356078030552...* 10^(-319).

Numerical values for $f(q) = (2/\pi)M_q$ follow. To get the data below, we used our relation (14) with the (naive—see text) choice $m(\omega) := 100$ and $L = 128$, the latter counting the degree to which the log-sin series is taken. (Note these next values are radically less precise than D. Bailey's equivalent values above; we cite the values below just to indicate what happens with given m, L parameters in the alternative, log-sin method of the text.)

```
f(0) = 0.49998879160937983591179672655936274390974036542331027985
12165103759527779130761632750121865916814102713490742096739287915...

f(1) = 0.24999999999999999999999999999999999999999999995284336715103517
65662884479522788965397366240873596443356521085266038103670362283 95181...

f(2) =  5.60419531008204410163672031862804512981728834486007439174481
20236110434619183624939067041592948643254628951630356042 11685... * 10^(-6)

f(3) =  4.71566328489648234337115520477211034602633759126403556 6434
7891 47339618963296377160... * 10^(-43).
```

As for the integrity of all the above data,

- All the approximations $(f(0), f(1), f(2), f(3))$ above are consistent with D. Bailey's values for the corresponding Mares integrals, to the displayed precision(s). We remind ourselves that D. Bailey's method is somewhat different—he used the original, log-cos scheme, with more powerful routines and machinery to get his extreme-precision values.

- D. Bailey's values above actually imply (recall Mares-integral/probability-density equivalence)

$$0 < \frac{1}{2} - f(0) - 2f(2) - 2f(4) < 10^{-500},$$

$$0 < \frac{1}{4} - f(1) - f(3) < 10^{-500},$$

all consistent with the sum rules in Theorem 2 and Corollary 2.

- Values for $f(5)$ and beyond have been quite difficult to obtain, and in view of the superexponential decay of f that we have proven, this difficulty makes sense. After all, we know as in-text that $f(2\pi) < 10^{-13680}$. Even with $f(5)$, it is not so much the computational complexity as the severe dynamic range of terms that hampers us. So we are still without a reasonably precise value for $f(5)$ and beyond. We do know from Corollary 1 that $f(5) < 10^{-1000}$, and D. Bailey has already inferred $f(5) < 10^{-500}$ numerically.

- It is fascinating that modern experimental mathematics still has not achieved a closed form for the canonical probability density $f(0) = (2/\pi)M_0$. The continued fraction displayed below shows at least that $f(0)$ is quite close to $1/2$, and is calculated from D. Bailey's extreme-precision value above for M_0:

{0, 2, 22304, 4, 1, 1, 2, 1, 25, 6, 9, 1, 1, 1, 2, 1, 2, 34, 4, 10, 29, 9, 1, 33, 1, 2, 1, 1, 6, 1, 2, 9, 21, 1, 12, 1, 6, 1, 9, 14, 9, 8, 5, 7, 1, 2, 27, 2, 1, 14, 5, 2, 1, 3, 1, 1, 1, 1, 5, 30, 2, 1, 1, 2, 2, 96, 1, 1, 12, 5, 1, 2, 3, 2, 2, 2, 1, 18, 7, 4, 1, 2, 1, 15, 1, 1, 2, 41, 1, 1, 17, 1, 35, 1, 307, 18, 14, 1, 134, 1, 13, 12, 1, 3, 33, 2, 1, 11, 1, 7, 5, 2, 1, 1, 19, 2, 27, 88, 11, 1, 7, 1, 1, 4, 11, 1, 2, 8, 1, 1, 2, 1, 1, 5, 1, 1, 1, 1, 17, 55, 1, 4, 2, 4, 3, 1, 1, 1, 1, 3, 5, 2, 88, 44, 53, 2, 7, 1, 2, 7, 4, 1, 1, 1, 33, 2, 1, 31, 6, 1, 1, 1, 4, 348, 3, 1, 1, 33, 1, 8, 1, 1, 5, 1, 79, 2, 7, 2, 1, 1, 2, 31, 1, 1, 9, 1, 3, 3, 2, 3, 5, 1, 1, 2, 2, 1, 2, 8, 2, 1, 1, 1, 1, 474, 29, 3, 1, 1, 4, 15, 1, 12, 1, 3, 1, 6, 1, 19, 1, 1, 1, 1, 1, 114, 1, 1, 2, 5, 2, 2, 2, 1, 9, 1, 12, 2, 1, 2, 3, 4, 1, 1, 2, 4, 1, 1, 2, 3, 1, 3, 3, 50, 1, 3, 9, 9, 1, 1, 1, 1, 2, 1, 5, 1, 1, 3, 1, 3, 1, 1, 1, 1, 1, 6, 2, 1, 1, 6, 1, 1, 1, 5, 1, 1, 1, 3, 1, 10, 7, 1, 1, 1, 1, 21, 1, 3, 4, 3, 1, 1, 5, 1, 18, 1, 3, 3, 1, 1, 30, 4, 1, 1, 8, 1, 1, 4, 1, 2, 1, 2, 6, 1, 1, 88, 5, 1, 1, 2, 3, 3, 4, 3, 1, 22, 6, 2, 1, 45, 1, 1, 27, 2, 1, 8, 3, 2, 1, 6, 3, 1, 1, 2, 1, 6, 1, 1, 3, 1, 1, 1, 3, 3, 4, 1, 2, 1, 21, 1, 1, 2, 1, 4, 1, 1, 511, 1, 1, 15, 4, 12, 1, 235, 4, 2, 10, 4, 1, 4, 12, 16, 7, 1, 1, 1, 3, 1, 1, 17, 1, 1, 1, 2, 6, 1, 4, 1, 6, 1, 3, 3, 2, 1, 2, 13, 1, 2, 1, 1, 1, 4, 5, 1, 1, 3, 1, 1, 4, 1, 2, 20, 897, 4, 4, 2, 13, 3, 13, 7, 52, 1, 1, 1, 1, 1, 1, 5, 2, 3, 1, ...}

10   Acknowledgments

The author thanks E. Bach, D. Bailey, R. Baillie, J. Borwein, and J. Buhler for valuable discussions and observations. T. Jenkins provided some fine ideas in regard to a physical walker model. A. Golden provided some key scholarship and computation. M. Trott was gracious to offer his *Mathematica* rendition of numerical-Mares integration. S. Wolfram offered some key insights a decade ago, as did O. Bonfim, long before this written work (and the moniker ROOF) was conceived. C. Pomerance provided some rigorous number-theoretical ideas relevant to the more exotic ROOF walks.

References

[1] E. Bach and J. Sorenson, "Computing prime harmonic sums," draft preprint, 2007.

[2] D. Bailey, private communication, 2007.

[3] D. Bailey, J. M. Borwein, N. Calkin, R. Girgensohn, R. Luke, and V. Moll, *Experimental Mathematics in Action*, A. K. Peters, Natick, Mass, 2007.

[4] D. Bailey, J. M. Borwein, V. Kapoor and E. Weisstein, "Ten Problems in Experimental Mathematics," *Amer. Math. Month.*, vol. 113, no. 6 (Jun 2006), pg. 481-409.

[5] Robert Baillie, D. Borwein, and Jonathan M. Borwein, "Surprising Sinc Sums and Integrals," *Amer. Math. Month.*, to appear, 2008.

[6] D. Borwein and J. M. Borwein, "Some remarkable properties of sinc and related integrals," *Ramanujan J.*, 5, 7390 (2001).

[7] J. M. Borwein and D. H. Bailey, *Mathematics by Experiment: Plausible Reasoning in the 21st Century*, A. K. Peters, Natick, Mass, 2004. Combined Interactive CD version, 2006.

[8] J. M. Borwein, D. H. Bailey and R. Girgensohn, *Experimentation in Mathematics: Computational Paths to Discovery*, A.K. Peters, Natick, Mass, 2004. Combined Interactive CD version, 2006.

[9] D. Borwein, J. M. Borwein and B. Mares, "Multi-variable sinc integrals and volumes of polyhedra," *Ramanujan J.*, 6 , 189208 (2002).

[10] R. Crandall, "Note on sinc-kernel sums and Poisson transformation," preprint, Jun 2007. Avaiable at `http://www.reed.edu/ crandall`

[11] Crandall, R. E. and Buhler, J. P., "On the evaluation of Euler sums," Experimental Mathematics, 3, 4, 275-285 (1995).

[12] R. Crandall and C. Pomerance, *Prime numbers: A computational perspective*, 2nd. Ed., Springer New York, 2002.

[13] P. Dusart, "The k-th Prime is Greater than $k(\ln k + \ln\ln k - 1)$ for $k \neq 2$," *Math. Comp.*, Vol. 68, No. 225. (Jan., 1999), pp. 411-415.

[14] K. E. Morrison, "Cosine products, Fourier transforms, and random sums," *Amer. Math. Month.*, 102, 716724 (1995).

[15] K. E. Morrison, "The final resting place of a fatigued random walker," preprint, 1998.

[16] Rosser and Schoenfeld, "Approximate formulas for some prime numbers, " *Illinois J. Math.*, 6 (1962), 64-94.

[17] B. Schmuland, "Random Harmonic Series," *Amer. Math. Month.*, 110, 407-416 (2003).

6. Unified algorithms for polylogarithm, *L*-series, and zeta variants

Discussion

This paper represents some twenty years of special-function algorithm development. The work dates back to when the present author, at NeXT, Inc., worked with educational institutions in regard to "workstation mathematics." Naturally this effort involved close ties with colleagues at Wolfram Research, Inc. who were courageously attempting to install arbitrary-precision calculation in the first versions of *Mathematica*. The dedication footnotes herein should speak for themselves.

This paper is not previously published.

Challenges

There is no end to the challenge list for polylogarithmic functions and their relatives. Some outstanding challenges are

1. Develop rapid convergent series for the Stieltjes constants, as foreshadowed in Section 7.

2. Work out a rigorous, extreme-precision, very fast rendition of the Riemann-splitting series in Section 3. (In this connection, see the Discussion preceding "Effective Laguerre asymptotics" in this volume.)

3. There is always the fabulous mystery of the Madelung constant, as in the ending section of the paper.

Source

Crandall, R.E., "Unified algorithms for polylogarithm, *L*-series, and zeta variants," (preprint 2012).

Unified algorithms for polylogarithm, L-series, and zeta variants

Richard E. Crandall[1]

March 2, 2012

In memory of gentle colleague Jerry Keiper
1953-1995

Abstract: We describe a general computational scheme for evaluation of a wide class of number-theoretical functions. We avoid asymptotic expansions in favor of manifestly convergent series that lend themselves naturally to rigorous error bounds. By employing three fundamental series algorithms we achieve a unified strategy to compute the various functions via parameter selection. This work amounts to a compendium of methods to establish extreme-precision results as typify modern experimental mathematics. A fortuitous byproduct of this unified approach is automatic analytic continuation over complex parameters. Another byproduct is a host of converging series for various fundamental constants.

---

[1]Center for Advanced Computation, Reed College. `crandall@reed.edu`

Contents

1 Motivation

This work began two decades ago, when the present author collaborated with J. Keiper—then of Wolfram Research—on optimal ways to calculate polylogarithmic entities, especially in view of wide parameter-range and analytic-continuation requirements as serious computational-software users would, and should expect.[2]

By "Unified" in this work's title, we refer to the notion that a few strong algorithms can cover a great many different kinds of function evaluations. Which algorithm to choose then depends upon such issues as memory, desired speed, and simplicity of coding. We shall be able thus to provide a unified picture of how to calculate these entities together with their relevant analytic continuations:

- Lerch transcendent $\Phi(z, s, a) := \sum_{n=0}^{\infty} \frac{z^n}{(n+a)^s}$.

- Riemann zeta function $\zeta(s) := \sum_{n=1}^{\infty} \frac{1}{n^s}$.

- Hurwitz zeta function $\zeta(s, a) := \sum_{n=0}^{\infty} \frac{1}{(n+a)^s}$.

- Periodic zeta function $E(s, x) := \sum_{n=1}^{\infty} \frac{e^{2\pi i n x}}{n^s}$.

- Polylogarithm function $\mathrm{Li}_s(z) := \sum_{n=1}^{\infty} \frac{z^n}{n^s}$.

- Polygamma functions $\psi_n(z) := \frac{d^{n+1}}{dz^{n+1}} \log \Gamma(z)$.

- Clausen function $\mathrm{Cl}_s(\theta) := \sum_{n=1}^{\infty} \frac{\sin(n\theta)}{n^s}$.

- MTW zeta functions $\mathcal{W}(s_1, \ldots, s_n) := \sum_{m_j \geq 1} \frac{1}{m_1^{s_1}} \cdots \frac{1}{m_{k-1}^{s_{k-1}}} \frac{1}{(m_1 + \cdots + m_k)^{s_k}}$.

- Multiple-zeta values $\overline{\zeta}(s_1, \ldots, s_n) := \sum_{k_1 > k_2 > \cdots > k_n} \frac{1}{k_1^{s_1}} \cdots \frac{1}{k_n^{s_n}}$.

- Epstein zeta functions (for matrix A) $Z_A(s; c, d) := \sum'_{n \in Z^D} \frac{e^{2\pi i c \cdot A n}}{|A n - d|^s}$.

- Various fundamental constants, such as Euler γ.

---

[2]Over these years I have been asked often by various researchers about such computational schemes, and thought it time now to write down the basics. Various ideas herein were carefully tested and implemented by Keiper, resulting in widely used *Mathematica* functions. (Any mistakes in exposition herein are not Jerry's.) Keiper was always fascinated by the Riemann zeta function, working on clever sidelines—such as the recursion scheme in Section 3.6—up until his untimely death. A superbly thoughtful obituary by S. Wolfram in Jerry's honor can be found at http://www.stephenwolfram.com/publications/other-pubs/95-keiper.html.

Often in computer systems, some or all of the above are calculated via Euler–Maclaurin expansions, which are of asymptotic character. As an alternative, we describe here how to use only convergent series. Just one reason for invoking series is that, typically, rigorous error bounds are accessible. We ultimately present three core algorithms for computation of the Lerch transcendent Φ—most everything else stems from that mighty function (the moniker "transcendent" is appropriate in the present context).

1.1 Acknowledgements and disclaimer

Over a long-term research project one hopes to enjoy substantial collegial support. The present author is happily indebted to pioneers in computation of functions: D. Bailey, D. Borwein, J. Borwein, P. Borwein, M. W. Coffey, P. Flajolet, A. Odlyzko, D. Platt, S. Plouffe, M. Rubinstein, J. Shallit, L. Vepstas. The pioneering work by R. Brent [16, 17] on fast evaluation is gratefully acknowledged. I. J. Zucker, expert in lattice sums, has over the years provided fascinating computational challenges. S. Wolfram has always provided encouragement for special-function investigation; he has suggested over the decades various of the lucrative pathways this work has followed. Virtually all of this research has involved intensive use of *Mathematica*; accordingly, the author thanks colleagues T. Gray, M. Trott, E. Weisstein, and Wolfram Research in general for all of the courtesy and aid over the years.

A disclaimer is in order here: Many of the ideas are already in the literature. To keep this paper at reasonable size, I endeavored to reference the key sources and discussions, so this work is not bibliographically perfect. Just my statement of a result should not be construed as implying it is an original thought. Importantly, my longtime colleague J. Buhler has conveyed many original thoughts now incorporated into this work.

2 Representations of the Lerch transcendent

It turns out that all of special functions listed above can be cast in terms of just one of them—the Lerch transcendent whose classical definition is (later we shall employ a modified transcendent we call $\overline{\Phi}$)

$$\Phi(z, s, a) := \sum_{n=0}^{\infty} \frac{z^n}{(n+a)^s}, \tag{1}$$

where under the constraints

$$|z| \leq 1, \quad a \notin \{0, -1, -2, \dots\}, \quad \Re(s) > 1 \tag{2}$$

Φ is absolutely convergent as a direct sum. Analytic continuation will eventually serve to relax constraints, but for the moment we shall stick to the absolute-convergence scenario.

5

The simplest instance of Lerch evaluation is the elementary sum

$$\Phi(z, 0, a) = \frac{1}{1-z}, \tag{3}$$

which when coupled with the formal relation

$$\Phi(z, s-1, a) = \left(a + z\frac{\partial}{\partial z}\right)\Phi(z, s, a) \tag{4}$$

yields all Lerch values at negative integer s as rational functions of z. However, for $z = 1$ the analytic continuation to nonpositive integer s simply gives a polynomial in the parameter a—that is, the operations of continuation and $z \to 1$ do not necessarily commute (see the discussion after relation (9)).

Another simple observation: The entire real line is a disjoint union of integer translates of the right-closed interval $(0, 1]$, $\Re(a)$ in (1) can always be relegated to said interval, as follows. One observes the elementary translation property for fixed positive integer m:

$$\Phi(z, s, a) = z^m\,\Phi(z, s, a+m) + \sum_{k=0}^{m-1}\frac{z^k}{(k+a)^s}. \tag{5}$$

Now a choice $m := |\lceil\Re(a)\rceil - 1|$ allows us to evaluate $\Phi(z, s, \overline{a})$ for some $\overline{a}$ having $\Re(\overline{a})) \in (0, 1]$.

Various other simple but useful relations include the doubling formula

$$\Phi(z, s, a) = \frac{1}{2^s}\Phi\left(z^2, s, \frac{a}{2}\right) + \frac{z}{2^s}\Phi\left(z^2, s, \frac{1+a}{2}\right),$$

obtained by using even/odd indices in the sum (1). This in turn implies

$$\Phi(z, s, a) - \Phi(-z, s, a) = \frac{z}{2^s}\Phi\left(z^2, s, \frac{a+1}{2}\right), \tag{6}$$

a form of doubling relation that is actually quite useful for certain computations (see Section 8).

Not so simple is the general functional relation discovered by Lerch [43] [40] [41]

$$\tag{7}$$

$$\frac{(2\pi)^s z^a}{\Gamma(s)}\Phi(z, 1-s, a) = e^{\frac{i\pi s}{2}}\Phi\left(e^{-2ia\pi}, s, \frac{\log z}{2\pi i}\right) + e^{2i\pi a - \frac{i\pi s}{2}}\Phi\left(e^{2ia\pi}, s, 1 - \frac{\log z}{2\pi i}\right),$$

valid over a wide range of complex parameters. Reference [40] has an interesting development of alternative functional relations involving combinations of Lerch-like functions.

For our present purposes, the primary advantage of this functional relation is in checking numerical schemes. Indeed, even though one may not know a closed form for say $\Phi(1, 5, 1/3)$, still the relation (7) should read $3.866 \cdots + 1025.9 \ldots i$ on both sides. In this way, a profound and beautiful theoretical result can be used in computations, to ensure self-consistency.

Conversely, one of our methods—the Riemann-splitting algorithm—implies by way of its very development the above functional relation.

2.1   Bernoulli-series representation of Lerch Φ

Under the stated constraints (2) one has a valid integral representation (with also $\Re(a) > 0$):

$$\Phi(z, s, a) = \frac{1}{\Gamma(s)} \int_0^\infty \frac{t^{s-1} e^{-at}}{1 - z e^{-t}} \, dt. \tag{8}$$

Our scheme is based on the observation that this integral can be cast in various, manifestly convergent, dual-series forms; moreover, a delicate branching chain based on parameter regions chooses the ideal series form thereby providing essentially invariant convergence rate across the various functions in the introduction—again, assuming bounded parameters.

We cite two important, classical expansions, both absolutely convergent under the given criteria on t:

$$\frac{e^{xt}}{e^t - 1} = \sum_{n \geq 0} B_n(x) \frac{t^{n-1}}{n!} \quad ; \quad |t| < 2\pi$$

$$\frac{2 e^{xt}}{e^t + 1} = \sum_{n \geq 0} E_n(x) \frac{t^n}{n!} \quad\quad ; \quad |t| < \pi,$$

where B_n, E_n are the Bernoulli, Euler polynomials, respectively. Incidentally the different criteria on $|t|$ here can be thought of in the following way: The Bernoulli-number series is essentially an expansion of $\mathrm{cosech}(t/2)$, while the Euler-number series is for $\mathrm{sech}(t/2)$. In the complex plane, the former has poles at $t = \pm 2m\pi$, while the latter has poles at $t = \pm m\pi$, and these poles constrain the radii of convergence.

To obtain what we call a Bernoulli master representation, we split the integral in (8) as $\int_0^\infty \rightarrow \int_0^\lambda + \int_\lambda^\infty$, where λ denotes a free parameter. Furthermore we invoke either Bernoulli- or Euler-polynomial expansions depending on the real part of the Lerch parameter z. This integral-splitting leads formally to such series as the following, where we assume $\Re(z) \in (1/2, 1]$, say:

$$\tag{9}$$

$$\Phi(z, s, a) = \frac{1}{\Gamma(s)} \sum_{n=0}^\infty \frac{\Gamma(s, \lambda(n + a)) z^n}{(n + a)^s} + \frac{z^{-a}}{\Gamma(s)} \sum_{m=0}^\infty \frac{B_m(1 - a)}{m!} \int_0^\lambda t^{s-1} (t - \log z)^{m-1} \, dt.$$

A similar construction involving Euler polynomials is straightforward, and applies best to say $\Re(z) \in [-1, -1/2)$. In this way a complete computational algorithm can be created for z on the closed unit disk. We exhibit this procedure as Algorithm 1.

Note that in the case of negative $\Re(z)$, the C_m in Algorithm 1 are all terminating hypergeometrics, so evaluations for such as z negative real are especially easy to implement.

Note also that when $z = 1$ and $s = -m = 0, -1, -2, \ldots$, the first sum in (9) vanishes and the second has a cancelled gamma-singularity, giving a closed-form analytic-continuation result

$$\Phi(1, -m, a) \;=\; -\frac{B_{m+1}(a)}{m + 1},$$

said value being a Hurwitz ζ evaluation $\zeta(-m, a)$.

Algorithm 1 Bernoulli-series algorithm for Lerch transcendent Φ.

This algorithm computes the classical Lerch transcendent $\Phi(z, s, a)$ for z on the complex unit disk, $|z| \leq 1$, and real $a \in (0, 1]$.

1) If($|z| \leq 1/2$) use parameter $\lambda := 0$, i.e. return the direct sum (1) which will be linearly convergent.

2) If $(s = -m \in (0, -1, -2, -3, \dots))$
     if($z == 1$) return

$$-\frac{B_{m+1}(a)}{m+1};$$

     else return the rational function of z determined by (3, 4);

3) If($\Re(z) \geq 0$) $p := -\log z$; else $p := -\log(-z)$;

4) $\lambda := 1 - p$;

5) Define coefficients C_m as follows:
     If($\Re(z) \geq 0$) {
         if($z \neq 1 + 0i$)

$$C_m := B_m(1-a)\frac{\lambda^s p^{m-1}}{s} \, {}_2F_1\left(-m+1, s; s+1; -\frac{\lambda}{p}\right);$$

     else

$$C_m := B_m(1-a)\frac{\lambda^{m-1+s}}{m-1+s};$$

     } else {
         if($z \neq -1 + 0i$)

$$C_m := \frac{1}{2}E_m(1-a)\frac{\lambda^s p^m}{s} \, {}_2F_1\left(-m, s; s+1; -\frac{\lambda}{p}\right);$$

     else

$$C_m := \frac{1}{2}E_m(1-a)\frac{\lambda^{m+s}}{m+s};$$

     }

5) return Φ as

$$\frac{1}{\Gamma(s)}\sum_{n=0}^{\infty}\frac{\Gamma(s, \lambda(n+a))\, z^n}{(n+a)^s} + \frac{e^{ap}}{\Gamma(s)}\sum_{m=0}^{\infty}C_m,$$

2.2   Bernoulli multisectioning

In Algorithm 1 and various algorithms to follow, we require perhaps a great many Bernoulli numbers. Realizing that $B_1 = -1/2$ but the rest of B_{odd} vanish, we can contemplate the power series

$$\frac{x^2}{\cosh x - 1} = -2 + \sum_{n \geq 0} \frac{2n-1}{(2n)!} B_{2n}\, x^{2n}$$

as an expedient for extracting Bernoulli numbers—-or even more stably, the numbers $B_{2n}/(2n)!$—to required precision. One simply takes enough power-series terms of cosh and performs Newton inversion of that finite cosh series. Further "multisectioning"—a technique pioneered by J. Buhler for the Bernoulli numbers (see the original treatise [20], which by now has been followed by several enhancement papers)—can reduce memory. In the more advanced setting, one calculates say all B_k with $k \bmod 8$ fixed, and does this separately for 4 values of k. The multisectioning technique is also discussed in the guise of "value recycling" in the calculation of large sets of $\zeta(2n)$ values [12], which values being required for many "rational zeta series."

Other special numbers appear in zeta-variant studies. For example the Nörlund numbers $B_n^{(n)}$, defined via generating function

$$\frac{t}{(1+t)\log(1+t)} = \sum_{n \geq 0} \frac{B_n^{(n)}}{n!} t^n \tag{10}$$

have entered naturally into studies on the Stieltjes constants [24, arXiv:1106.5146], a subject we discuss later in Section 7.1. Importantly, the Nörlund numbers *also* can be extracted via fast methods; one only need perform Newton inversion on the Taylor series for $\log(1+t)$.

2.3   Erdélyi-series representation for Lerch Φ

When s is not a positive integer, and for parameter $a \in (0, 1]$, $|\log z| < 2\pi$, one has the linearly convergent Erdélyi expansion [6]

$$z^a \Phi(z, s, a) = \sum_{n \geq 0} \zeta(s-n, a) \frac{\log^n z}{n!} + \Gamma(1-s)(-\log z)^{s-1}, \tag{11}$$

where appears the Hurwitz zeta function, itself an instance of Lerch, with direct sum (when convergent):

$$\zeta(s, a) := \sum_{k \geq 0} \frac{1}{(k+a)^s} = \Phi(1, s, a).$$

When s is a positive integer, adjustments must be made, being as the Erdélyi expansion's Γ singularity is neatly cancelled by one of the Hurwitz-ζ summands. An overall procedure for applying this Erdélyi form is as follows:

Algorithm 2 Erdélyi-series algorithm for Lerch transcendent Φ

This algorithm computes the classical Lerch transcendent $\Phi(z, s, a)$ for any complex s, with $a \in (0, 1]$, and any complex z in the region $|\log z| < 2\pi$ (which z-region certainly contains the annulus $|z| \in (e^{-2\pi}, 1]$).

1) If($|z| \leq 1/2$) return the direct sum (1) which will be linearly convergent.
2) If $(s = -m \in (0, -1, -2, -3, \dots))$
     if($z == 1$) return

$$-\frac{B_{m+1}(a)}{m+1};$$

     else return the rational function of z determined by (3, 4);
3) If(s is not a positive integer) return Φ as

$$z^{-a} \left(\sum_{n \geq 0} \zeta(s - n, a) \frac{\log^n z}{n!} \;+\; \Gamma(1 - s)(-\log z)^{s-1} \right).$$

4) Here, denote $s = k + 1$ for integer $k \geq 0$ and return Φ as

$$z^{-a} \left(\sum_{0 \leq n \neq k} \zeta(k + 1 - n, a) \frac{\log^n z}{n!} \;+\; \frac{\log^k z}{k!} \left(\psi^{(0)}(1 + k) - \psi^{(0)}(a) - \log(-\log z) \right) \right),$$

where $\psi^{(0)}$ is the standard digamma function.

2.4   Riemann-splitting representation for Lerch variant $\overline{\Phi}$

It turns out that one can forge a "master equation" for computation of the analytic continuation of a certain Lerch variant, call it $\overline{\Phi}$:

$$\overline{\Phi}(z, s, a) \;:=\; \sum_{n=0}^{\infty}{}' \frac{z^n}{((n + a)^2)^{s/2}}, \tag{12}$$

where $'$ on the sum indicates any denominator singularity is avoided. For $\Re(a) > 0$ this form $\overline{\Phi}$ is equivalent to the classical definition (1) for Φ. Our purpose in employing the square-then-half-power paradigm here is to allow transformation of integral representations such as this generalization of Riemann's ζ-function decomposition:

$$\sum_{n \in Z} \frac{z^n}{((n + a)^2)^{s/2}} \;=\; \frac{1}{\Gamma(s/2)} \int_0^{\infty} t^{s/2-1} \sum_{n \in Z} z^n e^{-(n+a)^2} \, dt. \tag{13}$$

The celebrated Riemann prescription is to split such an integral via $\int_0^\infty \to \int_0^\lambda + \int_\lambda^\infty$, then apply theta-function identities in the integrand's sum, eventually to yield a rapidly converging series with free parameter λ.

But how do we deal with the issue that $\overline{\Phi}$ is asking for a sum over nonnegative integer n, not $n \in Z$? One way is to use the following "magical" expedient—a phenomenon discovered by the present author at the start of this work in the late 1980s.[3] The idea is, a sum over integers on a specific half-line can be written

$$\sum_{S(n+a)>0} f(n) = \frac{1}{2}\sum_{n\in Z} f(n) + \frac{1}{2}\sum_{n\in Z} f(n)\, S(n+a),$$

where we define

$$S(z) := sign(\Re(z))$$

with $sign(0) := 0$. Now, the key is that, for the sake of Riemann-splitting, we can write simply (for nonzero complex ρ):

$$S(\rho) = \frac{\rho}{(\rho^2)^{1/2}},$$

and this allows the evaluation of the $\overline{\Phi}$ sum via *two* Riemann integrals.

Just as Riemann's original incomplete-gamma series for ζ is one way to establish the functional equation, so, too, is the current method for $\overline{\Phi}$; that is, we not only get a computational algorithm following, or we can prove such as (7). We omit the tedious details, opting instead to display the incomplete-gamma series in the next algorithm.

---

[3]It may well be that this expedient exists elsewhere in the literature; at any rate, this one simple trick opens up a whole world of computational analyticity.

Algorithm 3 Riemann-splitting algorithm for $\overline{\Phi}$ and its analytic continuation.

This algorithm computes $\overline{\Phi}(z, s, a)$—being for a wide class of parameters equivalent to the classical Φ—for any complex s, any complex z, and a in the complex region $\{\Re(z) \in [0, 1)\} \cup \{z = 1 + 0i\}$ (the translation relation (5) can be used for other a).

1) If($z == 0$) return $1/a^s$; Optionally: If($|z| \le 1/2$) return the direct sum (12) which will be linearly convergent.
2) If $(s = -m \in (0, -1, -2, -3, \dots))$
     if($z == 1$) return

$$-\frac{B_{m+1}(a)}{m+1};$$

     else return the rational function of z determined by (3, 4);
3) Choose a parameter λ, say $\lambda := \pi$ (but see the important discussion about the possibility of complex λ in Section 3);
4) Return the analytic continuation of $\overline{\Phi}$ as (note that for special cases of $\overline{\Phi}$, such as polylogarithms and zeta variants, the following incomplete-gamma decomposition can be significantly simplified):

$$
\overline{\Phi}(z, s, a) = -\frac{z^{-a}\lambda^{s/2}}{s\Gamma(s/2)}\delta_{a\in Z} + \frac{1}{s-1}\frac{\pi^{1/2}\lambda^{(s-1)/2}}{\Gamma(s/2)}\delta_{z=1} + \tag{14}
$$

$$
\frac{1}{2}\sum_{n\in Z}'\frac{z^n}{(A^2)^{s/2}}\left(\frac{\Gamma(s/2, \lambda A^2)}{\Gamma(s/2)} + \frac{\Gamma((s+1)/2, \lambda A^2)}{\Gamma((s+1)/2)}S(A)\right) +
$$

$$
\frac{\pi^{s-1/2}}{2z^a}\sum_{u\in Z}'\frac{e^{-2\pi iau}}{(U^2)^{(1-s)/2}}\left(\frac{\Gamma\left((1-s)/2, \frac{\pi^2}{\lambda}U^2\right)}{\Gamma(s/2)} + i\frac{\Gamma\left(1-s/2, \frac{\pi^2}{\lambda}U^2\right)}{\Gamma((s+1)/2)}S(U)\right),
$$

where in the first summation, $A := n + a$ and in the second summation $U := u + \frac{\log z}{2\pi i}$.

---

3 Riemann zeta function

Being as the Riemann ζ-function is defined for its literal-convergence region $\Re(s) > 1$ as

$$\zeta(s) := \sum_{n \geq 1} \frac{1}{n^s} = \Phi(1, s, 1) = \overline{\Phi}(1, s, 1),$$

the whole procedure behind Algorithms 1, 3 can be simplified. One may choose free parameter λ, and at the end of the algorithms one may return $\zeta(s)$ according to a Bernoulli series (requiring $|\lambda| < 2\pi$)

$$\zeta(s) = \frac{1}{\Gamma(s)} \sum_{n=1}^{\infty} \frac{\Gamma(s, \lambda n)}{n^s} + \frac{1}{\Gamma(s)} \sum_{m \geq 0} \frac{B_m}{m!} \frac{\lambda^{m+s-1}}{m+s-1}, \tag{15}$$

or in the case of Riemann splitting we have for general parameter λ

$$\zeta(s) = -\frac{\lambda^{s/2}}{s\Gamma(s/2)} + \frac{1}{s-1} \frac{\pi^{1/2} \lambda^{(s-1)/2}}{\Gamma(s/2)} + \tag{16}$$

$$\frac{1}{\Gamma(s/2)} \sum_{n=1}^{\infty} \left(\frac{\Gamma(s/2, \lambda n^2)}{n^s} + \pi^{s-1/2} \frac{\Gamma((1-s)/2, n^2 \pi^2 / \lambda)}{n^{1-s}} \right).$$

We can conveniently set $\lambda := \pi$ to obtain

$$\Gamma(s/2) \zeta(s) = \frac{\pi^{s/2}}{s(s-1)} + \sum_{n=1}^{\infty} \left(\frac{\Gamma(s/2, \pi n^2)}{n^s} + \pi^{s-1/2} \frac{\Gamma((1-s)/2, \pi n^2)}{n^{1-s}} \right). \tag{17}$$

This is essentially Riemann's classical prescription, and as we have intimated, leads to both the functional equation for ζ and a rapidly-convergent computational algorithm.

3.1 Riemann–Siegel formula and Dirichlet L-functions

The representation (16) is similar to the celebrated Riemann–Siegel expansion for $\zeta(s)$, in that the both incomplete-gamma summations have terms that decay significantly for large index n. We refer the reader to separate treatments of Riemann–Siegel expansions; practical exposition for the Riemann–Siegel construct is found in [26, 12]. There is also a modern treatment of Riemann–Siegel ideas by J. Reyna [48], such that the techniques can now also be invoked off the critical line.

Incidentally, a major application of efficient high-precision ζ evaluations is *analytic* counting of prime numbers [42, 26]. Evidently the prime count $\pi(10^{23})$ has been resolved in this way, namely

$$\pi\left(10^{23}\right) = 1925320391606803968923,$$

a recent result of D. J. Platt [45].

14

The analogy of Riemann–Siegel formulae with (16) can be rendered sharper if free parameter λ is allowed to have complex values—perhaps even s-dependent values. This promising idea is the focus of ongoing research [25]. In the Lerch connection specifically, see [2] and references therein. An excellent work of M. Rubinstein [49, Sections 3.3, 3.4] analyzes Riemann-splitting formulae similar to (16) above—along with the aforementioned complex freedom for free parameters; moreover, Rubinstein gives analogous formulae for Dirichlet *L*-function computations.

3.2   Riemann-zeta derivatives

In many of our computational formulae we require values of the Riemann zeta function—or its derivatives—at nonpositive integers. First, we know as in Section 2.1 that

$$\zeta(-m) \;=\; (-1)^m \frac{B_{m+1}}{m+1}$$

for $m = 0, 1, 2, \ldots$. From the functional equation

$$\zeta(s) \;=\; 2(2\pi)^{s-1} \sin \frac{\pi s}{2} \, \Gamma(1-s) \, \zeta(1-s)$$

we can extract derivatives

$$\zeta^{(1)}(0) \;=\; -\frac{1}{2}\log 2\pi, \quad \zeta^{(2)}(0) \;=\; \gamma_1 + \frac{\gamma^2}{2} - \frac{\pi^2}{24} - \frac{1}{2}(\log 2 + \log \pi)^2,$$

where γ_1 is a Stieltjes constant, and so on. For even $m = 2, 4, 6, \ldots$

$$\zeta^{(1)}(-m) := \frac{d}{ds}\zeta(s)\big|_{s=-m} \;=\; \frac{(-1)^{m/2} m!}{2^{m+1}\pi^m} \zeta(m+1),$$

for example

$$\zeta^{(1)}(-4) = \frac{3}{4}\frac{\zeta(5)}{\pi^4}.$$

For odd $m = 1, 3, 5 \ldots$, on the other hand,

$$\zeta^{(1)}(-m) \;=\; \zeta(-m)\left(\gamma \,+\, \log 2\pi \,-\, H_m - \frac{\zeta^{(1)}(m+1)}{\zeta(m+1)}\right),$$

for example

$$\zeta^{(1)}(-5) \;=\; \frac{15}{4\pi^6}\zeta^{(1)}(6) + \frac{137}{15120} - \frac{\gamma}{252} - \frac{1}{252}\log 2\pi.$$

Such machinations can be extended for higher derivatives, leading us to a

---

Principle: For an Erdélyi sum of the general form (for any integer s):

$$\sum_{m \geq 0}{}' \zeta^{(d)}(s-m) \, T_m,$$

where coefficient T_m has no ζ-component, all computations can refer to a set of derivatives $\{\zeta^{(d)}(s) \; : \; s = 2, 3, 4, \dots\}$. That is, no ζ values or derivatives are required for any integer argument < 2.

This principle shows that in some instances one really can avoid numerical differentiation. But there is more to be said: Even if one is also unwilling to adopt derivatives $\zeta^{(d)}(s = 2, 3, 4, \dots)$ as "fundamental constants," there is a way to generate such derivatives with still another series strategy. Let us describe the scenario for first derivatives. Take (15) with parameter $\lambda := 1$, multiply through by $\Gamma(s)$, then differentiate to obtain, for $\Re(s) > 1$,

$$\tag{18}$$

$$\zeta^{(1)}(s) \;=\; -\frac{\Gamma^{(1)}(s)}{\Gamma(s)} \zeta(s) + \frac{1}{\Gamma(s)} \sum_{n=1}^{\infty} \frac{\Gamma^{(1)}(s,n) - \Gamma(s,n) \log n}{n^s} - \frac{1}{\Gamma(s)} \sum_{m \geq 0} \frac{B_m}{m!} \frac{1}{(m+s-1)^2}.$$

It is relevant to point out here that there exist very simple and efficient schemes for evaluating ζ derivatives to moderate precision (see Section 7.6).

3.3   Gamma- and incomplete-gamma calculus

3.4 Γ and its derivatives

One of the best ways to calculate $\Gamma(s)$ itself is first to presume $\Re(s) > 0$ and use the Spouge formula with free real parameter $a > 0$:

$$\Gamma(1+z) \;=\; (z+a)^{z+1/2} e^{-z-a} \sqrt{2\pi} \left(c_0 + \sum_{k=1}^{\lceil a \rceil - 1} \frac{c_k(a)}{z+k} + E(z,a) \right),$$

where $c_0(a) := 1$ and the rest of the c coefficients are defined

$$c_k(a) \;:=\; \frac{1}{\sqrt{2\pi}} \frac{(-1)^{k-1}}{(k-1)!} (a-k)^{k-1/2} e^{-k+a}.$$

Remarkably, the relative-error term enjoys a very convenient bound [46]

$$|E(z,a)| \;<\; \frac{1}{\sqrt{a}(2\pi)^{a+1/2}},$$

so it is quite clear how to choose the free parameter a to achieve required precision. Some useful constants include the Taylor coefficients for Γ itself, from

$$\Gamma(1+x) \;=\; \sum_{n\geq 0} \Gamma^{(n)}(1)\frac{x^n}{n!},$$

which form is absolutely convergent for $|x| < 1$, the constants exemplified by:

$$\Gamma^{(0)}(1) \;=\; \Gamma(1) = 1,$$

$$\Gamma^{(1)}(1) \;=\; -\gamma,$$

$$\Gamma^{(2)}(1) \;=\; \gamma^2 - \gamma + \frac{\pi^2}{6},$$

$$\Gamma^{(3)}(1) \;=\; -2\gamma^3 + 9\gamma^2 - (\pi^2 + 6)\,\gamma + \frac{3}{2}\pi^2 - 4\zeta(3),$$

and so on. Such recondite forms can be deduced from relations involving polygamma functions, to which functions we shall later speak. We shall need these gamma-expansion constants in our work on general polylogarithm derivatives.

For limited argument ranges, there are other series of interest. One has two representation formulae (see [5, 12])

$$\log\Gamma(z) \;=\; -\log z \;+\; (z-1)(1-\gamma) \;+\; \sum_{n>1} \frac{\zeta(n)-1}{n}(1-z)^n, \tag{19}$$

$$\log\Gamma(x) \;=\; -\frac{1}{2}\log\frac{\sin(\pi x)}{\pi} + \frac{1}{2}\,(1-2x)\,(\gamma + \log 2\pi) \;+\; \frac{1}{\pi}\sum_{n\geq 1}\frac{\log n}{n}\sin(2\pi n x), \tag{20}$$

the first representation valid for $|z-1| < 2$, and the second for real $x \in (0,1)$. Interestingly, since the trigonometric sum in (20) is essentially a Clausen outer-derivative form, we can thus cast $\log\Gamma$, for suitable arguments, into a convergent series involving zeta derivatives. (See the remarks following representation (58) for more on this notion.)

3.5　Incomplete gamma and its outer derivatives

The incomplete gamma function $\Gamma(s,x)$ and its derivatives $\Gamma^{(d)}(s,x) := (\partial/\partial s)^d\,\Gamma(s,x)$ play a major role in our computational schemes. Defining

$$\Gamma(s,x) \;:=\; \int_x^\infty t^{s-1}e^{-t}\,dt, \tag{21}$$

there are standard recursion schemes connecting $\Gamma(s,x)$ and $\Gamma(s+1,x)$.

One representation of incomplete gamma is the ascending series

$$\Gamma(s,x) \;=\; \Gamma(s)\left(1 \;-\; x^s e^{-x}\sum_{k\ge 0}\frac{x^k}{\Gamma(s+k+1)}\right),\tag{22}$$

valid at least for $\Re(s) > 0$. Because we shall often need $\Gamma(0,x)$ in our developments, a special series applies, in the "exponential-integral" form

$$\Gamma(0,x) \;=\; -\gamma - \log x \;-\; \sum_{k\ge 1}(-1)^k\frac{x^k}{k!\,k}.\tag{23}$$

Then there is the general continued fraction

$$\Gamma(s,x) \;=\; \cfrac{x^s e^{-x}}{x+\cfrac{1-s}{1+\cfrac{1}{x+\cfrac{2-s}{1+\cfrac{2}{x+\cdots}}}}}\tag{24}$$

which now *is* valid for $s = 0$. The natural asymptotic expansion for general s is

$$\Gamma(s,x) \;\sim\; x^{s-1}e^{-x}\left(1 - \frac{1-s}{x} + \frac{(1-s)(2-s)}{x^2} \;-\;\cdots\right),$$

with error bounds not too hard to establish. Given the ascending series, continued fraction, and asymptotic forms, one may evaluate $\Gamma(s,x)$ to relative precision of D digits in $O(D^{1+\epsilon})$ operations. However, recall that one of our goals is to avoid asymptotics. We argue next that a manifestly converging (Laguerre) series can be used even for large $|x|$.

A remarkable, globally convergent (*not* merely asymptotic) expansion does exist for $\Gamma(s,x)$, namely

$$\Gamma(s,x) \;=\; x^s e^{-x}\sum_{n\ge 0}\frac{(1-s)_n}{(n+1)!}\frac{1}{L_n^{(-s)}(-x)L_{n+1}^{(-s)}(-x)},\tag{25}$$

where $L_n^{(-s)}$ here is an associated Laguerre polynomial. In fact, given that such forms as (25) are symbolically differentiable, we always have means to calculate any derivatives $\Gamma^{(d)}(s,x)$ *using a convergent series involving only elementary summands.*

We should observe that in various of our computational algorithms herein we need *sets* of values such as $\{\Gamma(0,n) \;:\; n = 1,2,3,\ldots\}$. It may be of value that the difference of two successive set members is

$$\Gamma(0,n+1) - \Gamma(0,n) \;:=\; \frac{e^{-n}}{n}\sum_{k\ge 0}\frac{(-1)^k}{n^k}\mu_k$$

where the $\mu_k := \int_0^1 y^k e^{-y}\,dy$ are elementary constants. This suggests the possibility of evaluating the given set of $\Gamma(0,n)$ values via recursion.

As for derivatives of incomplete gamma, here is a beautiful relation for calculating s-derivatives of the Laguerre entities $L_n^{(s)}(x)$, namely

$$\frac{\partial}{\partial s}L_n^{(s)}(x) \;=\; \sum_{m=0}^{n-1}\frac{L_m^{(s)}(x)}{n-m}.$$

The very existence of this identity establishes that every derivative of every order for any incomplete gamma at integer s can be given a converging series—one only has to symbolically differentiate (25).

In a 2008 work [9], effective (rigorous, with explicit big-O constants) are established for Laguerre polynomials. In fact,

$$L_n^{(-a)}(-z) \;\sim\; \frac{e^{-z/2}}{2\sqrt{\pi}}\,\frac{e^{2\sqrt{mz}}}{z^{1/4-a/2}\,m^{1/4+a/2}}\left(1+O\left(\frac{1}{m^{1/2}}\right)\right), \qquad (26)$$

which is certainly enough to infer a complexity bound for computing (25) to some prescribed number of digits.

We can in the present context think of the fundamental derivatives as the set

$$\{\Gamma^{(d)}(0,x)\;:\;d=0,1,2,3\dots\}.$$

This is because of another valuable recurrence, obtainable directly from the integral definition (21) via integration by parts:

$$\Gamma^{(d)}(s,x) \;=\; \int_x^\infty t^{s-1}\log^d t\,e^{-t}\,dt$$

$$=\; x^{s-1}\log^d x\,e^{-x} + (s-1)\Gamma^{(d)}(s-1,x) + d\,\Gamma^{(d-1)}(s-1,x).$$

So we are allowed reductions of any derivative $\Gamma^{(d)}(s,x)$ for positive integer s into a combination of $\Gamma^{(\delta)}(0,\cdot)$ terms with $\delta < d$. For example,

$$\Gamma^{(1)}(1,x) \;=\; e^{-x}\log x \;+\; \Gamma(0,x)$$

and

$$\Gamma^{(4)}(2,x) \;=\; ((x+1)\log x \;+\; 4)\log^3 x\,e^{-x} + 4\,\Gamma^{(3)}(0,x) + 12\,\Gamma^{(2)}(0,x).$$

3.6 Stark–Keiper and other "summation form" strategies

The Stark–Keiper formula reads, for positive integer N but arbitrary s,

$$\zeta(s,N) = \frac{1}{s-1}\sum_{k=1}^\infty\left(N+\frac{s-1}{k+1}\right)(-1)^k\binom{s+k-1}{k}\zeta(s+k,N),$$

where $\zeta(s,a)$ is the Hurwitz zeta function discussed in the next section. Interestingly, Keiper was able to use this relation for *recursive* computation of various zeta values. A code example of such recursion is exhibited in [30, p. 55].

There is another "summation form" for Riemann ζ

$$\zeta(s) = \lim_{N\to\infty} \frac{1}{2^{N-s+1} - 2^N} \sum_{k=0}^{2N-1} \frac{(-1)^k}{(k+1)^s} \left(\sum_{m=0}^{k-N} \binom{N}{m} - 2^N \right) \tag{27}$$

for which it is possible to give a rigorous error bound as a function of the cutoff N and s itself [7]. This and the Stark–Keiper scheme previous should always be considered for ζ calculations because the sums involved are extremely simple. (One drawback occurs when imaginary height t is very large, for $s := \sigma + it$, in which domain the prescriptions of the Riemann–Siegel class dominate.)

Continued fraction schemes for Riemann zeta and polylogarithms are found in [34]; and, we do not refer to the incomplete-gamma fractions—instead, those authors exhibit actual continued fractions for polylogarithmic entities (the fraction elements are quite complicated).

Later in the present treatment we discuss the Stieltjes expansion

$$\zeta(s) = \frac{1}{s-1} + \sum_{n\geq 0} (-1)^n \gamma_n \frac{(s-1)^n}{n!}, \tag{28}$$

where the γ_N—essentially Taylor coefficients—start with $\gamma_0 := \gamma$, the celebrated Euler constant. A similar expansion that yields certain closed-form coefficients is discussed by Vepstas in [53], namely

$$\zeta(s) = \frac{1}{s-1} + \sum_{n\geq 0} (-1)^n b_n \frac{[s]_n}{n!},$$

where here, $[s]_n$ denotes the (falling) Pochhammer symbol, $[s]_n := s(s-1)\cdots(s-n+1)$.

4   Hurwitz zeta function

The Hurwitz ζ-function is defined for its literal-convergence region $\Re(s) > 1$ as

$$\zeta(s,a) := \sum_{n\geq 0} \frac{1}{(n+a)^s} = \overline{\Phi}(1,s,a).$$

It is interesting to look first at Algorithm 1 and infer quickly that for $a \neq 0, -1, -2, \ldots$,

$$\zeta(s,a) = \frac{1}{\Gamma(s)} \sum_{n\geq 0} \frac{\Gamma(s,\lambda(n+a))}{(n+a)^s} + \frac{1}{\Gamma(s)} \sum_{m\geq 0} \frac{(-1)^m B_m(a)}{m!} \frac{\lambda^{m+s-1}}{m+s-1}, \tag{29}$$

20

with free parameter $\lambda \in [0, 2\pi)$. Note the interesting pole cancellation we have seen before: At nonpositive integers s the second sum's Γ prefactor has a pole neatly cancelled by the $m = 1 - s$ summand. Thus we obtain again the exact evaluations $\zeta(-m, a) = -B_{m+1}(a)/(m+1)$ for $m \in (0, 1, 2, \dots)$.

Algorithm 3 can be simplified for, say, $\Re(a) \in (0, 1]$, so including the instance $\zeta(s, 1) := \zeta(s)$, yielding

$$\zeta(s, a) = -\frac{\pi^{s/2}}{s\Gamma(s/2)}\delta_{a=1} + \frac{1}{s-1}\frac{\pi^s}{\Gamma(s/2)} + \tag{30}$$

$$\frac{1}{2}\sum_{n\in Z}'\frac{1}{(A^2)^{s/2}}\left(\frac{\Gamma(s/2, \pi A^2)}{\Gamma(s/2)} + \frac{\Gamma((s+1)/2, \pi A^2)}{\Gamma((s+1)/2)}S(A)\right) +$$

$$\pi^{s-1/2}\sum_{u=1}^{\infty}\frac{1}{u^{1-s}}\left(\frac{\Gamma((1-s)/2, \pi u^2)}{\Gamma(s/2)}\cos(2\pi au) + \frac{\Gamma(1-s/2, \pi u^2)}{\Gamma((s+1)/2)}\sin(2\pi au)\right),$$

where $A := n + a$ as in Algorithm 3.

4.1 Generalized *L*-series

A generalized *L*-series, which includes many Dirichlet series of theoretical interest, is

$$L_s(\chi) := \sum_{n\geq 1}\frac{\chi(n)}{n^s},$$

where $\chi(n)$ is periodic with period P. (No particular number-theoretical import is required here for the character χ—mere periodicity is enough.) We have in essence a finite number of sums, each over an arithmetic progression of n values, with the immediate Hurwitz-zeta decomposition following:

$$L_s(\chi) = \frac{1}{P^s}\sum_{k=1}^{P}\chi(k)\,\zeta\left(s, \frac{k}{P}\right).$$

5 Polylogarithms

The polylogarithm Li_s is defined—when convergent—as

$$\mathrm{Li}_s(z) := \sum_{n\geq 1}\frac{z^n}{n^s} = z\,\Phi(z, s, 1). \tag{31}$$

It turns out that Algorithm 2 can be quite useful in polylogarithm computations. We have, for s not a positive integer, and $|\log z| < 2\pi$, a convergent form

$$\mathrm{Li}_s(z) = \sum_{n\geq 0}\zeta(s-n)\frac{\log^n z}{n!} + \Gamma(1-s)(-\log z)^{s-1}. \tag{32}$$

21

However, when $s = k + 1$ for nonnegative integer k, a limiting procedure shows

$$\mathrm{Li}_s(z) = \sum_{0 \le n \ne k} \zeta(k + 1 - n) \frac{\log^n z}{n!} + \frac{\log^k z}{k!} \left(H_k - \log(-\log z)\right), \qquad (33)$$

where $H_k := \sum_{m=1}^{k} 1/m$ denotes the k-th harmonic number (with $H_0 := 0$). Later we explain the limiting procedure that gives (33); we shall also be able to compute, along such lines, polylogarithm derivatives in the s variable.

Algorithm 3 can again be simplified using $a := 1$, to yield a general polylogarithm analytic continuation as

$$
\begin{aligned}
\mathrm{Li}_s(z) = \; & -\frac{\lambda^{s/2}}{s\Gamma(s/2)} + \frac{1}{s-1} \frac{\pi^{1/2}\lambda^{(s-1)/2}}{\Gamma(s/2)} \delta_{z=1} + \qquad (34) \\
& \sum_{n=1}^{\infty} \frac{z^n}{n^s} \left(\frac{\Gamma(s/2, \lambda n^2)}{\Gamma(s/2)} \cosh(n \log z) + \frac{\Gamma((s+1)/2, \lambda n^2)}{\Gamma((s+1)/2)} \sinh(n \log z) \right) + \\
& \frac{\pi^{s-1/2}}{2} \sideset{}{'}\sum_{u \in Z} \frac{e^{-2\pi i a u}}{(U^2)^{(1-s)/2}} \left(\frac{\Gamma\left((1-s)/2, \frac{\pi^2}{\lambda}U^2\right)}{\Gamma(s/2)} + i \frac{\Gamma\left(1 - s/2, \frac{\pi^2}{\lambda}U^2\right)}{\Gamma((s+1)/2)} S(U) \right),
\end{aligned}
$$

where $U := u + \frac{\log z}{2\pi i}$.

5.1 Analytic properties of polylogarithms

Outside the open unit z-circle, there are two difficulties: First, there is no absolute convergence, and second, cuts in the complex plane must be carefully considered. So for example, it is known that

$$\mathrm{Li}_2\left(\frac{1}{2}\right) = \frac{\pi^2}{12} - \frac{1}{2} \log^2 2,$$

as may be verified numerically by direct summation of (31), with a precision gain of about 1 bit per summand. However, it is also standard that the analytic continuation has

$$\mathrm{Li}_2(2) = \frac{\pi^2}{4} - i\pi \log 2,$$

even though the sum (31) cannot be performed directly. Incidentally, all along the cut $z \in [1, \infty)$ there is a discontinuity in the correct analytic continuation, exemplified (for $\epsilon > 0$) by

$$\mathrm{Li}_2(2 + i\epsilon) = \frac{\pi^2}{4} + i\pi \log 2,$$

and in general

$$\mathrm{Disc}\,\mathrm{Li}_s(z) = 2\pi i \frac{\log^{s-1} z}{\Gamma(s)},$$

with $\Im(\mathrm{Li})$ always being split equally across the cut—thus we know exactly the imaginary part of any $\mathrm{Li}_n(z)$ on the real ray $z \in [1, \infty)$; said part is $(i/2)\mathrm{Disc}$. This discontinuity relation is quite useful in checking of any software.

There are relations that allow analytic continuation, namely [35] [44]:

$$\mathrm{Li}_s(z) + \mathrm{Li}_s(-z) = 2^{1-s}\mathrm{Li}_s(z^2),$$

true for all complex s, z, and for n integer, and complex z,

$$\mathrm{Li}_n(z) + (-1)^n\mathrm{Li}_n(1/z) = -\frac{(2\pi i)^n}{n!}B_n\left(\frac{\log z}{2\pi i}\right) - 2\pi i\Theta(z)\frac{\log^{n-1}z}{(n-1)!}, \qquad (35)$$

where B_n is the standard Bernoulli polynomial and Θ is a domain-dependent step function: $\Theta(z) := 1$, if $\Im(z) < 0$ or $z \in [1, \infty)$, else $\Theta = 0$. That is, the final term in this latest relation is included when and only when z is in the lower open half-plane union the real cut $[1, \infty)$.

Another useful relation is, for any complex z but for $n = 0, -1, -2, -3, \ldots$,

$$\mathrm{Li}_n(z) = (-n)!(-\log z)^{n-1} - \sum_{k=0}^{\infty}\frac{B_{k-n+1}}{k!(k-n+1)}\log^k z.$$

5.2 Polylogarithm-Hurwitz relation

One interesting option we do not explore in the present treatment is the functional relation between polylogarthims and the Hurwitz ζ:

$$(36)$$

$$\mathrm{Li}_s(z) := \frac{\Gamma(1-s)}{(2\pi)^{1-s}}\left(i^{1-s}\zeta\left(1-s, \frac{1}{2} + \frac{\log(-z)}{2\pi i}\right) + i^{s-1}\zeta\left(1-s, \frac{1}{2} - \frac{\log(-z)}{2\pi i}\right)\right).$$

One application of this functional relation is to render polylogarithms into a Bernoulli-series form. Indeed, the functional relation (36) together with (29) yields the following algorithm. Considering $\mathrm{Li}_s(e^{-x})$, define $\alpha := x/(2\pi i)$. If α is an integer, return $\zeta(s)$, else use, with free parameter $\lambda \in [0, 2\pi)$, and noting if necessary that α can now be restricted such that $\Re(\alpha) \in [0, 1)$, the representation

$$\mathrm{Li}_s(e^{-x}) := i^{1-s}\sum_{n\geq 0}\frac{\Gamma(1-s, \lambda(n+1-\alpha))}{(n+1-\alpha)^{1-s}} + i^{s-1}\sum_{n\geq 0}\frac{\Gamma(1-s, \lambda(n+\alpha))}{(n+\alpha)^{1-s}} - \qquad (37)$$

$$\pi\sum_{m\geq 0}\frac{(-i)^m\lambda^{m-s}}{m!}\mathrm{sinc}\left(\frac{\pi}{2}(m-s)\right)B_m(\alpha).$$

Refinements on this series representation are straightforward; e.g., use real parts when x, s are real. Note that s can be a nonnegative integer, courtesy of the sinc function—in fact, for even integer s the last sum with the B_m can be especially simple. Remarkably, one hereby obtains an algorithm for the general analytic continuation of $\mathrm{Li}_s(z)$ in the complex s-plane and for any complex z—a feature not present in some of our other computational algorithms (recall, as in Algorithm 2, there are restrictions on $|\log z|$).

5.3   Lerch–Hurwitz (periodic) zeta function

The function (periodic in real x)

$$E_s(x) \ := \ \sum_{n \geq 1} \frac{e^{2\pi i n x}}{n^s} \ = \ \Phi(e^{2\pi i x}, s, 1)$$

can also be thought of as a polylogarithm evaluation $\mathrm{Li}_s\left(e^{2\pi i x}\right)$. So E_s is susceptible to various of our algorithms. In particular, a straightforward way to calculate this periodic zeta function is to calculate an entity $\mathrm{Li}_s(e^{-2\pi i a})$ using (37) as-is.

6   Digamma and polygamma functions

A convergent scheme for the polygamma functions arises from the Hurwitz-zeta representation (29) and a definition of the digamma function

$$\psi^{(0)}(z) \ := \ \frac{d}{dz} \log \Gamma(z) = -\gamma \ + \ \sum_{n \geq 0} \left(\frac{1}{n+1} - \frac{1}{n+z} \right),$$

where we note the sum is a limiting form of the difference of two Hurwitz ζ functions. One infers a converging series ($\lambda \in (0, 2\pi)$ is the usual free parameter here)

$$(38)$$

$$\psi^{(0)}(z) \ = \ -\gamma - 1 - \log\left(1 - e^{-\lambda}\right) + \sum_{k=1}^{\infty} \left((-1)^k \frac{\lambda^k}{k!k} \left(B_k - B_k(z)\right) - \frac{e^{-\lambda(k+z-1)}}{k+z-1} \right),$$

then observes that higher-order polygammas, namely

$$\psi^{(n)}(z) \ := \ \frac{d^{n+1}}{dz^{n+1}} \log \Gamma(z)$$

are simply Hurwitz-zeta evaluations:

$$\psi^{(n)}(z) \ = \ (-1)^{n+1} \, n! \, \zeta(n+1, z)$$

for positive integer n. Thus we have established convergent series for all polygammas (there are standard recurrence relations that allow us to restrict argument a in $\psi^{(m)}(1+a)$ to be in $(0, 1]$).

7 Key fundamental constants

7.1 Euler and Stieltjes constants

We begin with the alluring historical expansion (28) involving the Stieltjes constants γ_N. Expanding (15) near $s = 1$ yields

$$\gamma_1 = \frac{\pi^2}{12} - \frac{1}{2}\gamma^2 + \sum_{n\geq 1}\frac{1}{n}\left(-\Gamma(0,n) + \frac{B_n}{n!\,n}\right) \tag{39}$$

Similar series accrue for the higher-index γ_N—one may employ the derivative machinery described at the end of Section 3.5 for $\Gamma^{(d)}(0,n)$ in such a context. It is of interest that the first sum in representation (39) can be put in the form

$$\sum_{n\geq 1}\frac{1}{n}\Gamma(0,n) = \int_0^{1/e}\frac{\log(1-t)}{t\log t}\,dt, \tag{40}$$

which in turn can be cast as a series involving the Nörlund numbers from (10).

On the other hand, we can use the incomplete-gamma expansion (17) for the case of the Riemann ζ to provide a computational formula for the Euler constant, as:

$$\gamma = -2 + \log 4\pi + 2\sum_{n=1}^{\infty}\left(\frac{\mathrm{erfc}\left(n\sqrt{\pi}\right)}{n} + \Gamma\left(0,n^2\pi\right)\right). \tag{41}$$

Based on the material in Section 3.5 we state

Principle: Any Stieltjes constant γ_N can be cast as a convergent two-dimensional sum involving elementary summands, such that D good digits require, for α some absolute constant, $O\left(D^\alpha\right)$ operations. (It is allowed that the implied big-O constant can be N-dependent.) It is natural to conjecture that α can be taken to be $1 + \epsilon$.

Incidentally, based on our previous discussion of incomplete-gamma evaluations, simply equating two different series forms (23, 25) gives us an immediate representation of Euler γ in rather simple terms of Laguerre polynomials L_n:

$$\gamma = -\frac{1}{e}\sum_{n\geq 0}\frac{1}{n+1}\frac{1}{L_n(-1)L_{n+1}(-1)} - \sum_{k\geq 1}\frac{(-1)^k}{k!\,k}. \tag{42}$$

Because of the subexponential growth of the Laguerre function, this series representation achieves D good digits of γ in $O(D^{2+\epsilon})$ operations. However, various acceleration techniques may well apply, especially as the summands of the Laguerre sum here are all rational. Indeed, the creation of the Laguerre denominators is not in itself expensive; one has an especially simple recurrence

$$L_k(-1) = 2L_{k-1}(-1) - \left(1 - \frac{1}{k}\right)L_{k-2}(-1),$$

and may create quickly each successive summand of (42). Note that the use of (41) together with the remarks of Section 3.5 should result in lower complexity. So (42) is not the fastest available scheme, but consistent with the above general principle on Stieltjes constants. It could well be, however, that these incomplete-gamma series for the γ_N turn out to be susceptible to binary-splitting or other acceleration techniques, so that current best complexity bounds for γ itself can perhaps be approached in this way. (See [18, 19, 54] for modern results on computing γ.)

7.2   The works of M. W. Coffey

As for convergence rate of such Stieltjes-constant schemes, the waters are deep. In regard to other series for the γ_n, see the excellent works of M. W. Coffey [22, 38, 39, 23, 24]. Coffey's algorithms in some cases also involve double sums, so all of these methods should eventually be assessed as to computational complexity. It is of interest that Coffey's analyses sometimes involve Meijer-G evaluations—equivalently in this context, high-order hypergeometrics—which functions amount to alternative ways to write incomplete-gamma terms. (Note that Coffey's papers are also relevant to other sections of the present work.) This complexity-of-series is a fascinating research problem; for one thing, the study of magnitude growth for the γ_n remains an imperfect science.

7.3   Glaisher–Kinkelin constant

One relation for the Glaisher–Kinkelin constant A is

$$A = (2\pi)^{1/12}e^{\gamma/12-\zeta'(2)/(2\pi^2)},$$

which can be cast in the form

$$\log A = \frac{1}{12}(1 + \log 2\pi) - \frac{1}{2\pi^2}\frac{d}{ds}\Gamma(s)\zeta(s)|_{s=2}.$$

Now we can exploit our Bernoulli-series form (15) for $\Gamma \cdot \zeta$, to infer

$$\frac{d}{ds}\Gamma(s)\zeta(s)|_{s=2} = \sum_{n\geq 1}\frac{1}{n^2}\left(((1+\lambda n)\log\lambda + 1)e^{-\lambda n} + \Gamma(0,\lambda n)\right) +$$

$$\sum_{m \geq 0} \frac{B_m}{m!} \lambda^{m+1} \left(\frac{\log \lambda}{m+1} - \frac{1}{(m+1)^2} \right)$$

An instance of this convergent series is afforded by setting $\lambda := \log 2$, whence

$$\log A = \frac{1}{24} + \frac{\log^2 2}{4\pi^2} - \frac{\log^2 2 \, \log \log 2}{4\pi^2} + \frac{1}{12} \log 2\pi - \frac{1}{24} \log \log 2 -$$

$$\frac{1}{2\pi^2} \sum_{n \geq 1} \left(\frac{\Gamma(0, n \log 2)}{n^2} + \frac{B_{n-1}}{n!} \log^n 2 \left(-\frac{1}{n} + \log \log 2 \right) \right).$$

300 summands of this series yield about 100 good decimal digits for $\log A$.

The above development for $\log A$ is to show how Bernoulli-series forms may be employed. There are somewhat more efficient series; for example, (19) may be used together with [50, (8)]:

$$\int_0^{1/2} \log \Gamma(x+1) \, dx = -\frac{1}{2} - \frac{7}{24} \log 2 + \frac{1}{4} \log \pi + \frac{3}{2} \log A$$

to effect a Glaisher–Kinkelin representation in terms of a rational zeta series, in the spirit of [12]. A goal of "one decimal digit of precision per series term" can thus be met in the form

$$\log A = \frac{265}{144} - \frac{1}{12}\gamma + \frac{571}{36} \log 2 - \frac{14}{3} \log 3 - \frac{5}{3} \log 5 - \frac{7}{3} \log 7 - \frac{1}{6} \log \pi +$$

$$\frac{1}{3} \sum_{n \geq 2} \frac{\left(\zeta(n) - 1 - \frac{1}{2^n} - \frac{1}{3^n} - \frac{1}{4^n} \right)(-1)^n}{n(n+1)2^n},$$

where, indeed, summands decay as 10^{-n}.

7.4   Khintchine constant

The celebrated Khintchine constant—the (almost everywhere) geometric mean of continued fraction elements, is given from the measure theory of fractions as

$$K_0 = \prod_{r \geq 1} \left(1 + \frac{1}{r(r+2)} \right)^{\log_2 r}.$$

Over the years, rapidly convergent expansions have accrued, such as [3]

$$(\log 2)(\log K_0) = \sum_{n \geq 1} \frac{\zeta(2n) - 1}{n} \sum_{k=1}^{2n-1} \frac{(-1)^{k+1}}{k},$$

a series that can be accelerated by "peeling off" terms from $\zeta(2n)$ and so involve Hurwitz-zeta values $\zeta(2n, N)$ for a chosen positive integer N.

Thus our various algorithms for computing ζ or Hurwitz-ζ can be brought to bear for the Khintchine constant. But there is more to be said. Another series for K_0 is

$$(\log 2)(\log K_0) = \log^2 2 + \text{Li}_2\left(-\frac{1}{2}\right) + \frac{1}{2}\sum_{k\geq 2}(-1)^k \text{Li}_2\left(\frac{4}{k^2}\right).$$

While this series converges rather slowly (1000 summands yield about 6 good decimals), there is the interesting research problem of somehow speeding up the polylogarithmic sum.

A modern, very efficient series is attributed to O. Pavlyk, who built on the work of Gosper [52], which series we paraphrase here as

$$(\log 2)(\log K_0) = -\frac{\pi^2}{6}(\gamma + \log 2 + \log \pi - 12\log A) +$$

$$2\sum_{k=2}^{\infty}(-1)^k \frac{\log k}{(k+2)k^{k+2}}\left(2^{k+1}\,_2F_1\left(1, k+2; k+3; -\frac{2}{k}\right) - \,_2F_1\left(1, k+2; k+3; -\frac{1}{k}\right)\right) +$$

$$2\sum_{k=2}^{\infty}\frac{(-1)^k(2^k-1)}{k+1}\left(\zeta^{(1)}(k+1) + \sum_{j=1}^{k-1}\frac{\log j}{j^{k+1}}\right),$$

where A is the aforementioned Glaisher–Kinkelin constant. Again we witness the appearance of previously studied entities, namely ζ derivatives.

Another Gosper relation is

$$(\log 2)(\log K_0) = \sum_{j\geq 2}\frac{(-1)^j(2-2^j)}{j}\zeta^{(1)}(j).$$

This series converges slowly, so again a good research problem is to somehow apply acceleration. It may be helpful to observe that this series may be transformed to

$$(\log 2)(\log K_0) = 2\gamma\log 2 - \sum_{j\geq 2}\frac{(-2)^j}{j}\eta^{(1)}(j). \tag{43}$$

7.5   The MRB constant

The M. R. Burns (MRB) constant is given by the attractive summation [51]:

$$B = \sum_{k\geq 1}(-1)^k\left(k^{\frac{1}{k}} - 1\right) = 0.1878596\ldots.$$

Remarkably, if we look longingly at our voluminous list of zeta variants and use a series for $k^{1/k} = e^{(\log k)/k}$, we obtain, at least formally, a true jewel of an expression:

$$B = -\sum_{m \geq 1} \frac{(-1)^m}{m!} \eta^{(m)}(m),$$

which series actually converges rather quickly (60 summands here yield > 100 correct decimal digits). It may turn out to be lucrative to employ not the eta derivatives $\eta^{(m)}(m)$ explicitly, instead to use a Taylor expansion of $\eta(d)$ around $d = 0$ to write, again formally:

$$B = \sum_{m \geq 1} \frac{c_m}{m!} \eta^{(m)}(0), \tag{44}$$

where the c coefficients are defined

$$c_j := \sum_{d=1}^{j} \binom{j}{d} (-1)^d d^{j-d}.$$

Interestingly, as pointed out by J. Borwein, $c_j/j!$ is the j-th Taylor coefficient in the expansion of e^{-xe^x} about $x = 0$. Though (44) converges rather slowly, it is conceivable that by focusing on the fixed argument 0 of the eta derivatives, we might uncover some algorithm advantages; see for example Section 3.2 for remarks on Riemann-zeta derivatives at argument 0.

7.6   Examples of alternating-series acceleration

The already classic paper [21] lays out historical and modern means of accelerating alternating series. The idea is that many a sum $S = \sum_n (-1)^n a_n$ can be restructured to converge—sometimes surprisingly—rapidly to its value S. As just one application of Algorithm 1 of [21], we can take our alternating relation (43) and run *Mathematica* code:

```
eta[s_] := (1 - 2^(1 - s)) Zeta[s];
s1 = 2 EulerGamma Log[2];
a[j_] := 1/(j + 2) 2^(j + 2) Derivative[1][eta][j + 2];
(* Now start accelerate algorithm for given a[0], a[1], ... *)
n = 130;   (* Number of a's involved. *)
d = (3 + Sqrt[8])^n; d = 1/2 (d + 1/d);
{b, c, s} = {-1, -d, 0};
Do[
   c = b - c;
   s = s + c a[k];
   b = (k + n) (k - n) b/((k + 1) (k + 1/2))
```

```
    , {k, 0, n - 1}
  ];
s2 = s/d;
(* s2 is now the desired sum of (-1)^k a[k]. *)
N[s1 - s2, 110]
```

Remarkably enough, the (only!) $n = 130$ evaluations of the η-derivative here yield 100 good decimal digits of the Khintchine constant K_0. A good many of our other alternating series throughout the present paper can likewise benefit from this specific acceleration technique.

Another example using the same algorithm—just different $a[]$ terms—is that of computing a ζ derivative. Consider that

$$\zeta^{(1)}(s) \;=\; \frac{\eta^{(1)}(s) - 2^{1-s}(\log 2)\,\zeta(s)}{1 - 2^{1-s}},$$

with

$$\eta^{(1)}(s) = -\sum_{n \geq 0} \frac{(-1)^n \log(n+1)}{(n+1)^s}.$$

Now we can apply the above coded algorithm to this alternating sum, to get for example $\zeta^{(1)}(2)$ to 100 good decimals with the same $n = 130$ total $a[j]$ terms involved. Note that even the $\zeta(2)$ appearing can be calculated in the same fashion, from the alternating $\eta(2)$. Incidentally, this example shows that $\log A$, where A is the Glaisher-Kinkelin constant, can be obtained quickly to 100 decimals, with a very short program involving only elementary-functions calls.

A good research challenge is to calculate in this accelerated fashion more general zeta variants, such as the MTW values discussed next.

8   MTW zeta values and polylogarithm derivatives

8.1   Rapidly convergent series representation

The Mordell–Tornheim–Witten (MTW) zeta function has a canonical form

$$\mathcal{W}(r,s,t) \;:=\; \sum_{m,n \geq 1} \frac{1}{m^r n^s (m+n)^t} \tag{45}$$

and has been the focus of much study in regard to symbolic identities, analyticity and so on. The literal sum converges over a certain r, s, t space, and can be analytically continued for all complex r, s, t—although there are singularities. We borrow the expedient of definite integration from Bailey, Borwein et al. [5] to represent $\mathcal{W}$ as

$$\mathcal{W}(r,s,t) \;=\; \frac{1}{\Gamma(t)} \int_0^1 \mathrm{Li}_r(z) \mathrm{Li}_s(z) (-\log z)^{t-1} \frac{dz}{z}. \tag{46}$$

By adroitly splitting $\int_0^1 \to \int_0^{1/e} + \int_{1/e}^1$, we may insert the series (32, 33) into the second integral, employing the evaluations

$$U_{p,q} := \int_{1/e}^1 \log^p z \log^q(-\log z)\frac{dz}{z}$$

$$= \frac{(-1)^{p+q}\, q!}{(p+1)^{q+1}},$$

and defining coefficients

$$A_p := \Gamma(1-p); \quad p \notin Z^+,$$

$$:= \frac{(-1)^{p-1}}{\Gamma(p)}H_{p-1}; \quad p \in Z^+,$$

where $H_k = \sum_{j=1}^k 1/j$ is the k-th harmonic number, with $H_0 := 0$. Similarly, defining

$$B_p := 0; \quad p \notin Z^+,$$

$$:= \frac{(-1)^p}{\Gamma(p)}; \quad p \in Z^+,$$

we obtain a series for $\mathcal{W}$ that is valid for any complex r, s, t, with $|\lambda| < 2\pi$:

$$(47)$$

$$\Gamma(t)\, \mathcal{W}(r,s,t) = \sum_{m,n\geq 1} \frac{\Gamma(t,(m+n)\lambda)}{m^r n^s (m+n)^t} +$$

$$\sum_{u,v\geq 0}{}' (-1)^{u+v}\frac{\zeta(r-u)\zeta(s-v)}{u!v!(u+v+t)}\lambda^{u+v+t} +$$

$$\sum_{q\geq 0}{}' (-1)^q\frac{\zeta(r-q)}{q!}\lambda^{s+q+t-1}\left(\frac{A_s + B_s\log\lambda}{(s+q+t-1)} - \frac{B_s}{(s+q+t-1)^2}\right) +$$

$$\sum_{q\geq 0}{}' (-1)^q\frac{\zeta(s-q)}{q!}\lambda^{r+q+t-1}\left(\frac{A_r + B_r\log\lambda}{(r+q+t-1)} - \frac{B_r}{(r+q+t-1)^2}\right) + \lambda^{r+s+t-2}\,\cdot$$

$$\left(\frac{(A_r + B_r\log\lambda)(A_s + B_s\log\lambda)}{r+s+t-2} - \frac{A_r B_s + A_s B_r + 2B_r B_s\log\lambda}{(r+s+t-2)^2} + \frac{2B_r B_s}{(r+s+t-2)^3}\right).$$

The notation $\sum'$ means that we *avoid* any $\zeta(1)$ evaluations entirely.

A fascinating result from this Erdélyi-series procedure: There are singularities of $\mathcal{W}$ at $r+s+t = 2$, with residue $\Gamma(1-r)\Gamma(1-s)/\Gamma(t)$ when said residue is finite. But there are *also* singularities whenever $s+t-1$ or $r+t-1$ is a nonpositive integer. Also, in the limit $t \to 0$ we see that the residual term is just the first $(u = v = 0)$ term of the u, v summation, and so $\mathcal{W}(r, s, 0) = \zeta(r)\zeta(s)$ is verified. As for testing such a complicated but converging representation,

1) One should obtain the same basic numerical answer over a range of free λ.

2) One might verify numerically the Zagier triangle identity

$$\mathcal{W}(r, s, t) = \mathcal{W}(r - 1, s, t + 1) + \mathcal{W}(r, s - 1, t + 1).$$

3) A typical numerical value for three noninteger r, s, t is

$$\mathcal{W}(\pi, \pi, \pi) \approx 0.121784932649073172392415831466446\ldots.$$

4) A typical evaluation near a pole is, for $d := 200001/300000$,

$$\mathcal{W}(d, d, d) = 529982.9016524962105\ldots.$$

5) Analytic-continuation values *are* obtainable outside of the domain of literal convergence of the defining sum for $\mathcal{W}$, e.g.

$$\mathcal{W}\left(-\frac{1}{2}, -\frac{1}{2}, 1\right) = 0.637833177149232816022942231906\underline{2}\ldots.$$

8.2 Multiple-zeta values

The twofold multiple-zeta function is

$$\overline{\zeta}(t, r) = \sum_{n > m \geq 1} \frac{1}{n^t} \frac{1}{m^r},$$

which is seen to be

$$= \mathcal{W}(r, 0, t).$$

Thus, the $s = 0$ case of series (47) is in play, and we get values such as

$$\overline{\zeta}\left(\frac{3}{2}, 1\right) = 4.681814411556227035692210279337199\underline{5}\ldots,$$

and for one of the still algebraically unresolved cases

$$\overline{\zeta}(2, 6) = 0.651565163715126904556463962090377\ldots.$$

In this connection see [28, 29]—the latter reference reveals a surprising efficiency in higher-dimensional sum calculations, which efficiency was later echoed in [14, Sec 7]. The pioneering algebraic work on such entities is exemplified in papers such as [4, 10] and references therein.

8.3   Alternating MTW variants

Alternating MTW sums are generally more numerically tractable, due to the existence of the alternating zeta function $\eta(s)$ defined below. The polylogarithmic instance of doubling relation (6) is

$$\mathrm{Li}_s(z) + \mathrm{Li}_s(-z) \;=\; \frac{1}{2^s}\mathrm{Li}_s(z^2),$$

which identity can be used in conjunction with (32) to establish an attractive, compact series (valid for $|\log z| < \pi$)

$$\mathrm{Li}_s(-z) \;=\; -\sum_{m \geq 0} \eta(s-m)\frac{\log^m z}{m!} \;, \tag{48}$$

where

$$\eta(s) \;:=\; \left(1 - 2^{1-s}\right)\zeta(s)$$

is a regular function; the $(s=1)$-pole of ζ has vanished, as $\eta(1) = \log 2$, so the sum (48) requires no singularity avoidance. Most important for certain analyses: The eta-series (48) can be differentiated with respect to the outer variable, say

$$\mathrm{Li}_s^{(k)}(-z) \;:=\; \left(\frac{\partial}{\partial s}\right)^k \mathrm{Li}_s(-z) \;=\; -\sum_{m \geq 0} \eta^{(k)}(s-m)\frac{\log^m z}{m!}.$$

We note in passing the derivatives

$$\eta^{(1)}(1) \;=\; \gamma \log 2 - \frac{1}{2}\log^2 2,$$

$$\eta^{(1)}(-2) \;=\; -\frac{1}{3}\log 2 - \frac{1}{4} + 3\log A,$$

where A is the Glaisher–Kinkelin constant; in general, though, one only needs a library of numerical Riemann-ζ derivatives to obtain η derivatives.

     An immediate byproduct of this eta-expansion is a rapidly convergent series for certain MTW zeta-variants. Consider for example the entity (here we echo the notation of Borwein et al. [5]):

$$\omega_{a,b,0}^{--}(r,s,t) \;:=\; \sum_{m,n \geq 1} (-1)^{m+n}\frac{\log^a m \,\log^b n}{m^r n^s (m+n)^t} \;,$$

which can also be written

$$= \frac{(-1)^{a+b}}{\Gamma(t)} \int_0^1 \mathrm{Li}_r^{(a)}(-z)\mathrm{Li}_s^{(b)}(-z)(-\log z)^{t-1} \frac{dz}{z} \;.$$

By again splitting $\int_0^1 \rightarrow \int_0^{1/e} + \int_{1/e}^1$, and employing the eta expansion (48) for the integral over $(1/e, 1)$, we can use evaluations

$$\int_0^{1/e} z^\alpha (-\log z)^\beta \, dz \;=\; \frac{1}{(1+\alpha)^\beta} \Gamma(1+\beta, 1+\alpha)$$

we arrive at a convergent series involving the η-function. An example of such series is the case $\omega_{0,0,0}^{--}$:

$$\tag{49}$$

$$\omega_{0,0,0}^{--}(r,s,t) \;=\; \sum_{m,n\geq 1} \frac{(-1)^{m+n}}{m^r n^s (m+n)^t} \frac{\Gamma(t, m+n)}{\Gamma(t)} \;-\; \frac{1}{\Gamma(t)} \sum_{j,k\geq 0} \frac{\eta(r-j)}{j!} \frac{\eta(s-k)}{k!} \frac{(-1)^{j+k}}{j+k+t}.$$

Note that for positive integer t, the incomplete gamma function here is elementary. For the case $r = s = t = 1$ we can give a limit on each summation index, say $m, n, j, k \leq 240$, to obtain

$$\omega_{0,0,0}^{--}(1,1,1) \;:=\; \sum_{m,n\geq 1} \frac{(-1)^{m+n}}{mn(m+n)}$$

$$= \; 0.30051422578989857134993454037786249769124657308512472044806788388354\ldots,$$

agreeing in fact with the $\omega_{0,0,0}^{--}(1,1,1) = \frac{1}{4}\zeta(3)$, a known evaluation [11].

Intriguing it is that certain apparently difficult sums can be computed more easily than one might expect. An exemplary calculation—using first derivatives of η—is

$$\omega_{1,1,0}^{--}(1,1,1) \;:=\; \sum_{m,n\geq 1} \frac{(-1)^{m+n} \log m \, \log n}{m \, n (m+n)}$$

$$= \; 0.0084654591832435660002204654483\ldots$$

8.4   More difficult ω-sums

The computational situation with the ω-sums is much more problematic when the alternating sign factor $(-1)^{m+n}$ is absent. Let us consider

$$\omega_{a,b,c}^{++}(r,s,t) \;:=\; \sum_{m,n\geq 1} \frac{\log^a m \, \log^b n \, \log^c(m+n)}{m^r n^s (m+n)^t}$$

$$= \; (-1)^{a+b+c} \frac{\partial^c}{\partial t^c} \frac{1}{\Gamma(t)} \int_0^1 \mathrm{Li}_r^{(a)}(z) \mathrm{Li}_s^{(b)}(z) (-\log z)^{t-1} \frac{dz}{z},$$

where now it will be recognized that the argument of the polylogarithms is not $-z$, but $+z$. This seemingly innocent alteration motivates us to work out the complete analytic expansion of polylogarithms near integer outer argument—a task to which we next turn.

9   Analytic expansion of Li_s for near-integer s

An intricate analysis of the Erdélyi representation (32) yields an analytic expansion of $\mathrm{Li}_s(z)$ for $s = k + 1 + \tau$, for k a fixed nonnegative integer and a new complex variable τ. It is convenient first to state the result informally:

Principle: For nonnegative integer k, the d-th derivative $\mathrm{Li}_{k+1}^{(d)}$ can be cast as

$$\mathrm{Li}_{k+1}^{(d)}(z) = (\text{series in powers of } \log z) + \log^k z \cdot (\text{degree-}(d+1) \text{ polynomial in } \log(-\log(z))).$$

Let us choose a nonnegative integer k and rewrite (32) with $s = k + 1 + \tau$:

$$
\mathrm{Li}_s(z) \;=\; \sum_{0 \le n \ne k} \zeta(k + 1 + \tau - n)\, \frac{\log^n z}{n!} \;+\; \left\{ \Gamma(-k - \tau)(-\log z)^{k+\tau} + \zeta(1 + \tau)\, \frac{\log^k z}{k!} \right\}.
\tag{50}
$$

Now there is no singularity avoidance whatever in the sum, so the τ-expansion of said sum can be written in terms of derivatives of ζ. We proceed to expand the entity in braces in powers of τ. Omitting the tedious details, the full τ-expansion is, for $|\log z| < 2\pi$ and $\tau \in [0, 1)$:

$$
\mathrm{Li}_{k+1+\tau}(z) \;=\; \sum_{0 \le n \ne k} \zeta(k + 1 + \tau - n)\frac{\log^n z}{n!} \;+\; \frac{\log^k z}{k!} \sum_{j=0}^{\infty} c_{k,j}(\mathcal{L})\, \tau^j,
\tag{51}
$$

with $\mathcal{L} := \log(-\log z)$ and the c coefficients involve the Stieltjes constants from (28):

$$
c_{k,j}(\mathcal{L}) \;=\; \frac{(-1)^j}{j!} \gamma_j - b_{k,j+1}(\mathcal{L}),
$$

for b coefficients defined in turn by

$$
b_{k,j}(\mathcal{L}) \;=\; \sum_{\substack{p+t+q=j \\ p,t,q \ge 0}} \frac{\mathcal{L}^p}{p!} \frac{\Gamma^{(t)}(1)}{t!} (-1)^{t+q} f_k(q),
$$

where $f_k(q)$ is the coefficient of x^q in

$$
\prod_{m=1}^{k} \frac{1}{1 + x/m},
$$

35

easily calculable via $f_{k,0} = 1$ and the recursion

$$f_{k,q} = \sum_{h=0}^{q} \frac{(-1)^h}{k^h} f_{k-1,q-h}.$$

The first instances are harmonic-number relations $f_{k,q} = -H_k$ and $f_{k,2} = \frac{1}{2}H_k^2 + \frac{1}{2}H_k^{(2)}$.

Recondite as this τ-expansion may be, the rewards are substantial. Indeed, one has

$$c_{k,0} = H_k - \mathcal{L},$$

giving immediately the Erdélyi expansion (33) at any positive integer $s = k+1$. Another instance: To get first derivatives $\mathrm{Li}_1^{(1)}(z)$ for $k = 0$ we shall need the exact evaluation

$$c_{0,1} = -\gamma_1 - \frac{1}{2}\gamma^2 - \frac{\pi^2}{12} - \gamma\mathcal{L} - \frac{1}{2}\mathcal{L}^2.$$

This begets a convergent expansion for first outer derivative (again for $|\log z| < 2\pi$):

$$\tag{52}$$

$$\mathrm{Li}_1^{(1)}(z) = \sum_{n \geq 1} \zeta^{(1)}(1-n)\frac{\log^n z}{n!} - \gamma_1 - \frac{1}{2}\gamma^2 - \frac{\pi^2}{12} - \gamma\log(-\log z) - \frac{1}{2}\log^2(-\log z).$$

As a sharp test of this result, a numerical value for 50 terms of the ζ'-series here works out as

$$\mathrm{Li}_1^{(1)}\left(\frac{1}{2}\right) = -0.17289680030426475956117932688102820711445969877714\ldots,$$

correct to the implied precision, when compared to different algorithms that simply take numerical derivatives of polylogarithms.

9.1   Convergent sums for the difficult ω values

Let us employ the above analytical results to attempt an exemplary series representation of

$$\omega_{1,1,0}^{++}(1,1,1) := \sum_{m,n \geq 1} \frac{\log m \log n}{mn(m+n)}$$

$$= \int_0^1 \mathrm{Li}_1^{(1)}(z)\mathrm{Li}_1^{(1)}(z)\,\frac{dz}{z},$$

again using notation and integral from Borwein et al. [5]. Combining results so far, we have a series form:

$$
\omega_{1,1,0}^{++}(1,1,1) = \beta + \sum_{m,n\geq 1}\frac{\log m\,\log n}{mn(m+n)}e^{-m-n} + \tag{53}
$$

$$
\sum_{m,n\geq 1}\frac{\zeta^{(1)}(1-m)}{m!}\frac{\zeta^{(1)}(1-n)}{n!}U_{m+n,0} -
$$

$$
\sum_{m\geq 1}\frac{\zeta^{(1)}(1-m)}{m!}\left(2\alpha\,U_{m,0}+2\gamma\,U_{m,1}+U_{m,2}\right),
$$

where constants are defined:

$$
\alpha := \gamma_1 + \frac{1}{2}\gamma^2 + \frac{\pi^2}{12},
$$

$$
\beta = 6 + 2\alpha + \alpha^2 - 6\gamma - 2\alpha\gamma + 2\gamma^2.
$$

Taking a limit of 375 on every summation index, we obtain a 140-digit value

$$
\omega_{1,1,0}^{++}(1,1,1) \approx
$$

4.30244762034222643331985179818698942752019447430436225569749833\
73406896458605708423834220669589660132906320814848672664367230\
39847882031570623622399525661106071826 04

in evident agreement with preliminary computations of Bailey, Borwein et al. [5].

9.2   Combining series with extreme-precision quadrature

There is an interesting way to achieve precise numerics via a combination of series and quadrature. Consider the MTW sums as in (45), but now 4-dimensional:

$$
\mathcal{W}(r,s,t,u) := \sum_{m,n,p\geq 1}\frac{1}{m^r n^s p^t (m+n+p)^u}. \tag{54}
$$

Clearly, a multidimensional sum of the type (47) can be generated, but it can become—for higher dimensions—more practical to use a simpler series together with quadrature. Indeed, it is not hard to derive

$$
\Gamma(u)\,\mathcal{W}(r,s,t,u) = \sum_{m,n,p\geq 1}\frac{\Gamma(u,\lambda(m+n+p))}{m^r n^s p^t (m+n+p)^u} + \tag{55}
$$

$$
\int_0^\lambda T^{u-1}\mathrm{Li}_r\left(e^{-T}\right)\mathrm{Li}_s\left(e^{-T}\right)\mathrm{Li}_t\left(e^{-T}\right)\,dT,
$$

valid for free parameter $\lambda \in [0, \infty)$. This wide parameter range also yields various theoretical truths; we obtain on the assignment $\lambda := \infty$ such cases as

$$\mathcal{W}(1,1,1,1) = 6\,\zeta(4),$$

$$\mathcal{W}(0,0,0,u) = \zeta(u) - \frac{3}{2}\zeta(u-1) + \frac{1}{2}\zeta(u),$$

and so on. As for the use of extreme-precision quadrature, taking $\lambda := 3$ and a limit of 80 on each summation index results in such as

$$\mathcal{W}(1,1,1,1) = 6.4939394022668291490960221792470074166485057115123614446097\ldots,$$

$$\mathcal{W}(2,2,2,2) = 0.2045567550777060275894820832804412223721321137466674838202\ldots,$$

$$\mathcal{W}(2,4,6,8) = 0.0001578532767602254319626298912297777035954013043090624119\ldots.$$

The first of these four approximations is consistent with $6\,\zeta(4)$.

Moreover, if one wants MTW outer derivatives, one may use such representations as (52) together with quadrature, to obtain, for free parameter $\lambda \in (e^{-2\pi}, 1]$,

$$
\begin{aligned}
\omega_{1,1,1,1}^{++++}(1,1,1,1) &:= \sum_{m,n,p \geq 1} \frac{\log m \, \log n \, \log p \, \log(m+n+p)}{m\,n\,p\,(m+n+p)} \qquad (56) \\
&= \int_0^1 \left(\mathrm{Li}_1^{(1)}(z)\right)^3 (\gamma + \log(-\log z)) \frac{dz}{z} \\
&= \int_\lambda^1 \left(\mathrm{Li}_1^{(1)}(z)\right)^3 (\gamma + \log(-\log z)) \frac{dz}{z} \; -
\end{aligned}
$$

$$
\sum_{m,n,p \geq 1} \frac{\log m \, \log n \, \log p}{m\,n\,p\,(m+n+p)} \left((\gamma + \log(-\log \lambda))\lambda^{m+n+p} \; + \; \Gamma(0, -(m+n+p)\log \lambda) \right)
$$

$$
= 393.95644190297417690269554547960277193912677747341826027083865333879\ldots.
$$

9.3   Alternative formulae for polylogarithm analysis

Another option for possible derivative-based analysis of polylogarithms and ω-sums is to invoke the identity

$$\mathrm{Li}_s(z) = \frac{z}{2} + \sum_{m \in Z} \frac{\Gamma(1-s, 2m\pi i - \log z)}{(2m\pi i - \log z)^{s-1}}.$$

It is suggested that if this sum be performed symmetrically—i.e. summed over $m \in [-M, +M]$ then $M \to \infty$, the relation is valid for all complex s, z [47].

9.4   Clausen function

A generalized Clausen function is

$$\mathrm{Cl}_s(x) \ := \ \sum_{n \geq 1} \frac{\sin nx}{n^s},$$

which can be formally put in Lerch–Hurwitz (complex-argument polylogarithmic) form

$$= \ \frac{1}{2i} \left(\mathrm{Li}_s \left(e^{ix} \right) - \mathrm{Li}_s \left(e^{-ix} \right) \right).$$

From the eta-decomposition (48) we quickly obtain (valid for $|\theta| < \pi$)

$$\mathrm{Cl}_s(\pi - \theta) \ = \ - \sum_{\text{odd } n > 0} \frac{\eta(s-n)}{n!} (-1)^{(n-1)/2} \, \theta^n \ . \tag{57}$$

For nonnegative integer k we know $\eta(-k) = (1 - 2^{k+1})(-1)^k \frac{B_{k+1}}{k+1}$, so when s is a positive integer we have

$$\mathrm{Cl}_s(\pi - \theta) \ = \ - \sum_{\text{odd } n \in [1, s-1]} \frac{\eta(s-n)}{n!} (-1)^{(n-1)/2} \, \theta^n \ +$$

$$\sum_{\text{odd } n \geq s+1} \left(1 - 2^{n-s+1}\right) (-1)^{(n-1)/2} \frac{B_{n+s-1}}{n!(n+s-1)} \, \theta^n.$$

We have split up the eta series in this way to signal the fact of eta being elementary at nonpositive integer arguments. For example, the Catalan constant G can be written

$$G \ = \ \mathrm{Cl}_2 \left(\frac{\pi}{2} \right) \ := \sum_{n \geq 1} \frac{\sin(\pi n/2)}{n^2} \ = \ \frac{1}{1^2} - \frac{1}{3^2} + \frac{1}{5^2} - \cdots$$

$$= \ \frac{\pi}{2} \log 2 \ + \sum_{\text{odd } n \geq 3} \left(2^{n-1} - 1\right) (-1)^{(n-1)/2} \frac{B_{n-1}}{n!(n-1)} \left(\frac{\pi}{2} \right)^n.$$

A more exotic example is

$$\mathrm{Cl}_8 \left(\frac{\pi}{2} \right) \ := \sum_{n \geq 1} \frac{\sin(\pi n/2)}{n^8} \ = \ \frac{1}{1^8} - \frac{1}{3^8} + \frac{1}{5^8} - \cdots$$

$$= \frac{\pi^5 \zeta(3)}{5120} - \frac{5\pi^3 \zeta(5)}{256} + \frac{63\pi \zeta(7)}{128} - \frac{\pi^7 \log 2}{645120} + \sum_{\text{odd } n \geq 9} \left(2^{n-7} - 1\right) (-1)^{(n-1)/2} \frac{B_{n-7}}{n!(n-7)} \left(\frac{\pi}{2} \right)^n.$$

For some θ the convergence of (57) may be slow, in which case an alternative series may be derived from the polylogarithm representation (33). Taking $s = 2$ for example, we find

$$\mathrm{Cl}_2(t) \;=\; t\left(1 - \frac{1}{2}\log t^2\right) + \sum_{\mathrm{odd}\ n \geq 3} (-1)^{(n-3)/2} \frac{B_{n-1}}{n!(n-1)} t^n,$$

valid for $|t| < 2\pi$, so one may switch between this series or (57) (with $s = 2$) to achieve rapid convergence over the useful range $t \in [-\pi, +\pi]$. Extension to arbitrary integer s is straightforward.

Incidentally, in [13] it is mentioned that a single Clausen series can also be accelerated using the duplication rule

$$\frac{1}{2}\mathrm{Cl}_2(2\theta) \;=\; \mathrm{Cl}_2(\theta) - \mathrm{Cl}_2(\pi - \theta).$$

In addition, those authors also mention yet another use of Bernoulli multisectioning to facilitate extreme-precision Clausen evaluation (see our Section 2.2). The importance of Clausen functions in modern experimental mathematics is exemplified by the questions raised in [31].

In the recent literature [5] there appear such entities as outer Clausen derivatives. We record here a certain, striking instance via our polylogarithm derivative form (52):

$$\mathrm{Cl}_1^{(1)}(\theta) \;=\; -\sum_{n \geq 1} \frac{\log n}{n} \sin(n\theta)$$

$$= \; \frac{\pi}{2}(\gamma + \log \theta) + \sum_{\mathrm{odd}\ n > 0} \zeta^{(1)}(1-n)(-1)^{(n-1)/2} \frac{\theta^n}{n!}, \qquad (58)$$

valid for real $\theta \in (0, 2\pi)$. It is fascinating that the $\log \Gamma$ forms (19, 20) now imply a direct connection between a zeta-sum and the zeta-derivative-sum here in (58). Indeed, denote by $\sum_\zeta$ the zeta-sum in (19), and by $\sum_{\zeta'}$ the derivative-sum in (58) with $\theta := 2\pi z$. Then

$$\sum_\zeta + \frac{1}{\pi}\sum_{\zeta'} \;=\; 1 - \gamma - z(1 + \log 2\pi) - \frac{1}{2}\log\frac{\sin \pi z}{\pi z},$$

which relation also means that we can cast the Clausen derivative in question in terms of a zeta-sum (devoid of zeta-derivatives). This connection is currently not fully understood; certainly, one would like to avoid zeta derivatives in sums, in favor of zeta values per se. (See [12] for various "recycling" methods that generate and process large sets of zeta values.)

10   Epstein zeta functions and lattice sums

For a higher-dimensional analogue of our algorithm developments, we consider a certain generalized (Epstein) zeta function of four parameters. Specifically, assume A is a D-by-

D, real, positive definite matrix, s is a complex number, and c, d are real vectors. We adopt the definition

$$Z_A(s; c, d) := {\sum_{n \in Z^D}}' \frac{e^{2\pi i c \cdot An}}{|An - d|^s}, \tag{59}$$

where the vector n runs over the full, D-dimensional integer lattice and the notation $\sum'$ means as usual that any singularities are to be ignored in the course of summation. As to the very existence of the sum, there will generally be some real threshold σ for which the constraint $\Re(s) > \sigma$ yields an absolutely convergent $\sum'$. Then we may infer the analytic continuation to $\Re(s) \leq \sigma$ by standard methods. We note that Z_A has a one-dimensional instance that coincides with the Riemann ζ function. For $A = 1$ (the identity matrix) we have

$$Z_1(s; 0, 0) = 2\zeta(s).$$

Just as with the Riemann zeta itself, there exist attractive closed-form evaluations of Z_A, in various dimensions D, as we present later.

Over the years, the sums Z_A have arisen in chemistry and physics research. Calculation of electrostatic energies of crystals use such sums in a natural way. In fact, in this context the matrix A is what characterizes the very structure of a periodic crystal, so that Z_A is related to physical energy. Interesting also is that the avoidance of the singularity in the Σ' is manifest in the physical scenario as avoidance of self-energy at an ionic center. There is a large and long-standing literature on such chemical application, grouped under such topical headings as "lattice sums," or "Madelung problem," and so on (see [27] for some survey material). In theoretical physics, vacuum energy—-or Casimir—calculations may also involve Epstein zeta functions [1]. The Epstein sums also figure into analytic number theory, for example in problems regarding sum-of-squares representations.

There are some interesting optimizations we can state right off. First, it will turn out to be computationally efficient to handle matrices of unit determinant. To this end we observe a scaling property, valid for positive real γ:

$$Z_A(s; c, d) = \gamma^{-s} Z_{A/\gamma}(s; \gamma c, d/\gamma). \tag{60}$$

Now the choice

$$\gamma = (\det A)^{1/D}$$

allows us to carry out calculations for the right-hand side of (60), with its matrix A/γ having unit determinant. In the common case where A is a diagonal matrix, the γ parameter is simply the geometric mean of the diagonal elements.

A second initial optimization is this: we need not evaluate Z_A for unconstrained c, d vectors, because one can always reduce them according to the integer lattice Z^D. Define the inverse transpose:

$$B = A^{-T}$$

41

and reduce c, d with respect to the Z^D lattice by:

$$c' = B((B^{-1}c) \bmod 1),$$

$$d' = A((A^{-1}d) \bmod 1),$$

where it is intended that each component of a vector be reduced (mod 1); i.e., so that each component is in the interval $[0, 1)$. It turns out, then, that under such reduction to the lattice we have

$$Z_A(s; c, d) = e^{2\pi i c' \cdot (d - d')} Z_A(s; c', d'). \tag{61}$$

In this way the magnitudes of the c, d parameters can be controlled during computation. Accordingly, we shall heretofore assume c, d have been *a priori* reduced. In particular, a c or d parameter now belongs to the Z^D lattice if and only if it is the zero vector.

10.1   Riemann splitting for Epstein zeta

Along the lines of Algorithm 3 a dimensionally generalized form for Epstein zeta function Z_A works out to be

$$
e^{-\pi i c \cdot d} \frac{\Gamma(s/2)}{\pi^{s/2}} Z_A(s; c, d) \quad = \quad \frac{\delta(c) \det B}{s/2 - D/2} - \frac{\delta(d)}{s/2} + \tag{62}
$$
$$
e^{-\pi i c \cdot d} {\sum_n}' \frac{\Gamma(\frac{s}{2}, \pi |An - d|^2) e^{2\pi i c \cdot An}}{(\pi |An - d|^2)^{s/2}} +
$$
$$
\frac{e^{\pi i c \cdot d}}{\det A} {\sum_k}' \frac{\Gamma(\frac{D-s}{2}, \pi |Bk - c|^2) e^{-2\pi i d \cdot Bk}}{(\pi |Bk - c|^2)^{\frac{D-s}{2}}}.
$$

This incomplete-gamma series together with our preceding optimization discussion suggests the following:

42

Algorithm 4 Riemann-splitting algorithm for Epstein zeta function $Z_A(s; c, d)$.

---

1) Invoke the reduction formulae (60, 61) as necessary, so that inputs A, c, d have these properties: That $\det A = 1$ and for $B = A^{-T}$ all components of $B^{-1}c, A^{-1}d$ lie in $[0, 1)$ via (mod 1) reduction to the integer lattice.

2) Next we handle the two possibilities $s = 0, D$. First check for the known evaluation at $s = 0$: If $s = 0$ return either -1 (in the case d vanishes), or return zero (in the case $d \neq 0$). Then check for possible singularity at $s = D$: If both $s = D$ and $c = 0$, return with indication of singularity, otherwise continue to step (3).

3) Choose a summation limit L (so that computational error will be, roughly speaking, $e^{-\pi L^2}$). Find also the least and greatest eigenvalues, respectively λ, μ, of A.

4) Sum the first series of (62) over each component $n_i \in [-L/\lambda, +L/\lambda]$, evaluating only those summands having $|An - d| \leq L$. Then sum the second series of (62) similarly, but using instead $k_i \in [-\mu L, +\mu L]$, resolving only those summands with $|Bk - c| \leq L$.

5) Using the rest of the terms in (62), and any reduction formulae that were invoked in step(1), return an evaluation of the original Z_A.

10.2 Exact theoretical evaluations as sharp algorithm tests

There exist a host of closed-form results and known evaluations of the Epstein zetas. These results are extremely useful—if not indispensable—for proper testing of any computational engine that is supposed to be evaluating Z_A. Happily, there are known evaluations for different classes of parameters; e.g., d or c or both $\neq 0$, and so on. In what follows, we remind ourselves that c, d parameters are assumed reduced to the Z^D lattice, as was explained for reduction relation (61).

First there is the functional equation, which we can infer from the Riemann strategy. From (62) it is evident that the entity:

$$\Lambda_A(s; c, d) = \sqrt{\det A}\, e^{-\pi i c \cdot d} \frac{\Gamma(s/2)}{\pi^{s/2}} Z_A(s; c, d)$$

is invariant under a specific parametric transformation. Indeed,

$$\Lambda_A(s; c, d) = \Lambda_B(D - s; -d, c).$$

This functional equation is of course a higher-dimensional analogue of the celebrated relation for $\zeta(s)$.

An exact evaluation at $s = 0$ also follows from (62); namely:

$$Z_A(0; c, d) = -\delta(d),$$

43

from which the well known evaluation $\zeta(0) = -1/2$ follows in the one-dimensional setting. Likewise, the fact of a pole at $s = D$ depends, as we see in (62), on whether c vanishes: there is a pole at $s = D$ if and only if $c = 0$, and the residue is easy to infer as $2\pi^{D/2}(\det B)/\Gamma(D/2)$.

A beautiful class of closed-form evaluations has arisen over the years in the art of applying Z_A in the sciences. Many such relations arise in the theory of Jacobi theta-functions; often combinatorics and number theory are involved in fascinating ways. Here and elsewhere, 1_D means either the D-dimensional identity matrix, or a D-vector consisting of all 1's as appropriate by context; while α_D denotes a D-vector $\{\alpha, \alpha, ..., \alpha\}$. In addition, we define some specific L-series which figure into some of the closed-form evaluations:

$$\beta(s) = 1^{-s} - 3^{-s} + 5^{-s} - ...,$$

$$\eta(s) = (1 - 2^{1-s})\zeta(s) = 1^{-s} - 2^{-s} + 3^{-s} - ...,$$

$$\lambda(s) = (1 - 2^{-s})\zeta(s) = 1 + 3^{-s} + 5^{-s} + ...,$$

$$L_{-3}(s) = 1^{-s} - 2^{-s} + 4^{-s} - 5^{-s} + 7^{-s} - ...,$$

$$L_{-8}(s) = 1^{-s} + 3^{-s} - 5^{-s} - 7^{-s} + 9^{-s} + 11^{-s} - ...$$

$$L_{+8}(s) = 1^{-s} - 3^{-s} - 5^{-s} + 7^{-s} + 9^{-s} - 11^{-s} - ...$$

Some of these series themselves enjoy exact evaluation, such as $\beta(1) = \pi/4$ and the analytic continuation evaluation $\beta(0) = 1/2$.

Various exact evaluations are known for $D = 2$ dimensions, for example

$$Z_{1_2}(s; 0_2, 0_2) = \sum_{m,n \in Z} {}'\frac{1}{(m^2 + n^2)^{s/2}} = 4\zeta(s/2)\beta(s/2),$$

$$Z_{1_2}(s; (\tfrac{1}{2})_2, 0_2) = \sum_{m,n \in Z} {}'\frac{(-1)^{m+n}}{(m^2 + n^2)^{s/2}} = -4\eta(s/2)\beta(s/2).$$

Note that the second zeta above is a "Madelung constant" for $D = 2$, i.e. the potential energy of the origin charge in a certain 2-dimensional charge lattice. About the physically genuine, 3-dimensional Madelung constant we shall have more to say later. For the moment, we note that 2-dimensional zeta evaluations have been taken yet further, for example the Madelung problem on an hexagonal lattice has been solved, in the sense of exact evaluation. The relevant hexagonal sum is taken to be:

$$H(s) = \sum_{m,n \in Z} {}'\frac{\left(\frac{n-m+1}{3}\right)}{((n + m/2)^2 + 3(m/2)^2)^{s/2}},$$

where $\left(\frac{x}{3}\right)$ is the Legendre symbol, which runs $1, -1, 0, 1, -1, 0, ...$ as x runs $1, 2, 3, 4, 5, 6, ...$. Now $H(x)$ can be written as a superposition of four complex Epstein zetas. However, it is argued in [8] that one can instead evaluate just two (real) zetas:

$$H(s) = \frac{1 - 3^{1-s/2}}{2}(2Z_A(s; 0_2, 0_2) - Z_A(s; (\tfrac{1}{2})_2, 0_2)),$$

with matrix

$$A = \{\{1, 1/2\}, \{0, \sqrt{3}/2\}\},$$

and by such means arrive at the exact value

$$H(s) = 3(1 - 3^{1-s/2})\zeta(s/2)L_{-3}(s/2),$$

so that the "Madelung" value is $H(1)$, while the case $s = 2$ comes down to the attractive result $H(2) = \pi \ln 3\sqrt{3}$. Beyond this, but still in two dimensions, sums with denominators $(m^2 + |d|n^2)^{s/2}$ have been evaluated, through adroit application of number-theoretical and transform techniques, for certain classes of discriminant d [37]. An example of the beauty of some of the known two-dimensional Epstein zeta evaluations is Zucker's sum:

$$\sum{}' \frac{(-1)^{m+n}}{(4m-1)^2 + 13(4n-1)^2} = \frac{\pi}{16\sqrt{13}} \log \frac{(1+\sqrt{2})^3}{5+\sqrt{26}}.$$

Further dimensional generalizations include the following collection for $D = 4$, starting with an attractive chain of evaluations [56] [37] [32]:

$$\log 2 = -\frac{1}{4}Z_{1_4}(2, (\tfrac{1}{2})_4, 0_4) = -\frac{1}{4} \sum_{\{a,b,c,d\}\in Z^4} \frac{(-1)^{a+b+c+d}}{a^2+b^2+c^2+d^2}$$

$$= -\frac{\pi}{2} - Z_{1_4}(2; ((\tfrac{1}{2})_3, 0), 0_4)$$

$$= -\frac{1}{2}Z_{1_4}(2; ((\tfrac{1}{2})_2, 0_2), 0_4)$$

$$= \frac{\pi}{2} - Z_{1_4}(2; (\tfrac{1}{2}, 0_3), 0_4)$$

$$= -\frac{1}{8}Z_{1_4}(2; 0_4, 0_4).$$

Some of which being instances of the following:

$$Z_{1_4}(s; (\tfrac{1}{2}, 0_3), 0_4) = 4\beta(s/2)\beta(s/2-1) - 2^{3-s}\eta(s/2)\eta(s/2-1),$$

$$Z_{1_4}(s; ((\tfrac{1}{2})_3, 0), 0_4) = -4\beta(s/2)\beta(s/2-1) - 2^{3-s}\eta(s/2)\eta(s/2-1),$$

$$Z_{1_4}(s; 0_4, ((\tfrac{1}{2})_3, 0)) = 2^s(\lambda(s/2)\lambda(s/2-1) - \beta(s/2)\beta(s/2-1)),$$

$$Z_{1_4}(s; 0_4, (\tfrac{1}{2}, 0_3)) = 2^s(\lambda(s/2)\lambda(s/2 - 1) + \beta(s/2)\beta(s/2 - 1)),$$

$$Z_{1_4}(s; ((\tfrac{1}{2})_3, 0), (0_3, \tfrac{1}{2})) = 2^s(L_{-8}(s/2)L_{-8}(s/2 - 1) + L_{+8}(s/2)L_{+8}(s/2 - 1)),$$

$$Z_{1_4}(s; (\tfrac{1}{2}, 0_3), (0, (\tfrac{1}{2})_3)) = 2^s(L_{-8}(s/2)L_{-8}(s/2 - 1) - L_{+8}(s/2)L_{+8}(s/2 - 1)),$$

Then there are some isolated evaluations in even higher dimension, such as the beautiful 6-dimensional result of Zucker:

$$\sum{}' \frac{(-1)^{a+b+c+d+e+f}}{(a^2 + 3b^2 + 3c^2 + 3d^2 + 3e^2 + 3f^2)^{3/2}} = -\frac{2\pi\zeta(3)}{3\sqrt{3}} - \frac{4\pi^3 \log 2}{81\sqrt{3}},$$

and an 8-dimensional result:

$$Z_{1_8}(8; (\tfrac{1}{2})_8, 0_8) = \sum_{n \in Z^8}{}' \frac{(-1)^{n_1+...+n_8}}{(n_1^2 + ... + n_8^2)^4} = -\frac{8\pi^4 \log 2}{45}.$$

For the remainder of this section we focus on $D = 3$, which cases include evaluations applicable in the physical sciences.

10.3   Madelung constant

The celebrated Madelung constant of chemistry and physics is a 3-dimensional construct

$$M = Z_{1_3}(1; (\tfrac{1}{2})_3, 0_3) = \sum_{(x,y,z) \in Z^3}{}' \frac{(-1)^{x+y+z}}{\sqrt{x^2 + y^2 + z^2}},$$

which has never been cast into any convenient closed form.[4] It is of interest that this 3-dimensional setting is in many ways more difficult than any of the 2,4,8-dimensional settings previous. This discrepancy could be thought of as a relative paucity of relations for odd powers of Jacobi theta functions.[5] It should be remarked, however, that literally dozens of three-dimensional Epstein zetas admit of closed-form evaluation, as in [57] [58], yet none of these is precisely the Madelung M. Nevertheless, using the method of the present treatment—which amounts to a long-known "Ewald expansion" for such a crystal energy—one is able to obtain numerical values such as

$$M \sim -1.74756459463318219063621203554439740348516143662474175815282535076\ldots$$

---

[4]There is an attempt to do so in [27], where a few closed-form terms come rather close to the known M value.

[5]Although, the present author does use in [27] the compelling Andrews identity for θ^3, and such a procedure results in a *finite-domain definite integral* for M.

in a short time, and, in practice, carry out such numerics to thousands of decimal places. Incidentally one reason for attaining such precision would be somewhat more than recreational: The Madelung constant enjoys relations with other Z_A evaluations, for example the following nontrivial connections are known [55] [58] [33]:

$$M = \frac{3}{\pi} Z_{1_3}\left(2; \left(\left(\tfrac{1}{2}\right)_2, 0\right), 0_3\right) = \frac{3}{\pi} \sum_{(x,y,z) \in Z^3}{}' \frac{(-1)^{x+y}}{x^2 + y^2 + z^2}$$

$$= \frac{3}{\pi} \sum_{(x,y,z) \in Z^3}{}' \frac{(-1)^{x+y+z}}{x^2 + 2y^2 + 2z^2}$$

$$= \frac{6}{\pi} \sum_{(x,y,z) \in Z^3}{}' \frac{(-1)^{x}}{x^2 + y^2 + 2z^2}$$

$$= \frac{1}{\pi} Z_{1_3}\left(2; 0_3, \left(\tfrac{1}{2}\right)_3\right) = \frac{4}{\pi} \sum_{(x,y,z) \in (2Z+1)^3} \frac{1}{x^2 + y^2 + z^2},$$

this last form defined over all odd triples (and, as always, as an analytic continuation since the literal sum does not converge).

There are other interesting approaches to precise estimation of M. Indeed, J. Buhler and R. Crandall [32] showed that one need not necessarily invoke Riemann splitting to achieve a rapidly convergent series. For example,

$$M = -\lambda + \sum_{\vec{v} \in Z^3}{}' \frac{(-1)^{\sum v_i}}{|\vec{v}|}(1 - \tanh \lambda|\vec{v}|) + \frac{2\pi}{\lambda} \sum_{\vec{u} \in O^3} \frac{\operatorname{cosech}(\pi^2|\vec{u}|/(2\lambda))}{|\vec{u}|},$$

where O^3 denotes the odd-integers 3-dimensional lattice, and λ is a free parameter. It follows that

$$M = \lim_{\lambda \to \infty} \left(-\lambda + \frac{2\pi}{\lambda} S\left(\frac{\pi}{2\lambda}\right)\right)$$

where

$$S(x) := \sum_{\vec{u} \in O^3} \frac{\operatorname{cosech}(\pi x|\vec{u}|)}{|\vec{u}|}.$$

Through the efforts of S. Tyagi and I. J. Zucker, S values are known exactly for certain arguments. The deepest known such result is Zucker's $S(1/\sqrt{24})$, which gives a Madelung estimate as the argument of lim above,

$$M \approx -\pi\sqrt{6} + \frac{\sqrt{\tfrac{1}{6}\left(3\sqrt{2} - 6\sqrt{3} + 6\sqrt{6}\right)}\,\Gamma\left(\tfrac{1}{8}\right)^2}{\pi\,\Gamma\left(\tfrac{1}{4}\right)} = -1.74756\ldots,$$

47

which is correct to the implied precision. For some ruminations on hyperclosed numbers and how the mighty Madelung might fit into the universe of fundamental constants, see [15].

One of the mysteries in regard to the Madelung M is the lack of closed-form representation *in spite of* the existence of closed forms for related entities. There is, for example, the attractive relation:

$$\sideset{}{'}\sum \frac{3(-1)^x + 3(-1)^{x+y} + (-1)^{x+y+z}}{(x^2 + y^2 + z^2)^{3/2}} = -4\pi \log 2,$$

which can be cast as an identity involving $Z_{1_3}(3, b/2, 0_3)$ for a certain three binary vectors b.

In any test suite for numerical evaluation, all the above are useful checking relations. We have seen – for the hexagonal lattice sum $H(s)$–a nontrivial A matrix. Another case of nontrivial A, in fact for the matrix $A = diag(\{2, \sqrt{2}, \sqrt{2}\})$, is:

$$Z_A(s, (\frac{1}{4}, 0_2), (\frac{1}{2}, 0_2)) = \sum_{(x,y,z)\in Z^3} \frac{(-1)^m}{((2x - \frac{1}{2})^2 + 2y^2 + 2z^2)^{s/2}} = 2^s L_{-8}(s - 1).$$

One may also test further some nontrivial instances of c, d parameters such as:

$$Z_{1_3}(s, (\frac{1}{2}, 0_2), (0_2, \frac{1}{2})) = \sum_{(x,y,z)\in Z^3} \frac{(-1)^x}{(x^2 + y^2 + (z - \frac{1}{2})^2)^{s/2}} = 2^{s+1}\beta(s - 1).$$

In particular, setting $s = 2$ yields the Catalan constant $G = \beta(2) = 1 - 1/3^2 + 1/5^2 - 1/7^2 + ...$ as:

$$G = \frac{1}{16} Z_{1_3}(3, (\frac{1}{2}, 0_2), (0_2, \frac{1}{2})),$$

while for $s = 1$ we would obtain the exact potential energy at a certain point within a peculiar, alternating charge-plane crystal. Another example of nontrivial—but physically meaningful—parameters is the following attractive evaluation of [36], which can be thought of as the potential energy at the point $(x, y, z) = (\frac{1}{6}, \frac{1}{6}, \frac{1}{6})$ within the (sodium chloride) lattice of the basic Madelung problem:

$$Z_{1_3}(1, (\frac{1}{2})_3, (\frac{1}{6})_3) = \sum_{(x,y,z)\in Z^3} \frac{(-1)^{x+y+z}}{\sqrt{(x - \frac{1}{6})^2 + (y - \frac{1}{6})^2 + (z - \frac{1}{6})^2}} = \sqrt{3}.$$

Of course, this last formula is *not* the most efficient way to calculate the square root of three!

References

[1] Ambjorn J and Wolfram S, "Properties of the Vacuum I. Mechanical and Thermodynamic," Annals of Physics, 147, 1 (1983).

[2] E. Balanzario and J. Sanchez-Ortiz, "Riemann–Siegel Formula for the Lerch Zeta Function," Math. Comp., 29 Nov 2011,
http://www.ams.org/journals/mcom/0000-000-00/
S0025-5718-2011-02566-4/S0025-5718-2011-02566-4.pdf

[3] D. Bailey, J. Borwein, and R. Crandall "On the Khintchine Constant," *Math. Comp.*, **66**, 417–431 (1997).

[4] D. Bailey, J. Borwein, and R. Girgensohn, "Experimental evaluation of Euler sums," Exp. Math. 3, Issue, 17-30 (2004).

[5] D. Bailey, D. Borwein, and J. Borwein, "On Eulerian Log-Gamma Integrals and Tornheim–Witten Zeta Functions," preprint (2012).

[6] H. Bateman and A. Erdélyi, *Higher Transcendental Functions*, Vol. I, New York: McGraw-Hill, 1952.

[7] D. Borwein, P. Borwein, and A. Jakinovski, "An efficient algorithm for the Riemann Zeta function," CECM preprint server,
http://www.cecm.sfu.ca/preprints/1995pp.html, CECM-95-043. (1995)

[8] Borwein J and Borwein P, *Pi and the AGM*, John Wiley & Sons, New York (1997).

[9] D. Borwein, J. Borwein, and R. Crandall, "Effective Laguerre Asymptotics," SIAM J. Numerical Anal., 6, 285-3312 (2008).

[10] J. Borwein, D. Bailey, and R. Girgensohn, *Experimentation in mathematics: computational paths to discovery*, AK Peters (2004).

[11] J. Borwein, personal communication (2012).

[12] J. Borwein, D. Bradley, and R. Crandall, "Computational strategies for the Riemann zeta function," J. Comp. App. Math., 121, 247-296 (2000).

[13] J. Borwein and D. Broadhurst, "Determinations of rational Dedekind-zeta invariants of hyperbolic manifolds and Feynman knots and links," preprint (19 Nov 1998).
arXiv:hep-th/9811173v1

[14] J. Borwein, D. Bradley, D. Broadhurst, P. Lisonek, "Special Values of Multiple Polylogarithms," (8 Oct 1999).
http://arxiv.org/abs/math/9910045

[15] J.Borwein and R.Crandall, "Closed forms: what they are and why we care," Notices Amer. Math. Soc. (2010, to appear). Reprinted in similar form: J. Borwein and P. Borwein, *Experimental and computational mathematics: Selected writings*, PSIpress (2010).

[16] R. P. Brent and P. Zimmermann, *Modern Computer Arithmetic*, Cambridge Monographs on Computational and Applied Mathematics (No. 18), Cambridge University Press (2010).

[17] R. P. Brent, "Multiple-precision zero-finding methods and the complexity of elementary function evaluation, in Analytic Computational Complexity," (edited by J. F. Traub), Academic Press, New York, 51-176 (1975).

[18] R. P. Brent, personal communication (2012).

[19] R. P. Brent, "Some new algorithms for high-precision computation of Euler's constant," at http://maths.anu.edu.au/ brent/pub/pub049.html

[20] J. Buhler, R. Crandall, and R. Sompolski, "Irregular primes to one million," Math. Comp., Vol. 59, no. 200, pp. 717-722 (1992).

[21] H. Cohen, F. Villegas, and D. Zagier, "Convergence Acceleration of Alternating Series," *Exp. Math.*, 9, 3-12 (2000).

[22] M. W. Coffey, "On some series representations of the Hurwitz zeta function," J. Comp. Appl. Math., 216, 297-305 (2008).

[23] M. W. Coffey, "Series representations for the Stieltjes constants," Rocky Mtn. J. Math., to appear (2011).

[24] M. W. Coffey, arXiv papers:
arXiv:1106.5148 "Hypergeometric summation representations of the Stieltjes constants"
arXiv:1106.5147 "Sums of alternating products of Riemann zeta values and solution of a Monthly problem"
arXiv:1106.5146 "Series representations of the Riemann and Hurwitz zeta functions and series and integral representations of the first Stieltjes constant"
arXiv:1101.5722 "Evaluation of some second moment and other integrals for the Riemann, Hurwitz, and Lerch zeta functions"
arXiv:1101.4257 "Fractional part integral representation for derivatives of a function related to $\ln \Gamma(x+1)$"
arXiv:1008.0040 "Integral and series representations of the digamma and polygamma functions"
arXiv:1008.0037 "Double series expression for the Stieltjes constants"

[25] R. Crandall, "Alternatives to the Riemann–Siegel formula," manuscript (2011).

[26] R. Crandall and C. Pomerance, *Prime numbers: A computational perspective*, 2nd Ed., Springer, New York (2002).

[27] Crandall, R. E. "New representations for the Madelung constant," Exp. Math. 8:4, 367-379 (1999).

[28] R. Crandall and J. Buhler, "On the evaluation of Euler sums," Exp. Math., 3:4, 275-285 (1995).

[29] R. Crandall. "Fast evaluation of multiple zeta sums," Math. Comp., 67, 223, 1163-1172 (1996).

[30] R. Crandall, *Topics in Advanced Scientific Computation*, Springer-Verlag, New York (1996).

[31] Crandall, R., "Theory of log-rational integrals," Contemp. Math., Vol. 517, pp. 127–142 (2010).

[32] J. Buhler and R. Crandall, "Elementary function expansions for Madelung constants," J. Phys. A.: Math. Gen, 5497-5510 (1986).

[33] R. Crandall and J. Delord, "The potential within a crystal lattice," J. Phys. A.: Math. Gen, 20, 2279-2292 (1987).

[34] D. Cvijovic and J. Klinowski, "Continued-Fraction Expansions for the Riemann Zeta Function and Polylogarithms," Proc. Amer. Math. Soc., 125, 2543-2550 (1997).

[35] A. Erdélyi, et al., *Higher Transcendental Functions*, Vol. 1, New York: Krieger.

[36] P. Forrester and M. Glasser, "Some new lattice sums including an exact result for the electrostatic potential within the NaCl lattice," J. Phys. A.: Math. Gen. 15, 911-914 (1982).

[37] M. Glasser M and I. J. Zucker, "Lattice Sums," Theor. Chem.: Adv. Persp., 5, 67-139 (1980).

[38] C. Knessl and M. Coffey, "An Asymptotic Form for the Stieltjes Constants $\gamma_k(a)$ and for a Sum $S_\gamma(n)$ Appearing Under the Li Criterion," Math. Comp. 276, 2197-2217 (2011).

[39] C. Knessl and M. Coffey, "An Effective Asymptotic Formula for the Stieltjes Constants," Math. Comp. 273, 379-386 (2011).

[40] J. C. Lagarias and W-C Winnie Li, "The Lerch Zeta Function I: Zeta Integrals," http://arxiv.org/pdf/1005.4712v2

51

[41] J. C. Lagarias and W-C Winnie Li, "The Lerch Zeta Function II: Analytic Continuation,"
http://arxiv.org/abs/1005.4967

[42] J. Lagarias and A. Odlyzko, "Computing $\pi(x)$: an analytic method," J. Algorithms, 8:173–191 (1987).

[43] A. Laurincikas and R. Garunkstis, *The Lerch Zeta Function*, Kluwer, 2002.

[44] L. Lewin, *Polylogarithms and Associated functions*, North Holland (1981).

[45] D. J. Platt, "Computing Degree 1 L-functions Rigorously," Thesis, Univ. Bristol (2011).

[46] G. Pugh, "An Analysis of the Lanczos Gamma Approximation," Thesis, Univ. British Columbia (2004).

[47] http://en.wikipedia.org/wiki/Polylogarithm

[48] J. Reyna, "High Precision Computation of Riemann's Zeta Function by the Riemann–Siegel Formula," Math. Comp. 80, 274, 995-1009 (2011).

[49] M. Rubinstein, "Computational Methods and Experiments in Analytic Number Theory," (2004).
http://arxiv.org/pdf/math/0412181v1.pdf

[50] E. Weestein (2012).
http://mathworld.wolfram.com/LogGammaFunction.htm

[51] E. Weisstein (2012).
http://mathworld.wolfram.com/MRBConstant.html

[52] E. Weisstein (2012).
http://mathworld.wolfram.com/KhinchinsConstant.html

[53] L. Vepstas, "Notes Relating to Newton Series for the Riemann Zeta Function," manuscript (2006).
http://linas.org/math/norlund-l-func.pdf

[54] A. Yee, "Euler–Mascheroni Constant—116 million digits on a laptop," manuscript (2009).
http://numberworld.org/euler116m.html

[55] I. J. Zucker, "Functional equations for poly-dimensional zeta functions and the evaluation of Madelung constants," J. Phys. A.: Math. Gen., 9, 499-505 (1976).

[56] I. J. Zucker, "Some Infinite Series of Exponential and Hyperbolic Functions," SIAM J. Math., 15, 2, 406-413 (1984).

[57] I. J. Zucker, "A systematic way of converting infinite series into infinite products," J. Phys. A.:Math. Gen., 20, L13-L17 (1987).

[58] I. J. Zucker, "Further relations amongst infinite series and products: II. The evaluation of three-dimensional lattice sums," J. Phys. A.:Math. Gen., 23, 117-132 (1990).

Part III

Physics, biology, epidemics, and physiology

"Since first seeing neurons clearly over a century ago, humans have been struck by the beauty of these tree-like objects. Perhaps the species' recent evolutionary stint as arboreal primates has hard-wired us to love the sight of trees and forests! Thus, as we begin to see not only the brain's trees, but its vastly intricate and diverse forests, we may find that the most beautiful sights of all really are within."

—Stephen J. Smith

1. The potential within a crystal lattice

Discussion

This paper is—to this author—a satisfying mix of intuitive physics and algorithmic notions. I recall the day when my eminent colleague J. F. Delord explained that the elusive—still mysterious— Madelung constant is the potential energy of a *single* positive charge residing at the center of a unit, grounded conducting cube. That is, the infinite NaCl lattice serves to create six faces of dead-zero potential on such a cube. Section 4 of the enclosed paper outlines such intuitive arguments.

As for algorithmic issues, formulae (6.2)-(6.8) and (1.3) cover the most general periodic crystal as discussed in Section 2: Crystal nomenclature. The formulae will yield generalized Madelung constants to high precision. (See also the section on lattice sums in "Unified algorithms for polylogarithms, *L*-series, and zeta variants," elsewhere in this collection.)

Challenges

Many challenges abound. The Madelung constant M remains unresolved in anything like a closed form. The power series (7.7) has not been studied enough; note that it represents essentially a perturbed-quartic potential, and (7.9) can perhaps be made more precise.

Source

R. Crandall, and J. Delord, "The potential within a crystal lattice," *J. Phys. A: Math. Gen.*, **20**, 2279–2292 (1987).

J. Phys. A: Math. Gen. **20** (1987) 2279-2292. Printed in the UK

The potential within a crystal lattice

R E Crandall and J F Delord

Department of Physics, Reed College, Portland OR 97202, USA

Received 19 August 1986

Abstract. The generalised Madelung problem, that of finding the spatial electric potential within a crystal lattice, is often approached via analytic function theory. But new results can be achieved through careful application of standard theorems from classical electrostatics. In this way we show that the celebrated Madelung constant for the NaCl crystal can be rigorously bounded through symmetry arguments devoid of summations. We derive new expansions for Madelung constants of general crystals: an absolutely convergent 'sine' series which has the advantage of non-alternating summands, and an 'exponential' series which agrees in the simple cubic cases with the 'cosech' expansions of previous authors. Finally, we show how the NaCl Coulomb singularity may be removed to yield a regular power series expansion for the potential well at the origin.

1. Introduction

The electrostatic potential at a vector x within a periodic point structured crystal is given formally by a sum over Coulomb terms

$$V(x) = \sum_p q(p)|p - x|^{-1} \qquad (1.1)$$

where p generally denotes the location of point charge $q(p)$. We shall call this simply the *crystal potential*. A closely related construct describes the potential well in which a particular charge resides. This *Madelung potential*, indexed by some point charge location r, is given formally by a sum in which the self-potential of the reference charge is removed:

$$V_r(x) = V(x) - q(r)|r - x|^{-1}. \qquad (1.2)$$

The electrostatic energy binding a particular site to the rest of the crystal is thus $q(r)V_r(r)$. This potential can be used to determine the electrostatic potential energy of the crystal. When the crystal is structured generally as a periodic lattice with identical charge assemblies at each lattice site, a *Madelung constant* can be determined by summing and normalising the potential energies associated with a single cell:

$$M = v^{-1} \sum_r q(r)V_r(r) \qquad (1.3)$$

where v is the volume of a lattice cell and r runs over the charge sites associated with one cell. This definition of the Madelung constant is consistent with traditional dimensionless ones (Sherman 1932), provided some lattice constant, such as unit nearest-neighbour separation, is enforced. We discuss later the relation of M to total cell energy and other crystal energy measures.

0305-4470/87/092279 + 14$02.50   © 1987 IOP Publishing Ltd

In the case of the simple cubic crystal NaCl we adopt the common convention that nearest-neighbour separation is unity. For this and other highly symmetric crystals, M turns out to be the potential energy in which a unit origin charge resides. The Madelung constant in this case is just $V_0(0)$, and can be written out as a Coulomb sum over integer triplets, with the origin taken as reference site:

$$M_{\text{NaCl}} = \sum_{m,n,p \neq 0} (-1)^{m+n+p} (m^2 + n^2 + p^2)^{-1/2}. \qquad (1.4)$$

The difficulty in performing this and similar sums is well known, having occupied various researchers for more than seventy years (Madelung 1918, Ewald 1921, Born and Goppert-Mayer 1933, Born and Huang 1954, Glasser and Zucker 1980). Numerical values can be computed from transformations of (1.4) having exponentially decaying summands involving only elementary functions (Hautot 1975, Zucker 1976). General crystal sums suitable for computer work will appear later in this treatment. We compute the NaCl case as

$$M_{\text{NaCl}} = -1.747\,564\,594\,633\,182\,190\,636$$

$$212\,035\,544\,397\,403\,485\,161\,436\,624\,741\,758\,152\,672 \ldots \qquad (1.5)$$

presumed correct to 50, and possibly 60, places.

A simple thought experiment using fundamental electrostatic principles can relegate the NaCl Madelung problem to the confines of a finite space. We shall find that a bound on the Madelung constant M for this problem

$$-2 < M < -(4/3)^{1/2} \qquad (1.6)$$

can be obtained 'in one's head', without recourse to infinite summations. Using the principles in conjunction with a previous result for the NaCl crystal potential, namely (Forrester and Glasser 1982)

$$V(\tfrac{1}{6}, \tfrac{1}{6}, \tfrac{1}{6}) = \sqrt{3} \qquad (1.7)$$

one can obtain, again with virtually no calculation, a fair estimate

$$M \sim -\sqrt{3} \qquad (1.8)$$

which is in error by only one per cent of the value (1.5).

Using such principles with proper precision we can deduce some striking exact relations. We shall show that *for any t in the interval* $(0,1)$, the Madelung constant M for the NaCl crystal can be obtained from a non-alternating absolutely convergent series

$$8\pi^{-3} \sum_{r \in O^3} \sin^2(\pi rt/2)/r^4 = t + Mt^2/2 \qquad (1.9)$$

where the sum is taken over all distances r from the origin to odd integer triplets. We still do not know just how much one may discover about M by exploiting the free parameter t. But certainly this kind of series, with its lack of alternating signs, can be used to provide rigorous bounds tighter than (1.6).

We eventually arrive at efficient expansions for the various electrostatic quantities. These expansions are reminiscent of previous work (Zucker 1975, Glasser and Zucker 1980) on cubic crystals but can handle general crystals. In particular we derive the spatial behaviour of the potential well $V_0(x)$ in which the (positive) origin charge of NaCl sits. We provide formal expressions for the coefficients of the axial expansion

$$V_0(x, 0, 0) = M + ax^4 + bx^6 + \ldots . \qquad (1.10)$$

It is interesting that there is no quadratic term, as will follow again from electrostatic principles, so that the origin charge of NaCl sees at least a quartic, not a harmonic electrostatic well.

We have used such expansions in computer programs which will compute general crystal potentials and Madelung constants. Besides generating values such as (1.5), we have confirmed previous work (Hajj 1972) and discovered, by numerical accident, some provable theorems concerning exact potential values. For the CsCl crystal, an exact value similar to the previous (1.7) for NaCl can be given at a particular point in space:

$$V_{CsCl}(\tfrac{1}{4}, 0, 0) = 2. \tag{1.11}$$

We were given pause upon receiving the computed value 1.999 999 999 . . ., and indeed the exact value can be proved. There is also a similar result for a crystal consisting of parallel planes of like charge, i.e. the charge at (m, n, p) is given by $(-1)^m$. For this case it turns out that

$$V(0, 0, \tfrac{1}{2}) = 2. \tag{1.12}$$

A method for establishing such particular potential values depends upon previous applications of Jacobi theta functions, as discussed in § 7.

2. Crystal nomenclature

Our general crystal is taken to be built upon a periodic Bravais site lattice generated by replicated non-coplanar basis vectors A_1, A_2, A_3 such that the position vector for a lattice site (not necessarily a charge site) has three components

$$p_i = m_1 A_{1i} + m_2 A_{2i} + m_3 A_{3i} \qquad m_j \in Z. \tag{2.1}$$

Denote by **A** the matrix collection of all nine basis components. These components can be thought of either as spatial lengths, or dimensionless multiples of some lattice constant. Then for three columns m of integers we can write the general site column vector as

$$p = Am \qquad m \in Z^3. \tag{2.2}$$

It should be noted that the volume of one Bravais cell is given by the determinant, det **A**. It will be convenient to define the reciprocal crystal matrix whose basis components are essentially those of the reciprocal lattice:

$$B = A^{-T} \tag{2.3}$$

where −T refers to inverse transpose. Having defined the lattice, it remains to place at every site an assembly of point charges. We shall assume a neutral crystal, where at each site p we assemble n charges q_j at positions determined by *distinct* fixed offset vectors d_j:

$$p + d_j \qquad j = 0, \ldots n - 1 \tag{2.4}$$

where the neutrality condition is

$$\sum q_j = 0. \tag{2.5}$$

The general crystal is thus defined by the quantities

$$A, n, \quad \{d_j : j = 0, \ldots, n-1\}, \quad \{q_j : j = 0, \ldots, n-1\} \tag{2.6}$$

together with the constraint (2.5) and the distinctness of the d_j. Examples of this nomenclature for common crystal structures are

CsCl: $n = 2, d_0 = 0, d_1 = (\frac{1}{2}, \frac{1}{2}, \frac{1}{2}), q_0 = 1, q_1 = -1$ (2.7)

$$A = \begin{pmatrix} 1 & 0 & 0 \\ 0 & 1 & 0 \\ 0 & 0 & 1 \end{pmatrix}$$

ZnS: $n = 2, d_0 = 0, d_1 = (\frac{1}{2}, \frac{1}{2}, \frac{1}{2}), q_0 = 1, q_1 = -1$

$$A = \begin{pmatrix} 0 & 1 & 1 \\ 1 & 0 & 1 \\ 1 & 1 & 0 \end{pmatrix}$$

NaCl: $n = 2, d_0 = 0, d_1 = (1, 1, 1), q_0 = 1, q_1 = -1$

$$A = \begin{pmatrix} 0 & 1 & 1 \\ 1 & 0 & 1 \\ 1 & 1 & 0 \end{pmatrix}$$

Equivalent form:

$n = 8, \{d_j: n = 0, \ldots, 7\}$ = ordered set of eight binary triplets

$\{q_j: n = 0, \ldots, 7\} = \{+1, -1, -1, +1, -1, +1, +1, -1\}$

$$A = \begin{pmatrix} 2 & 0 & 0 \\ 0 & 2 & 0 \\ 0 & 0 & 2 \end{pmatrix}.$$

The second form for NaCl may seem unwieldy, but the simplicity of the **A** matrix represents a certain advantage in the analysis.

The extension of the ideas herein to handle charge distributions, such as electron orbitals combined with point nuclei, is straightforward. Though our treatment does not go beyond the generality of the point-charge crystal, we shall indicate in what follows those junctures where charge distributions would be appropriate to introduce.

We shall often perform summations where some singularity must be avoided. A symbol such as

$$\sum_{m \in S}{}'$$ (2.8)

means to sum over all vectors m in the set S except to allow only finite summands (because of the $'$ superscript). Note that $'$ does *not* mean simply to avoid the zero element of S. A simple example of this singularity removal is as follows. Let

$$g(u) = \sum_{m \in Z}{}' (m - u)^{-2}$$ (2.9)

be defined for real u, and denote by $f(u)$ the same sum but without the $'$ superscript. Now almost everywhere $f = g$, but whereas $f(u)$ diverges for any integer value of u, the sum for g becomes just $\pi^2/3$ at such u, because the diverging term 0^{-2} is to be removed.

Potential within a crystal lattice   2283

At several points in our analysis we shall use the Poisson transformation method, which has enjoyed wide use in lattice summation problems (Hautot 1975, Zucker 1976). For suitable complex valued functions f defined on R^3, the transformation is

$$\sum_{m \in Z^3} f(m) = \sum_{u \in Z^3} \int f(m) \exp(2\pi i m u) \, d^3 m. \tag{2.10}$$

The integral is performed over a continuous 3-vector m even though the left-hand sum is over integer triplets. Here and elsewhere, a form such as mu is interpreted as a dot product.

3. Electrostatics

The above nomenclature will now be used to establish various electrostatic quantities. The *crystal potential* at any point x is obtained by summing Coulomb terms over each charge in the crystal, knowing that the vector from x to the jth charge at site (m_1, m_2, m_3) is $Am + d_j - x$:

$$V(x) = \sum_{m \in Z^3} \sum_{j=0}^{n-1} q_j |Am + d_j - x|^{-1}. \tag{3.1}$$

The question of convergence naturally arises, since for no crystal is the sum absolutely convergent. We do not handle such questions here, referring the reader to other work on convergence theorems over polyhedral domains (Borwein *et al* 1985) and Coulombic convergence corrections for spherical domains (Crandall and Buhler 1987b). At this point we indicate that if the charges were more general than mere points, the relevant spatial integrals must be used in place of the q_j. This crystal potential V diverges naturally as the position x approaches any charge location $r = Am + d_j$, with behaviour

$$V(x) \sim q_j |r - x|^{-1}. \tag{3.2}$$

It is now evident that the Madelung potential (1.2) at any site is just the sum (3.1) with the first summation given a prime superscript. Accordingly, the energy density for the crystal is

$$M = (\det A)^{-1} {\sum_{m \in Z^3}}' \sum_{k=0}^{n-1} \sum_{j=0}^{n-1} q_k q_j |Am + d_j - d_k|^{-1}$$

$$= (\det A)^{-1} \sum_{k=0}^{n-1} M_k \tag{3.3}$$

where $M_k = q_k V_{d_k}(d_k)$ represents the building energy for the kth charge of the cell. Expression (1.4) for the NaCl constant can be obtained directly from (3.3) together with the second definition in (2.7) for the crystal.

The constant M is not to be confused with the total electrostatic energy per cell, which is given by

$$U = \tfrac{1}{2} {\sum_{k=0}^{n-1}}' q_k V_{d_k}(d_k) \tag{3.4}$$

$$= \tfrac{1}{2} \sum_{k=0}^{n-1} M_k \tag{3.5}$$

2284   *R E Crandall and J F Delord*

with the factor of $\frac{1}{2}$ arising from pairwise combinations for the Coulomb energy. The energy U depends via the **A** matrix upon the definition of the unit cell, whereas M is independent of the choice of **A** as long as the crystal structure is unchanged. Yet another measure of crystal energy is Sherman's (1932) per ion measure, namely $E = (2*U)/n$. For example, across the two definitions of the same NaCl crystal in (2.7), U varies but M and E do not. Whether to use U, E or $M = 2U/\det \mathbf{A} = nE/\det \mathbf{A}$ depends upon context. For NaCl, using the first definition (2.7), we have $n = \det \mathbf{A} = 2$, so $M = U = E$ in this case. If the second representation for NaCl is adopted, however, the cell has become larger since $n = \det \mathbf{A} = 8$, and we have $M = E = U/4$. For general crystals, one may avoid the issue of what measure is best by reverting to calculation of the separate 'Madelung constants' M_k to settle any energy problem involving the crystal.

The electrostatic analysis begins with consideration of the exact charge density of the crystal, keeping in mind the potential law:

$$\nabla^2 V(x) = -4\pi\rho(x). \tag{3.6}$$

The charge density is given formally by placing delta functions at all sites:

$$\rho(x) = \sum_j q_j \sum_{m \in Z^3} \delta^3(\mathbf{A}m + d_j - x). \tag{3.7}$$

This can be immediately Poisson transformed to give

$$\rho(x) = (\det \mathbf{A})^{-1} \sum_j q_j \sum_{u \in Z^3} \exp[-2\pi\mathrm{i}(\mathbf{B}u)(d_j - x)]. \tag{3.8}$$

Note that if the charges were spatially distributed and not mere points, then the transform can still be performed in principle, with a Fourier transform of the site charge density appearing in (3.8).

Now the Poisson equation (3.6) can be solved by finding that summation whose Laplacian in x is proportional to (3.8). The difficulty is that in the resulting sum for V a denominator $|\mathbf{B}u|^2$ causes the u sum to diverge at $u = (0, 0, 0)$. This can be traced to the requirement of crystal neutrality: the constraint (2.5) on the total charge actually removes this divergence. A way to solve this is to place a primed sum as the second summation in (3.8). This can be done with impunity, again because of the constraint $\Sigma q_j = 0$.

Yet another equivalent way to cancel the infinity is to note that (3.6) can only be solved up to the ambiguity of an additive constant for V. As the 'infinite constant' obtained by direct solution is devoid of any crystal structure details, it can safely be removed. In any case we obtain a legitimate expression for the crystal potential:

$$V(x) = (\pi \det \mathbf{A})^{-1} \sum_j q_j \sum_{u \in Z^3}' |\mathbf{B}u|^{-2} \exp[-2\pi\mathrm{i}(\mathbf{B}u)(d_j - x)]. \tag{3.9}$$

This is valid for the general crystal, and forms the basis of many useful expansions. Such formulae are often derived in the literature by analytic continuation of Epstein zeta functions (Emersleben 1923), of which (3.1) and (3.9) are complementary special cases. Such analytic methods give us (3.9) from (3.1) directly, but already in passing through the electrostatic considerations we see a portent of simple physical arguments to come. For the moment, we write out some examples of (3.9). For NaCl one obtains

$$V_{\mathrm{NaCl}}(x) = 4\pi^{-1} \sum_{u \in O^3} \exp(\pi\mathrm{i}ux)/u^2. \tag{3.10}$$

Potential within a crystal lattice   2285

This being a sum over odd triplets, the prime notation has been safely removed. For the CsCl crystal, the potential is

$$V_{CsCl}(x) = 2\pi^{-1} \sum_{u_1+u_2+u_3 \, odd} \exp(2\pi i u x)/u^2 \qquad (3.11)$$

where, again, the prime notation is not required since the u origin is ruled out.

We shall make use of another principle of electrostatics as embodied in the Green theorem (Griffiths 1981) that for suitable functions V and W defined on and inside a closed surface S,

$$\int_{volume} (V\nabla^2 W - W\nabla^2 V) \, d^3 R = \int_{surface} (V\nabla W - W\nabla V) \, d^2 R. \qquad (3.12)$$

If V describes an electrostatic potential within a sphere S which happens to be devoid of interior and surface charge, then the choice $W = 1/R$ gives the familiar result that the value of V at the centre of S equals the average of V over the surface of S. Recall that the Madelung potential V_r referring to charge site r is computed by ignoring $q(r)$, so we conclude that V_r may in general be computed as an average over a sufficiently small sphere centred at r. This principle will be used to derive formulae such as (1.9).

4. Direct bounds

Symmetry principles are directly applicable to simple crystals such as NaCl. As suggested by (1.4) or (2.7) the charge at a lattice point m is

$$q(m) = (-1)^{m_1+m_2+m_3}. \qquad (4.1)$$

It is easy to see that each of the infinite planes $x = \frac{1}{2}$, $y = \frac{1}{2}$, or $z = \frac{1}{2}$ is a surface of zero potential, for charge values are antisymmetric when reflected through such planes. Looked at from the point of view of the expansion (3.10), this phenomenon follows from the observation that the real part of $\exp(\pi i k/2)$ vanishes for odd integers k. Therefore the unit cube centred at the origin with vertices $(\pm\frac{1}{2}, \pm\frac{1}{2}, \pm\frac{1}{2})$ is a closed surface of zero potential. This 'neutral cube' therefore represents a finite domain with known boundary conditions for which a complete solution to the electrostatic problem would yield also the Madelung constant for NaCl.

The constant M_{NaCl} can now be estimated in many ways. One has only to solve this problem: a solitary $+1$ charge resides at the centre of a grounded unit cube. What is the electrostatic energy of the configuration? This will be the Madelung constant. An equivalent question is: what is the interior surface charge density of the cube? Let σ represent said density. Then clearly the surface integral of σ will balance the origin charge:

$$\int_{cube} \sigma \, d^2 a = -1 \qquad (4.2)$$

while the Madelung constant will be the potential seen by the origin charge

$$\int_{cube} (\sigma/r) \, d^2 a = M. \qquad (4.3)$$

Evidently, then, M is bounded by the extreme possibilities for the integral (4.3), namely

$$-1/r_{min} < M < -1/r_{max} \qquad (4.4)$$

where r_{min}, r_{max} are the minimum and maximum distances, respectively, from the origin to the cube surface. Immediate calculation of these radii gives the bound (1.6).

A second way of analysing the neutral cube problem is to attempt to solve the Laplace equation for the Madelung potential

$$\nabla^2 V_0(\boldsymbol{x}) = 0 \tag{4.5}$$

subject to the boundary condition that the potential at a point on the cube surface is $1/r$. If one attempts an eigenfunction expansion solution, one obtains the sum (3.10) or a transformation of same. But there are numerical methods for solving the Laplace equation. Performing a standard neighbour-relaxation pass of a $31 \times 31 \times 31$ point cube yields a stable estimate at the origin of

$$M_{NaCl} \sim -1.748\,79\ldots. \tag{4.6}$$

Unfortunately such a method does not converge nearly as fast as the best exponential sums for M.

Still another approximation method for NaCl is to recall the sphere-averaging theorem discussed after (3.12). Note on the basis of (1.7) and the coordinate symmetry of the crystal that the eight points $(\pm\frac{1}{6}, \pm\frac{1}{6}, \pm\frac{1}{6})$ all sit at the potential $\sqrt{3}$. It is reasonable to estimate that the crystal potential on the sphere of radius $1/\sqrt{12}$, which passes through all eight points, is roughly $\sqrt{3}$. Thus for points $\boldsymbol{x}$ lying on that sphere, the Madelung potential is roughly

$$V_0(\boldsymbol{x}) \sim \sqrt{3} - |\boldsymbol{x}|^{-1} = -\sqrt{3} \tag{4.7}$$

and we obtain the estimate (1.8) for M_{NaCl}.

It is interesting that these arguments for the NaCl crystal appear not to extend readily to any other structure, not even the simple CsCl crystal. As in (2.7) we assume a positive origin charge. Analysis of the problem indicates that there is no finite neutral surface enclosing this charge. All that we have been able to show is that the crystal potential for CsCl vanishes on the six squares (thought of as wire frames and hence one dimensional):

$$x = \pm\tfrac{1}{4} \qquad |y| + |z| = \tfrac{1}{2} \tag{4.8}$$

with the remaining four obtained by permuting coordinates. This follows from symmetry considerations applied to the sum (3.11). It is true that all points on all squares lie on the Wigner–Seitz polyhedron for the crystal, but alas that polyhedron is not a neutral surface. The result (1.11) was prompted by the numerical results obtained in the course of the futile search for a closed neutral surface.

The NaCl estimate arising from (4.3) can be made more precise through more knowledge of the surface density σ. Later results, such as formula (7.1) for the exact Coulombic potential, yield an exact description of the density on the neutral cube surface, where we may take $x = \frac{1}{2}$, y and z ranging from $-\frac{1}{2}$ to $+\frac{1}{2}$:

$$\sigma(y, z) = -\frac{1}{2} \sum_{n,p\,\text{odd}} \text{sech}\ \pi R/2 \cos \pi ny \cos \pi pz \tag{4.9}$$

where $R = (n^2 + p^2)^{1/2}$. Application of the integral in (4.3) yields then a representation for the Madelung constant reminiscent of the Benson–Mackenzie sech^2 form (Glasser and Zucker 1980).

Potential within a crystal lattice   2287

5. Sine series for general crystals

We may apply the sphere-averaging principle, as a special case of (3.12), to obtain an expression for the general Madelung potential at a charge site *r*. In fact, if *S* is a sphere centred at *r* with radius *t*, and there is no other point charge in *S*, then the Madelung potential is given by the surface average:

$$V_r(r) = (4\pi t^2)^{-1} \int_S V_r(x) \, \mathrm{d}^2 x. \tag{5.1}$$

Inserting the expansion (3.9) for the crystal potential we find

$$V_r(r) = -q(r)/t + (2\pi^2 t \det \mathbf{A})^{-1} \sum_j q_j \sum_{u \in Z^3}{}' \frac{\exp[-2\pi \mathrm{i}(\mathbf{B}u)(d_j - r)] \sin 2\pi t |\mathbf{B}u|}{|\mathbf{B}u|^3}. \tag{5.2}$$

This 'sine series' can be used to calculate Madelung constants via (1.3), although faster convergence can be realised by exploiting the freedom to choose *t* as long as *t* lies between 0 and the distance from site *r* to nearest neighbour. Formal integration of (5.2) yields an absolutely convergent series:

$$tq(r) + V_r(r)t^2/2 = (2\pi^3 \det \mathbf{A})^{-1} \sum_j q_j \sum_{u \in Z^3}{}' \frac{\exp[-2\pi \mathrm{i}(\mathbf{B}u)(d_j - r)] \sin^2 \pi t |\mathbf{B}u|}{|\mathbf{B}u|^4}. \tag{5.3}$$

The formal integration has not been justified, but (5.3) can be verified independently by Poisson transformation of the inverse quartic sum. The Madelung constant may now be obtained from the definition (1.3). Define, for a 3-vector $m \in BZ^3$, the Fourier transform of one charge assembly

$$h(m) = \sum_j q_j \exp(2\pi \mathrm{i} m d_j). \tag{5.4}$$

For highly symmetric crystals, *h* tends to have a simple evaluation. For the second definition of NaCl (2.7), $h(m) = 8$ if $m \in BO^3$, otherwise $h(m) = 0$. Then for any *t* such that

$$0 \leqslant t < \min_{j \neq k} |d_j - d_k| \tag{5.5}$$

the Madelung constant *M* satisfies

$$t \det \mathbf{B} \sum_k q_k^2 + Mt^2/2 = (2\pi^3 \det \mathbf{A}^2)^{-1} \sum_{u \in Z^3}{}' \frac{|h(\mathbf{B}u)|^2 \sin^2 \pi t |\mathbf{B}u|}{|\mathbf{B}u|^4}. \tag{5.6}$$

This general formula has the advantage of non-alternating summands. This means that during calculations one may always use the current partial sum to establish a bound on *M*. It is tempting to integrate further, perhaps even arbitrarily, exploiting to the fullest the free parameter *t*. This idea is best demonstrated by analysing the special case of the NaCl crystal. In that case, (5.2) reduces to

$$\pi^2(1 + tM) = 4 \sum_{u \in O^3} \sin \pi t u / u^3 \tag{5.7}$$

valid for $0 < t < 1$. It may seem at first that (5.7) is impossible, since the limiting value of the right-hand sum appears to vanish as *t* approaches zero. This is not so—the sum approaches $\pi^2/4$ in the small *t* limit, a phenomenon not uncommon in the world of Fourier series. Integration of (5.7) gives the claim (1.9), which is the NaCl form of (5.6).

By manipulating further integrals of (5.7) one can derive some interesting expressions, such as

$$\pi^4(Mt^3/4 + 3t^2/16) = \sum_{u \in O^3} \sin^3 \pi t u / u^5 \tag{5.8}$$

valid for all $0 \leq t < \frac{1}{3}$. This sine series converges yet more rapidly than (1.9), but we do not yet know how much information about the Madelung constant can be gleaned with such techniques. Still, one advantage of the general crystal expansion (5.6) is that, in spite of its formidable appearance and non-optimal convergence, it is the simplest formula we know of for reasonable computer calculation. Much more convergent formulae we describe later are harder to program.

One may take advantage also of the positive definiteness of the summands in (1.9). If we take $t = 1/\sqrt{3}$, for example, then (1.9) gives on the basis of the first summand alone

$$M > 6(64\pi^{-3}/9 - 1/\sqrt{3}) \sim -2.088 \ldots \tag{5.9}$$

and for the other direction,

$$t + Mt^2/2 < 8\pi^{-3} \sum_{u \in O^3} u^{-4} \tag{5.10}$$

which when extremised in t gives $M < -1.318 \ldots$. Of course, tighter bounds result upon more complicated analysis of the summands of (1.9).

Bounds for general crystals can likewise be obtained from (5.6). For lower bounds, one may choose an integer triplet such that $t = (2|\mathbf{B}u|)^{-1}$ satisfies the constraint (5.5). Then the right-hand side of (5.6) can be bounded below by a finite set of summands in the spirit of (5.9).

6. Exponential series for general crystals

Previous authors have found rapidly convergent 'cosech' series for the Madelung constants of cubic crystals (Hautot 1975, Zucker 1975, Glasser and Zucker 1980). We have extended this work to the general case, starting with the crystal potential (3.9). Transformation of this potential to an exponential sum is somewhat tedious, involving a straightforward geometrical series identity valid for positive $y < 1$:

$$f(y) = \sum_{q \in Z} \exp[2\pi(\mathrm{i}t(q + y) - s|q + y|)]$$

$$= \frac{\sinh(2\pi sy) \exp[2\pi \mathrm{i}t(y - 1)] + \sinh[2\pi s(1 - y)] \exp(2\pi \mathrm{i}ty)}{2(\sinh^2 \pi s + \sin^2 \pi t)} \tag{6.1}$$

where s, t are positive reals. The function f can be continued for negative $y > -1$ from the relation $f(y) = f^*(-y)$. The idea is to use this one-dimensional sum for just one of the components of the u triplet in (3.9), to obtain a two-dimensional sum having the exponential convergence properties.

One hesitates to carry out the fundamentally asymmetrical calculations, especially when the lattice-defining matrix A enjoys no particular symmetry. But once a general formula is obtained, computer programs can take care of the numerical aspect. To this end we have worked out the generalised two-dimensional sum which agrees with previous authors for the simple cubic structures. The result is as follows. Define a matrix $\mathbf{D}$ and, for a given 3-vector x, define a 3-vector h by

$$\mathbf{D} = \mathbf{B}^T\mathbf{B}$$
$$h = \mathbf{B}x. \tag{6.2}$$

Now choose some component of h, say h_i, which is non-zero. There will be at least one such since the crystal potential will turn out to diverge if $h = (0, 0, 0)$. Define a positive quadratic form R^2 acting on a select pair of components of a 3-vector m by

$$(R(m))^2 = \sum_{j, k \neq i} (D_{jk}/D_{ii} - D_{ij}D_{ik}/D_{ii}^2)m_j m_k. \tag{6.3}$$

The two components of m which appear in this form will be the summation indices for the final two-dimensional potential sum. Next, define 2-vectors c, H having components

$$c = (D_{ij}/D_{ii}, D_{ik}/D_{ii})$$
$$H = (h_j, h_k) \tag{6.4}$$

where j, k are chosen here such that i, j, k is a cyclic ordering of $1, 2, 3$. Now consider the function defined for 3-vectors x:

$$J(x) = (\pi \det \mathbf{A})^{-1} \sum_{u \in Z^3}{}' \exp[2\pi i(\mathbf{B}u)x]|\mathbf{B}u|^{-2}. \tag{6.5}$$

This function can be Poisson transformed via the identity (6.1) to give

$$D_{ii} \det \mathbf{A} J(x) = \pi(2h_i^2 - 2|h_i| + \tfrac{1}{3}) \tag{6.6}$$

$$+ \sum_{N \in Z_2}{}' \frac{\sinh(2\pi R|h_i|)\cos\{2\pi N[\mathbf{H} - (|h_i| - 1)\,\mathrm{sgn}(h_i)c]\} + \sinh[2\pi R(1 - |h_i|)]\cos[2\pi N(\mathbf{H} - ch_i)]}{2R(\sin^2 \pi Nc + \sinh^2 \pi R)}$$

where $R = R(N_1, N_2)$. This expansion is valid for all x such that $0 < |h_i| < 1$. The final step in calculating the crystal potential is to observe that

$$V(x) = \sum_{j=0}^{n-1} q_j J(x - d_j). \tag{6.7}$$

In spite of the cumbersome notation, it should be clear that assignments (6.2)-(6.7) can be programmed with knowledge of the crystal parameters in (2.6). It is also evident that the summation in (6.6) has terms decreasing exponentially in either component of N. One may compute the crystal potential efficiently in this way, noting that in (6.7) the vector x should be taken to lie in the origin cell without loss of generality.

To obtain the Madelung potential in reference to a site d_k lying in the origin cell, one may subtract from $V(x)$ the Coulomb term $q_k|x - d_k|^{-1}$. If, however, one wishes to compute the limiting value of the Madelung potential as x approaches d_k, it is necessary to perform further analysis. We shall not pursue the analytic theory for the J function here, but it is possible to establish an analytic continuation of J for zero argument (Zucker 1976, Glasser and Zucker 1980, Crandall and Buhler 1987a), giving a workable formula for the Madelung potential at site d_k

$$V_{d_k}(d_k) = \lim_{x \to d_k} (V(x) - q_k|x - d_k|^{-1})$$

$$= \sum_j{}' q_j J(d_k - d_j) + q_k \sum_\delta{}' J(\mathbf{A}\delta/2) \tag{6.8}$$

where δ runs over binary triples. Note that the prime superscript for the first sum rules out $j = k$, while the prime in the second rules out $\delta = (0, 0, 0)$. Thus the sums have $n - 1$ and 7 terms, respectively.

General calculation of crystal potentials and Madelung constants is now possible with (6.2)-(6.8) and (1.3). We have used programs which accept numerical input of the general parameters (2.6) and have obtained precision such as that of (1.5) in this fashion. In addition, one may obtain the 'cosech' series formulae previously discovered (Hautot 1975) by algebraic manipulation of (6.8) and (6.6).

The Madelung constant itself, which arises from (6.8) via (1.3), can be obtained in certain cases by appeal to a different method of subtracting the singularity in the limit of site approach. This is the method which will yield a power series expansion for the NaCl Madelung potential, and to which we now turn.

7. The problem of the Coulomb singularity

A method of handling the Coulomb singularity without recourse to analytic function theory is most easily demonstrated for the NaCl structure. The method yields a regular power series expansion for the Madelung potential near the origin. From the general formulae (6.6) and (6.7) for the crystal potential, one may deduce that for NaCl

$$V(x, y, z) = 2 \sum_{n, p \,\text{odd}} \frac{\sinh \pi R(1/2 - |x|) \cos \pi n y \cos \pi p z}{R \cosh \pi R/2} \tag{7.1}$$

where $R = (n^2 + p^2)^{1/2}$. This expansion is valid everywhere within the $(0, 0, 0)$ lattice cell. Note that this exponentially convergent expression reveals immediately many of the properties we have discussed. For example, if any coordinate x, y, or z is $\pm\frac{1}{2}$ then $V = 0$. It is not so clear that V is completely symmetric under permutation of the coordinates, but such is true on the basis of the original sum (1.1).

The Coulomb singularity is evident upon the observation that if (x, y, z) represents a charge site, then the sinh/cosh ratio is not exponentially damped and the sum diverges. To obtain the Madelung potential, therefore, we must find an effective way to subtract from the sum (7.1) a Coulomb term. This can be effected with a useful Poisson identity. Let w be the 2-vector (y, z). Then

$$2 \sum_{u \in O^2} u^{-1} \exp(-\pi x |u| + i \pi u w) = (x^2 + w^2)^{-1} + \sum_{\substack{a \in Z^2 \\ a \neq (0,0)}} (-1)^{a_1 + a_2} (x^2 + |a + w|^2)^{-1/2}. \tag{7.2}$$

The first term on the right-hand side is a Coulomb term which has been separated off the right-hand sum; hence the omission of the a origin. If we identify the 2-vector u with the pair (n, p) from (7.1) we obtain an expression for the Madelung potential referred to the NaCl origin:

$$V_0(x, y, z) = \sum_{\substack{a \in Z^2 \\ a \neq (0, 0)}} (-1)^{a_1 + a_2} (x^2 + |a + w|^2)^{-1/2} - 4 \sum_{n, p \,\text{odd}} \frac{\cosh \pi R x \cos \pi n y \cos \pi p z}{R[1 + \exp(\pi R)]}. \tag{7.3}$$

Evaluating this expression at the NaCl origin yields a previously known expansion for the Madelung constant (Hautot 1975):

$$M_{\text{NaCl}} = 4(1 - \sqrt{2}) \zeta(\tfrac{1}{2}) \beta(\tfrac{1}{2}) - 4 \sum_{n, p \,\text{odd}} R^{-1} [1 + \exp(\pi R)]^{-1}. \tag{7.4}$$

Here, the Riemann zeta and beta functions appear due to the identity (Glasser 1973)

$$\sideset{}{'}\sum_{a \in Z^2} (-1)^{a_1+a_2} a^{-2s} = 4\zeta(s)\beta(s)(-1+2^{1-s}). \tag{7.5}$$

But (7.3) contains information also about the regular behaviour of the Madelung potential near the origin. In fact, one may use (7.5) and a binomial square root expansion to deduce an axial series. Define the coefficients

$$C_k = 4(-\tfrac{1}{2})^k (2k-1)!!(-1+2^{1/2-k})\zeta(k+\tfrac{1}{2})\beta(k+\tfrac{1}{2})/k!$$

$$-\frac{4\pi^{2k}}{(2k)!} \sum_{n,p \text{ odd}} R^{2k-i}[1+\exp(\pi R)]^{-1}. \tag{7.6}$$

Then the Madelung potential along the x axis can be written in a power series

$$V_0(x,0,0) = \sum_{k=0}^{\infty} C_k x^{2k}$$

$$= -1.747\,564\,59 - 3.578\,582 x^4 - 0.989\,4995 x^6$$

$$-2.942\,158 x^8 - 1.010\,713 x^{10} - 2.914\,156 x^{12} - 1.171\,4335 x^{14} - \dots \tag{7.7}$$

The coefficient C_0 is just M_{NaCl} as in (7.4). The coefficient C_1 is interesting: it must vanish. This is because $\nabla^2 V$ must vanish everywhere, and formally equals $3C_1$ at the origin. This in turn provides an interesting identity when we evaluate (7.6) for $k=1$:

$$\zeta(\tfrac{3}{2})\beta(\tfrac{3}{2}) = \pi^2(1-1/\sqrt{2})^{-1} \sum_{n=2 \bmod 8} r_2(n)\sqrt{n}[1+\exp(\pi\sqrt{n})]^{-1} \tag{7.8}$$

where $r_2(n)$ is the number of representations of integer n as the sum of two squares. This result for the zeta-beta product at $\tfrac{3}{2}$ has been derived previously (Zucker 1984).

The coefficients C_k for $k>1$ are all negative and tend to the value -2. This asymptotic value -2 is interesting in that it suggests a rough model for the NaCl Madelung potential in the form

$$V_0(x,0,0) \sim M - 2x^4 - 2x^6 - \dots$$

$$= M - 2x^4/(1-x^2) \tag{7.9}$$

which, although only an approximate formula, shows the proper quartic behaviour as well as the required Coulomb poles at $x = \pm 1$.

As a final note we refer to the exact special values (1.11) and (1.12). One may follow the arguments of Forrester and Glasser (1982) and Glasser and Zucker (1980) using Jacobi identities (Whittaker and Watson 1973) to obtain some useful formulae, for example:

$$\sum_{m,n,p \in Z} (-1)^{n+p}[(m-\tfrac{1}{4})^2 + n^2/2 + p^2/2]^{-s} = 16^s \beta(2s-1) \tag{7.10}$$

$$\sum_{m,n,p \in Z} (-1)^m[m^2 + n^2 + (p-\tfrac{1}{2})^2]^{-s} = 2^{2s+1}\beta(2s-1). \tag{7.11}$$

Expression (7.10) is an equivalent form of the CsCl crystal potential when $s = \tfrac{1}{2}$. Since $\beta(0) = \tfrac{1}{2}$, we immediately obtain $V_{\text{CsCl}}(\tfrac{1}{4},0,0) = 2$. For the crystal defined

$$n=2; \; d_0 = (0,0,0); \; d_1 = (1,0,0)$$

$$q_0 = +1; \; q_1 = -1$$

$$\mathbf{A} = \begin{pmatrix} 2 & 0 & 0 \\ 0 & 1 & 0 \\ 0 & 0 & 1 \end{pmatrix} \tag{7.12}$$

2292   *R E Crandall and J F Delord*

which amounts to parallel planes of like charge, the crystal potential is given by (7.11) as

$$V(0, 0, \tfrac{1}{2}) = 2. \tag{7.13}$$

As mentioned in § 1, our computer programs, which use the general prescription (6.2)-(6.8) with (1.3), verify these exact results and the previous (1.7) numerically. One good feature of exact potential results is that they may act as a sharp test of any programming effort.

Acknowledgments

The authors would like to thank P Cuaz, J P Buhler, R Mayer, D Griffiths and N Wheeler for valuable discussions during the course of this work. We are grateful to a referee for providing a proof of the CsCl theorem (1.11).

References

Born M and Goppert-Mayer M 1933 *Handb. Phys.* **24** 708-14
Born M and Huang K 1954 *Dynamical Theory of Crystal Lattices* (Oxford: Oxford University Press)
Borwein D, Borwein J and Taylor K 1985 *J. Math. Phys.* **26** 2999-3009
Crandall R and Buhler J 1987a *J. Phys. A: Math. Gen.* to be published
Crandall R, Buhler J and Delord J 1987b in preparation
Emersleben 1923 *Phys. Z.* **24** 73, 97
Ewald P 1921 *Ann. Phys., Lpz.* **64** 253
Forrester P and Glasser M 1982 *J. Phys. A: Math. Gen.* **15** 911-4
Glasser M 1973 *J. Math. Phys.* **14** 409-13, 701-3
Glasser M and Zucker I 1980 *Theoretical Chemistry: Advances and Perspectives* vol 5 (New York: Academic) pp 67-139
Griffiths D 1981 *Introduction to Electrodynamics* (Englewood Cliffs, NJ: Prentice-Hall)
Hajj F 1972 *J. Chem. Phys.* **56** 891-8
Hautot A 1975 *J. Phys. A: Math. Gen.* **8** 853-62
Madelung E 1918 *Phys. Z.* **19** 524
Sherman J 1932 *Chem. Rev.* **11** 93
Whittaker E and Watson G 1973 *Modern Analysis* (Cambridge: Cambridge University Press) 4th edn, p 470
Zucker I 1975 *J. Phys. A: Math. Gen.* **8** 1734-45
—— 1976 *J. Phys. A: Math. Gen.* **9** 499-505
—— 1984 *SIAM. Math. Anal.* **15** 406-13

2. The fractal character of space–time epidemics

Discussion

It was in the early 2000's at the Center for Advanced Computation at Reed College that I first encountered the wonderful equations and methods of mathematical epidemiology. We worked with the DOD on bioterror issues, and with Multnomah County Health Department on disease-strategy issues. Various algorithmic ideas arose, and this paper—actually just a set of notes for a 2007 conference—reflects some such notions:

1. There is the algorithm idea of determining disease severity *after* a conflagration (disease) has passed, by looking at the fractal distribution of survivors. Figure 2 in the paper is essentially a "forest fire" model in which surviving trees enjoy measurable fractal dimension, which dimension being strongly correlated to disease (fire) parameters.

2. The notion of conflagration "wavefront," as in Figure 1 of the paper, can actually be quantified, with expansion velocity derivable from disease parameters.

3. As intimated in Figure 4, there are algorithm ideas in regard to vaccination strategy. One relevant question is: Given limited vaccine or vaccination resources, exactly where in the population should vaccination occur? Initial experiments suggest that one should apply services—whether vaccination or other forms of medical intervention—at the most "fractal-like" boundaries, because that is where local expansion velocity is most extreme.

Challenges

Item (3) above is certainly of high importance. Another challenge is to properly model urban-infection scenarios, possibly arriving at some algorithms for minimizing disease spread based on daily-commuter dynamics. (At one point, our research effort uncovered quantitative predictions of how decreasing the number of typical work-days to 4/week would affect disease spread.)

Source

R. Crandall, "The fractal character of space–time epidemics," NOLTA Conference on the Applications of Nonlinear Theory to Infectious Disease Dynamics, (presentation 2007).

2007 International Symposium on Nonlinear Theory and its Applications
NOLTA'07, Vancouver, Canada, September 16-19, 2007

NOLTA'07

The fractal character of space-time epidemics

R. Crandall[1]

This brief is an extended abstract for a presentation at the NOLTA 2007 conference on the"Applications of Nonlinear Theory to Infectious Disease Dynamics."

From purely-temporal to space-time models

WIth space-time models being our eventual focus, let us first mention spatially independent models, so that inclusion of spatial dependence can be imagined as an extension. The standard susceptible-infected-recovered (SIR) model (for origins, see [3]) amounts to a system of coupled differential equations:

$$\partial S/\partial t \;=\; -\beta I S, \quad \partial I/\partial t = \beta I S - \gamma I, \quad \partial R/\partial t = \gamma I. \tag{1}$$

In the simplest SIR scenario, the susceptible, infected, and recovered populations—respectively $S(t), I(t), R(t)$—are functions only of time t, while β, γ are positive real coupling constants. It is a striking fact that one can give an exact relation for $S(\infty)$; namely

$$S(\infty) - \frac{\gamma}{\beta} \log \frac{S(\infty)}{S(0)} = S(0) + I(0), \tag{2}$$

It is not hard to show, given only $\beta, \gamma, S(0) > 0$, that a positive solution $S(\infty)$ always exists.

Our first connection to the fractal character of survival sets is this observation: That in the continuum model above, there is a nonzero final survivor density. It will turn out that in models with a characteristic *velocity* of disease propagation, the asymptotic survivor set often has fractal properties. The easiest way to extend a spatially independent model to involve both space and time is to introduce enough fundamental constants to forge a "natural speed," i.e. a number with dimensions of space/time. We need to admit right off that the existence of such a speed does not necessarily mean the main disease phenomenon is an infection moving with that speed. However, indeed, in some space-time models an obvious propagation at predicted speed is just what happens.[2] So, one way to obtain a relatively trivial space-time model is to assume an infection "jumps" a distance δ every time quantum τ. Such a model might be a forest fire of burning tress (of population $I(\vec{r}, t)$) amongst healthy trees (of population $S(\vec{r}, t)$), where $\vec{r}$ is spatial position in a 2-dimensional

---

[1]Center for Advanced Computation, Reed College, Portland OR USA 97202

[2]For example, the speed of propagation of rabies has been modeled in just this way [5]

forest. Such a theory—which must include some form of probabilistic, or mean-field manner of a tree catching fire—will end up with a characteristic speed δ/τ.

However, in some ways a more natural—or shall we say fitting—way to create a velocity parameter is *to allow the infecteds to spatially diffuse*. To that end, consider a "spatialized" SRI model:

$$\partial S/\partial t \;=\; -\beta IS, \quad \partial I/\partial t = d\,S\nabla^2 I + \beta IS - \gamma I, \quad \partial R/\partial t = \gamma I. \tag{3}$$

Here d is a diffusion constant, as might appear in the classical diffusion equation $\partial p/\partial t = d\nabla^2 p$. It turns out that this model has a natural speed $V \sim \sqrt{4d(\beta S_\infty - \gamma)}$, where S_∞ is simply the asymptotically constant population of susceptibles at spatial infinity (see [1], [5]).

Explanatory figures

- Figure 1 shows 2-dimensional propagation of a disease front in the continuum space-time model (3).

- Figure 2 shows the results of a discrete solving of the equations (3) above, and another view (for different parameters) of the asymptotic survivor set, again in the discrete picture.

- Figure 3 shows a typical parameter-dimension contour for a 2-dimensional model of the type for Figure 2. (Plot taken from [4].)

- Figure 4 shows a USA-influenza model [1] [2] that employs space-time equations similar to (3) above, but also including vaccination terms.

- Figure 5 shows the modeling of a county (Multnomah County, Oregon) population, in preparation for assessing discrete solutions and hence survivor sets, especially in regard to prevention measures such as quarantine and vaccination.

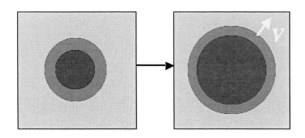

Figure 1: Infection (conflagration, fire, etc.) in red, propagates with asymptotic speed V as given in-text. This continuum-differential model only reveals a low density of surviving susceptibles (dark green center).

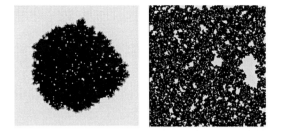

Figure 2: Left-side of figure is the state (outward-propagating infection ring, a few inside survivors, etc.). Right-side of figure is a closeup of a fractal survivor set of (measured) dimension 1.67.

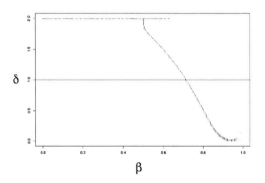

Figure 3: Typical plot of fractal dimension δ of a survivor set vs. an infection parameter β. The points of the curve were all taken empirically, and show remarkably small variance—thus suggesting an underlying theory of fractal dimension. Note that by tuning β one can obtain dimension $\delta > 1$ or < 1 (the middle, horizontal line is dimension 1).

Figure 4: USA-influenza model [1] [2] in action, showing the Eastern seaboard in (a typical) December (left), and in (a typical) April (right), where reddish = infection and bluish = recovered. This model certainly exhibits fractal phenomena, but as the model is highly complex—also involving age classes and vaccination—research is first needed at a more fundamental level, with fewer parameters. Ideally, though, fractal studies should lead to vaccination strategies, for reasons attendant on the fractal nature of infection boundaries.

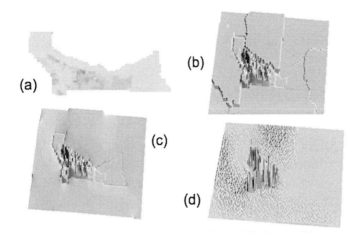

Figure 5: Another scenario that needs research on fractal character: A metropoltian-urban population with dynamical mixing. Part (a) is a population-density plot for Multnomah County, OR. Part (b) is a multilevel plot (only flat densities are known for surrounding counties). Part (c) is a custom spline for smoothing the multilevel data from (b). Part (d) is the commute model at high-noon-workday. Such plots are useful in developing a theory of the fractal character of infection boundaries and survivor pockets.

Acknowledgements

We thank M. Elman and A. Sullivan for their epidemiological expertise and research advice, and also the Multnomah County Department of Health for precise age-class sample population data.

References

[1] E. Cahill, R. Crandall, L. Rude, A. Sullivan, "Space-time influenza model with demographic, mobility, and vaccine parameters," Proc. 5th Ann. Hawaii Intl. Conf. on Math., Stat., and related fields, 2006 `http://academic.reed.edu/epi/papers/USAfluFINAL.pdf`

[2] G. Crandall and R. Crandall, USAfluModel v1.4, `http://academic.reed.edu/epi` (2005).

[3] W. Kermack and A. McKendrick, "A contribution to the mathematical theory of epidemics," *Proc. Roy. Soc,. London*, 115:700-721 (1927).

[4] B. O'Donnell, B. McPhail, D. Tran and P. Carlisle, "Computing the SIR Model: Fractal dimension leads to vaccination strategies," Sep. 2004 `http://academic.reed.edu/epi`

[5] J. Murray, *Mathematical biology*, Springer–Verlag (1989).

3. Mathematical signatures as tools for visual dysfunction

Discussion

Colleagues R. Weleber and S. Gillespie came to me and asked if there could be an algorithm for assessing eye disease based on sensitivity data—the "hill of vision." The hill of vision is a plot *not* of the retina itself but the retina's sensitivity. Figure 3 in this paper shows healthy retina (top left), a retina having retinitis pigmentosa (top center), and one with macular dystrophy (top right). Clearly these hill-of-vision plots differ greatly, with red (flat basin) regions corresponding to very low sensitivity.

The entire treatment describes algorithms for classifying eye disease. There are also surface-spline algorithms described, because the data that Dr. Weleber accumulated is at discrete points rather widely separated in some cases.

Challenges

An algorithm challenge is to perfect such work, and expect a computer-reported diagnosis of any disease. A conceptual/theoretical challenge is to explain why healthy eyes have "corrugated" hills of vision, as in top-left plot of Figure 3. In other words, even the healthiest eyes (I am told one can expect optimal eye health around age 13 in humans) have rippled hills of vision, not smooth. Could it be that in evolutionary terms the rippled sensitivity allows finer detection of moving objects, or danger, or ... ?

Source

R. Crandall, S. Gillespie, and R. Weleber, "Mathematical signatures as tools for visual dysfunction," (2009).
`http://www.perfscipress.com/papers/Signatures5\_psipress.pdf`

PSIpress
perfscipress.com
21 nov 09: C

Mathematical signatures as diagnostic tools for visual dysfunction

R. Crandall, S. Gillespie, and R. Weleber

Abstract: Herein we describe methods for data splining and signature analysis on retinal-sensitivity data. We propose four signatures, called Mercator-, Fourier-, Bessel-, and fractal-signatures, that we believe are candidates for detection and classification of retinal pathology. A prototype application VFMA (Visual-field modeling and analysis) that implements these ideas is described. One of our experimental discoveries is that the normal retina is, in a quantifiable sense, nonsmooth, in fact in said sense less smooth than diseased eyes.

1   Introduction

The visual field is that physical space, measured in degrees eccentric to fixation, to which the human eye can perceive images and detect motion. The Hill of Vision is a 3-dimensional construct of the visual field extent, with the added representative of visual sensitivity at each point of eccentricity. H. Traquair first introduced the concept of the Hill of Vision in 1927 with his description of the visual field as "an island of vision or hill of vision surrounded by a sea of blindness." [4] The visual field and its 3-dimensional representation, the Hill of Vision, become abnormal in many diseases and disorders that affect the visual system, extending from the retina, through the optic nerve, chiasm, optic tract, and radiation to the primary and secondary visual cortical regions where the field of vision and images are reconstructed and interpreted by the brain.

Modeling of the Hill of Vision is a major goal for computerized analysis of the visual field in health and disease. We have created a method of modeling and analysis of the Hill of Vision that will be useful for early detection, characterization, and quantification of visual defects. Specifically, we introduce mathematical signatures—essentially condensations of all the Hill of Vision measurements. A reasonable analogy is that of an EKG, with which one ends up with a condensed "signature" of cardiac function; indeed our signatures do appear as "strip-charts," the mathematical issue being that a signature is generally of lower dimension than the complete data. An application has been developed that implements the signature algorithms and methodology described in this work.

2   Hill-of-vision representation

2.1   Data-gathering methods

The Hill of Vision (Figure 1) can be assessed by either moving test targets, called kinetic perimetry, or through presentation of static stimuli of specific sizes with increasing brightness until the subject perceives the light, a technique termed static perimetry. We performed full-field static perimetry on normal subjects and patients with diseases affecting the visual field using an Octopus 101 perimeter (Haag-Streit, Inc., Koeniz, Switzerland) and radially designed, centrally condensed custom grids of up to 187 test locations extending from fixation to 80° temporally, 78° inferiorly, 56° nasally and 56° superiorly, and targets of Goldmann Size III (0.47° diameter) presented on a 10 cd/m^2 background. Thresholds were determined using the new German Adaptive Threshold Estimation (GATE) strategy [3] and were converted to differential luminance sensitivity (DLS) using published transformations [2]. These DLS values were used to model the Hill of Vision and to generate mathematical signatures describing the surface of this three-dimension structure.

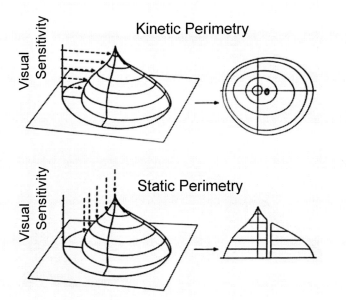

Figure 1: Perimetry nomenclature. The present treatment involves exclusively static perimetry, whereby sensitivity at specific retinal locations is determined as the threshold of perception.

2.2   Coordinate-system nomenclature

Our research into the efficacy of mathematical signature for detection of retinal pathology starts with sensitivity data. First, we shall carefully define nomenclature. Because most of the mathematical signature analysis involves *planar* calculations in standard polar coordinates (r, ϕ), we employ a specific transformation from the spherical retinal surface to a planar circle.

Referring to Figure 2, the (ideal) retinal hemisphere is modeled via standard spherical coordinates $(R, \theta, \phi) = (1, \theta, \phi)$, and so exactly two angular parameters define a physical location on the retina. As Figure 2 shows, a location is mapped to a pair (r, ϕ), with the important caveat that $r \in [0, 1]$ and is a *geodesic* distance along the surface of the hemisphere.[1] Though it can be somewhat confusing to use planar coordinates when the original dataset lives on a curved surface, it turns out to be quite convenient to use the final pair (r, ϕ) for the algebra of signatures. For example, the Bessel signature we have developed can therefore use Bessel arguments of standard, polar form.

In summary, what we shall call the *signature coordinate system* is defined by the transfor-

---

[1]In some of our graphical displays, r runs over $[0°, 90°]$ instead of $[0, 1]$. But we choose the unit-disk scaling, namely $r \in [0, 1]$ for actual signature calculations.

mation:

$$(R = 1, \theta, \phi) \ \rightarrow \ (r, \phi) := \left(\frac{2\theta}{\pi}, \phi \right),$$

so that the radius r in the resulting planar *signature space* runs over $[0, 1]$ while the azimuth runs over $(-\pi, \pi]$.

2.3 Area-sensitivity calculus

A caution is relevant here: When using the planar-polar system (right-hand view in Figure 2), measurement of arbitrary areas is nontrivial, and can be counter-intuitive. Imagine we have a "footprint" shape—a closed contour drawn in a planar-polar system—and that on the full circle we have a dbm-sensitivity function, say $D(r, \phi)$ but with D vanishing outside the footprint. In many instances we are interested in the total integrated D, but *integrated over the original retinal surface*, i.e. over the back-transformed "footprint." This total area-sensitivity in units of dB-steradians might seem to be something like

$$S \stackrel{?}{=} \int \int D(r, \phi) \ r \ dr \ d\phi,$$

but this is unphysical, incorrect. One must take into account the Jacobian curvature factor due to the transformation from spherical to planar-polar, with the correct integral given by

$$S = \int_{-\pi}^{\pi} d\phi \int_0^1 dr \ D(r, \phi) \frac{\pi}{2} \sin\left(\frac{\pi}{2} r \right).$$

Rather than give a detailed derivation of this, we shall simply point out that there are easy, mnemonic cases. For example, day D is a constant all retinal-surface points, and so is the same constant also in the transformed planar-polar picture. Then the correct integration yields

$$S = 2\pi D \int_0^1 \frac{\pi}{2} \sin\left(\frac{\pi}{2} r \right) \ dr = 2\pi D,$$

which is correct, in being 2π steradians (for a full hemisphere) times the constant sensitivity. When we use such estimates, we report them in units of dB-steradians.

Another example of area-sensitivy integration is in order here, primarily because said example suggests yet another caution. Say we again have a constant $D(r, \phi) = D$ but only over a small circle, origin-centered and of radius, say, $r = 1/2$ in signature coordinates. The correct integral is $\pi D / \sqrt{2}$; indeed, the polar cap that comes down to polar angle $\pi/4$ does subtend exactly $\pi/\sqrt{2}$ steradians. However, here is a counter-intuitive aspect of such "cap" transformations: A circle-cap sitting on the original, physical retinal sphere is *not* generally a circle in the planar signature space. (Our example here of a cap of radius $r = 1/2$ does give a circular cap upon back-transformation, but that is because such a cap is pole-centered to begin with.) The primary detail a researcher needs for circle-cap deformations is the angle formula

$$\cos \Omega = \cos \theta \cos \theta_0 + \sin \theta \sin \theta_0 \cos(\phi - \phi_0),$$

expressing the exact subtended angle Ω between two spherical points $(1, \theta_0, \phi_0)$ and $1, \theta, \phi)$. If (θ_0, ϕ_0) is the direction of the center of a circle-cap on the original sphere, then our signature-space substitution $\theta \to \pi r/2$ yields from the $\cos \Omega$ formula an actual polar-coordinates equation for an (r, ϕ) path in signature space, and—unless $\theta_0 = 0$, which is the special case of a pole-centered cap—we do *not* obtain a circle in the signature plane, rather a peculiar kind of oval.

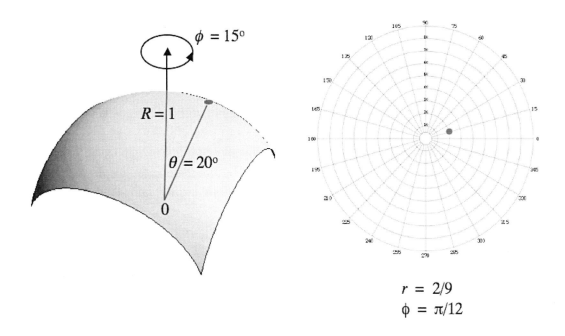

Figure 2: Pictorial of spherical $\to$ signature coordinate transformation. Spherical coordinates (R, θ, ϕ) for a retinal surface assume $R = 1$, with θ, ϕ being the standard polar angle and azimuth. A point of sensitivity measurement (red dot on left-hand surface above) is at, say, $(R, \theta, \phi) = (1, 20^o, 15^o)$ on the sphere. The right-hand figure has the measurement point located within a unit circle, at *planar* polar coordinates (that we call signature coordinates) $r = 2/9$, corresponding to $2\theta/180$, and the azimuth in algebraic form, $\phi = 15/360 \cdot 2\pi = \pi/12$.

2.4   Hill-of-vision graphical modeling

Referring to Figure 3, the steps for visualization of the Hill of Vision—namely the sensitivity profile on the retina—are as follows:

1. Load data points (anywhere from a few, to hundreds of such), as plotted in the upper strip of Figure 3, the plotting being for 2-dimensional signature space as previously defined. The three left-to-right plots are for NO, RP, and MD patients (see caption to Figure 3, and also our Discussion section).

2. Interpret these discrete points as constraints on a spline surface, and show each such surface as in the second strip of Figure 2. (All of our visualization to date uses the thin-plate spline (TPS) described later.)

3. As embodied in the VFMA application (see Appendix), allow for rotation/inspection of the spline surface, as in the remaining 3-view strips of Figure 3.

It is important to note that the black dots—i.e. data positions—each have respective data values. One can think of these as "spikes" coming out of the figure page, with heights given by dbm sensitivity measurement. (Moreover, it is also reasonable to imagine these spikes as "hairs" all grown normal to the retinal sphere, with hair length again being sensitivity value.) Regardless of the nature of the spline, or the metaphor for imagining sensitivity values, a crucial observation is that the spline surface—say the upper-right view of Figure 2—is *neither* a picture of the retina nor a sensitivity map directly on the spherical retina; it is a picture of sensitivity in signature space, and must be warped somewhat by the curvature Jacobian to represent on-retina sensitivity. In a word: The 3-dimensional spline at upper-right of Figure 2 shows a function defined on the unit disk in 2-dimensions, and this is the basis for all of our signature calculations.

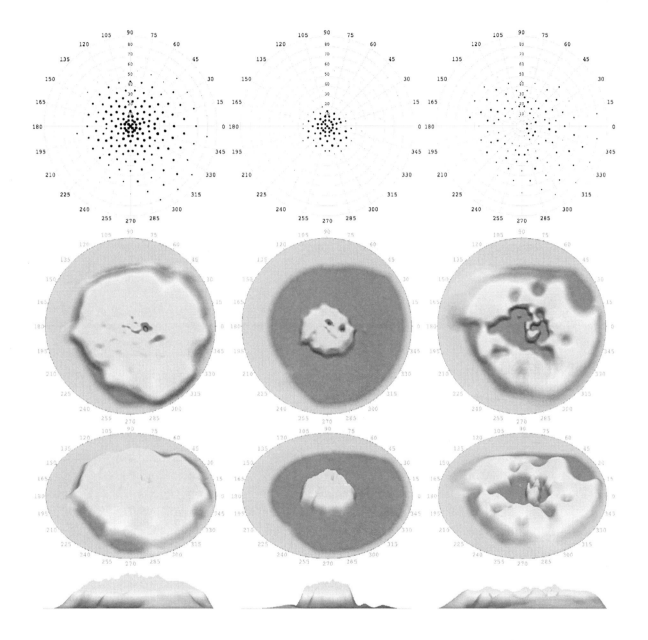

Figure 3: Hill-of-Vision representation. The data plotted in signature coordinates (upper three views) are, respectively left-to-right, from a normal subject (NO), a patient with retinitis pigmentosa (RP), and a patient with macular dystrophy (MD). A thin-plate spline (TPS) is used to infer a smooth surface (second row, left-to right for NO, RP, MD) as viewed from atop. The other variously tilted views are obtained via manual control in the VFMA application (see Appendix). See our Discussion section for precise historical reference of the patient data used.

3   Splines

3.1   Infinitely differentiable splines

The task at hand is: Given N input data vectors in D dimensions with given real evaluations, namely a set

$$(\vec{r}_k = (r_{0,k}, \ldots r_{D-1,k}) : k \in [0, N-1])$$

and an evaluation set

$$(I(\vec{r}_k) : k \in [0, N-1]),$$

establish a smooth, real-valued spline function $S(\vec{r})$ such that for all $k \in [0, N-1]$ we have

$$S(\vec{r}_k) = I(\vec{r}_k).$$

That is to say, we want a smooth function that agrees with I at each of the N input points.

3.2   Nuclear splines

One possible spline design runs as follows.[2] We should admit right off that the nuclear spline we are about to decribe is *not* used in the VFMA software (see Appendix) and so is not used for signatures reported in the present work. Instead, our results arise from thin-plate spline (TPS) described in various references such as [1]. Define a (nonnegative) radial-potential function $V(|\vec{r}|)$. It is essential for spline coherence (agreement on the input set) that V diverges at $\vec{r} = \vec{0}$; however, in practice we simply force V to be very large at said origin. Now define a spline function S on general D-dimensional vectors $\vec{t}$ by

$$S\left(\vec{t}\right) := \frac{\sum_{k=0}^{N-1} V\left(|\vec{t} - \vec{r}_k|\right) I(\vec{r}_k)}{\sum_{k=0}^{N-1} V\left(|\vec{t} - \vec{r}_k|\right)}.$$

The way the potential works is, for vector $\vec{t}$ coinciding with an input vector r_k, we expect, due to the divergence of $V(0)$, the surviving term in the ratio of sums above is just $I(\vec{r}_k)$.

We found that a computationally practical choice for the potential function is

$$V(R) = \frac{e^{-LR}}{(R^2 + \epsilon)^{\alpha}},$$

where ϵ is very small. Note that when $\alpha := 1/2, \epsilon := 0$ this is the radial Yukawa potential of nuclear physics. Thus we shall refer to potentials of this exponential form "Yukawa-class" potentials and presume freedom of choice for L, ϵ, α.

---

[2]It is not hard to show that there are many infinitely-differentiable functions that coincide with the N input-vector evaluations; in fact there are infinitely many splines that coincide with a *countably infinite* collection of input evaluations. Herein we are looking at just a few spline designs.

It is instructive here to provide an explicit graphical example: Data heights on the 2-dimensional unit circle. Figure 4 shows choices of parameters for the potential V and the resulting spline surfaces. In each case the input data are 2-dimensional evaluations, and so have

$$D = 2,$$

$$N := 21,$$

$$\{\{\vec{r}_0, I(\vec{r}_0)\}, \{\vec{r}_2, I(\vec{r}_2)\}, \ldots, \{\vec{r}_{20}, I(\vec{r}_{20})\}\} =$$

$$
\begin{pmatrix}
1 & 0 & 0 \\
\cos\left(\frac{\pi}{8}\right) & \sin\left(\frac{\pi}{8}\right) & 0 \\
\frac{1}{\sqrt{2}} & \frac{1}{\sqrt{2}} & 0 \\
\cos\left(\frac{3\pi}{8}\right) & \sin\left(\frac{3\pi}{8}\right) & 0 \\
0 & 1 & 0 \\
\cos\left(\frac{5\pi}{8}\right) & \sin\left(\frac{5\pi}{8}\right) & 0 \\
-\frac{1}{\sqrt{2}} & \frac{1}{\sqrt{2}} & 0 \\
\cos\left(\frac{7\pi}{8}\right) & \sin\left(\frac{7\pi}{8}\right) & 0 \\
-1 & 0 & 0 \\
\cos\left(\frac{9\pi}{8}\right) & \sin\left(\frac{9\pi}{8}\right) & 0 \\
-\frac{1}{\sqrt{2}} & -\frac{1}{\sqrt{2}} & 0 \\
\cos\left(\frac{11\pi}{8}\right) & \sin\left(\frac{11\pi}{8}\right) & 0 \\
0 & -1 & 0 \\
\cos\left(\frac{13\pi}{8}\right) & \sin\left(\frac{13\pi}{8}\right) & 0 \\
\frac{1}{\sqrt{2}} & -\frac{1}{\sqrt{2}} & 0 \\
\cos\left(\frac{15\pi}{8}\right) & \sin\left(\frac{15\pi}{8}\right) & 0 \\
1 & 0 & 0 \\
\frac{1}{2} & 0 & 1 \\
0 & \frac{1}{2} & 4 \\
-\frac{1}{2} & 0 & 3 \\
0 & -\frac{1}{2} & 2
\end{pmatrix}
$$

Thus for example, the very last array entry, namely $\{0, -1/2, 2\}$ means that the value of the I-function at $(x, y) = (0, -1/2)$ is 2. In fact our input data above has exactly four points with nonzero evaluations $(1, 2, 3, 4)$ in the indicated order. *However, it is important that we have intentionally placed a necklace of points around the unit circle, all having the value 0, to force in this particular demonstration the boundary condition that our spline wants to vanish on the unit circle.* Other boundary conditions can just as easily be specified by throwing appropriate input data into the set.

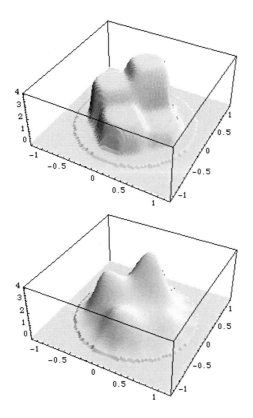

Figure 4: Two splines passing (very nearly) through heights $1, 2, 3, 4$ at given input positions. Note that we have artificially forced S to be slightly negative outside the unit circle, to show the effects of the forced boundary zeros. The top spline surface uses a Yukawa form with $L := 5, \epsilon := 0.00005, \alpha := 3$. The bottom spline surface uses $L := 2, \epsilon := 0.00005, \alpha := 1$.

3.3   Thin plate spline (TPS)

Another spline technique found in the literature is thin plate splines, introduced in 1976 by J. Duchon. Thin plate spline smoothing is inspired by the modeling of thin metal plates in elasticity theory. This spline is well established in the literature; we refer the reader to []. We need not supply an exemplary figure for the TPS, because Figures 2-5 already show the TPS in action. Just as with the nuclear spline design, the TPS has extra input; namely, the retinal-sensitivity data, and a necklace of zeros, to force near-vanishing of the TPS at circle's edge.

4   Mathematical signatures

4.1   Nonlinear-mercator signature

Our first method for generating a 2-dimensional signature for a spline surface is to create a nonlinear-Mercator projection. By this we mean the classical Mercator map projection *except* the lines of "latitude" are nonlinearly warped so that every sector has equal area. Then, a Mercator plot can be vertically integrated to yield a rotationally invariant 1-dimensional signature, or "strip-chart."

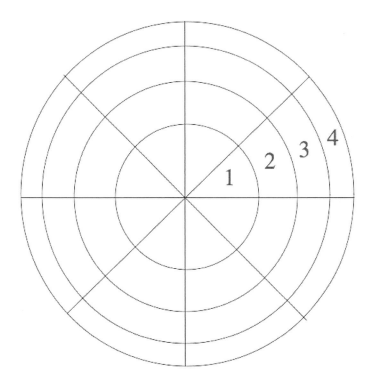

Figure 5: The nonlinear-Mercator transformation decimates the unit disk into lines of latitude (radius) and longitude (azimuthal angle), except that discrete radii are used in such a way that all incremental areas are equal. In the figure shown, we have decimated the unit disk into 32 regions, where sectors 1,2,3,4–and therefore all 32 sectors—all have the same area.

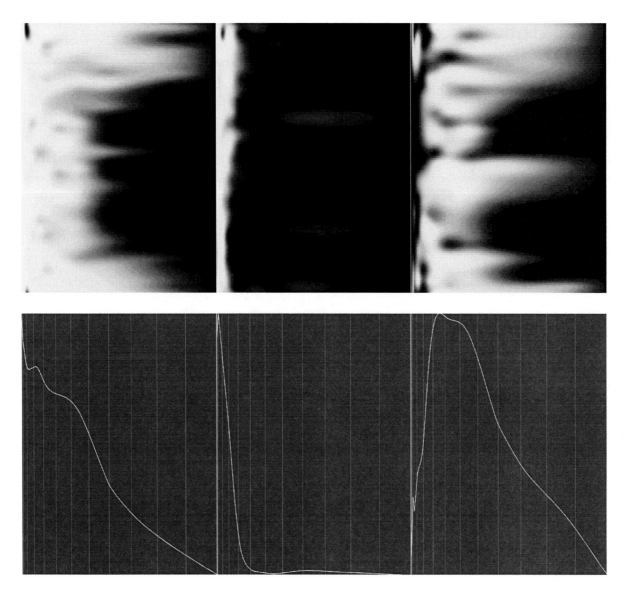

Figure 6: Mercator-signature results for NO, RP, MD patients, left-ro-right respectively. The maximum amplitudes of the normalized signature curves are, from left-to-right respectively, 30.0, 22.3, 13.0. The signature plots are created via columnar summation of the upper, 2-dimensional Mercator plots.

4.2   Radial-Fourier signature

A third signature involves taking the 2-dimensional FFT of the aforementioned nonlinear-Mercator 2-dimensional signature.

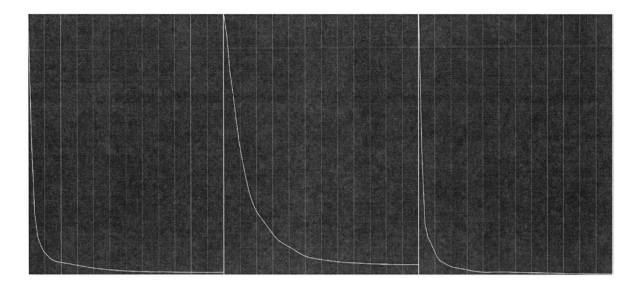

Figure 7: Radial-Fourier signatures for NO, RP, and MD patients, left-ro-right respectively. TThe signature curves are created via columnar summation of the relevant 2-dimensional power-spectrum plot. These are normalized signatures; the actual signature maxima are, for NO, RP, MD respectively, 4340, 743, 3163; therefore the normal patient (left-most panel) has the largest signature excursion.

4.3 Bessel signature

One reason to create a spline surface is to be able to invoke numerical integration algorithms on discrete data sets.

An example runs as follows. A general expansion for splined unit-circle data—now we speak of *radial* coordinates (r, ϕ) and a spline function $S(r, \phi)$—is the Bessel expansion valid for cases where S vanishes on the unit circle:

$$S(r, \phi) \;=\; \sum_{m=-\infty}^{\infty} \sum_{k=1}^{\infty} s_{mk} \, J_m(rz_{mk}) e^{im\phi}.$$

Here, z_{mk} denotes the k-th positive zero of the Bessel function J_m. Note that for real-valued splines S, one may further avoid negative m by noting $J_m = (-1)^m J_{-m}$ and observing $\cos\phi, \sin\phi$ terms. In any case the formal inversion that yields the actaul coefficients s_{nk} is

$$s_{nk} = \frac{1}{\pi} \frac{1}{(J_n'(z_{nk}))^2} \int_0^{2\pi} d\phi \int_0^1 r \, dr \; S(r, \phi) J_n(rz_{nk}) e^{-in\phi}.$$

The strategy, then, to provide a "signature" as a set of expansion coefficients (s_{nk}) for a discrete data input set is

- Create a spline S from the input data, making sure to use a "necklace" of zero-points around the unit circle so that S itself vanishes or nearly so on the unit rim.

- Choose a cutoff parameter M such that coefficient s_{nk} indices will run over $n \in [-M, M]$, $k \in [1, M]$, say. (Again, converting to pure-real algebra can allow n to be only in $[0, M]$; however the complex arithmetic is more elegant and understandable after all.)

- Perform numerical integration to obtain the $M(2M+1) = O(M^2)$ coefficients s_{nk}.

- Check that the spline surface reconstructs well via the (now finite) Bessel sum.

- Report the *signature* of the original data as the set $(|s_{nk}|^2)$.

Note that the signature elements $|s_{nk}|^2$ amount to a kind of *power spectrum* under Bessel decomposition. In general, all rotates of some data set in the unit circle should exhibit equivalent signatures, and this should be tested.

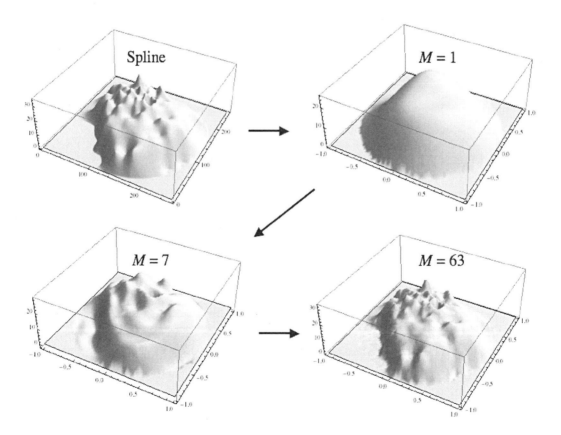

Figure 8: In the Bessel-signature strategy, splined retinal data in signature space (upper-left view) is broken down into $(2M+1)$ Bessel functions. For $M = 1$ (upper-right view) the Bessel functions add to a shape only vaguely reminiscent of the original spline. The sum of the relevant 15 Bessel terms for $M = 7$ (lower-left view) is still yields insufficient resolution,

but for $M = 63$, the sum of the 127 Bessel terms can be seen (lower-right view) to give excellent agreement with the original spline. The whole strategy is based on the idea that the *coefficients* in the Bessel superpositions are themselves the "Bessel signature" for the given retina.

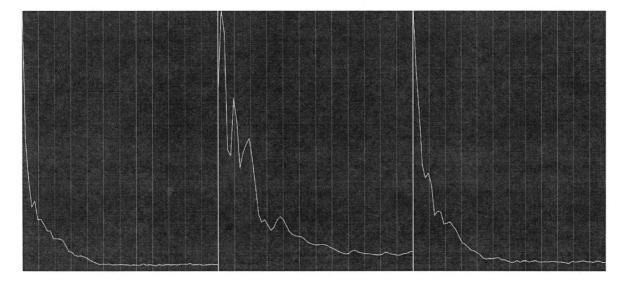

Figure 9: Bessel-signature results for NO, RP, and MD patients, left-ro-right respectively. These signature are obtained by columnar summation of $\log |s_{jk}|$ from the relevant 2-dimensional Bessel-coefficient matrix. Again these signatures are visually normalized, the actual respective maxima being: 0.81, 0.23, 0.58.

4.4   Fractal signature

Our fourth proposed signature we call a "fractal signature," although perhaps more accurate would be "fluctuation signature." The idea here actually arises in computer graphics; namely, a lunar landscape, say, of dimension near 3 is jagged—i.e. attempting, if you will, to be space-filling. On the other hand, a smoothly undulating lowlands, like an almost flat meadow, will have dimension closer to 2.

So, our fractal measure uses the standard box-dimension counting from computational physics, in order to assess the approximate fractal dimension of the spline surface.[3]

Referring to Figure 10, the plot has, for 3-dimensional boxes of side ϵ, a horizontal axis $\log(1/\epsilon)$, and a vertical axis $\#(\epsilon)$, the latter being the number of ϵ-boxes that contain spline points. (For this fractal measurement, we are using the spline on a 256×256 grid, so there

---

[3]Of course, fractal dimension is properly defined only for infinite sets with certain recursive properties. In this sense, "fluctuation signature" is perhaps more appropriate; however, graphics modelers know that the irregularities of "fractal mountain ranges" and so on are measured in such a way—and an analogy with the Hill of Vision is evident.

are 2^{16} points that can occupy boxes.)

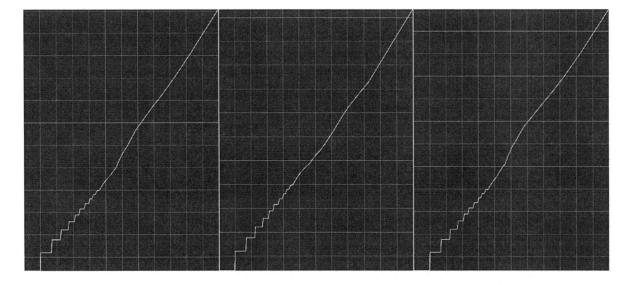

Figure 10: Fractal-signature results for NO, RP, and MD patients, left-to-right respectively. Though these plots look very similar, the dimension Δ varies in subtle ways, sometimes not by much; however, there does appear to be a systematic trend over normal vs. diseased patients (see our Discussion section). The actual dimensions, left-to-right, are: $\Delta = 2.63, 2.58, 2.53$.

5   Discussion

We give here the historical names of the actual data files used in our signature plots.

- Normal (NP), (j, s 124-439 2008.01.03 W187 Gi III OS shown as OD),

- Retinitis pigmentosa (RP), (ad.087.2007.11.13 W187 Gi III OD),

- Stargardt Macular Dystrophy (MD), (h, c 2009.04.27 Stargardt Dys W164 Gi III OS shown as OD).

Though the present treatment is mainly to *define* the mathematical signatures, we already can see—by inspecting the figures of the present treatment and a number of other signature examples not shown—some qualitative trends that suggest the utility of said signatures. Some immediate observations are:

- For the Mercator signature, normal (NO) eyes tend to give essentially linear-descending ramps, while diseased eyes tend to have very different. See Figure 6.

- For the Fourier signature, normal (NO) eyes tend to have exponentially decaying appearance, as do macular dystrophy (MD) eyes. However, the maximum amplitude for MD is reduced. See Figure 7.

- For the Bessel signature, NO eyes typically exhibit a damped form, as do MD eyes; yet again, the diseased case has smaller maximum amplitude. See Figure 9. Note the interesting fact that a Bessel signature that is a pure spike (technically, a delta-function at far left) would mean that the Hill of Vision is actually the zeroth-order Bessel function $aJ_0(br)$ for some constants a, b. That this is never experimentally the case leads to our discovery that in normal (NO) situations, a certain fluctuation in the Hill of Vision is expected. See next item on this very phenomenon.

- For the fractal signature—see Figure 10—there is no visually obvious difference between normal and diseased eyes. However, it is a striking fact that *normal (NO) situations tend to have highest fractal dimension* Δ. It may even be the case that, subtle as Δ turns out to be experimentally, the Δ values are sorted by disease. If so, that would turn out to be an important diagnostic measure. The interesting theoretical question is: Does the normal retina develop spatial sensitivity fluctuation as an artifact, or as an advantage? We are saying that, based on our evidence, the normal retina is nonsmooth in a quantifiable sense.

Further research should involve classification schemes for deciding on which disease, or what disease aspect, is at hand, based on *combining* of various signatures. For example, if the fractal measure gives an imperfect sort of disease categories, there is always the possibility of combining this with, say, Bessel or Mercator signature, to end up with what is sometimes called in classification theory as a "voting scheme."

6   Acknowledgements

The authors thank D. Chalmers and S. Kounitski for their aid in this research. U. Schiefer and J. Pätzold (University of Tübingen, Germany) provided use of their GATE software for data acquisition. Funding was, in part, from Hear See Hope, the Grousbeck Family Foundation, and Foundation Fighting Blindness.

7   Appendix: VFMA application notes

VFMA, an application for Visual Field Modeling and Analysis, was developed by the authors to serve as a platform for the design and testing of algorithms to generate 3-dimensional models of the Hill of Vision for the entire field of vision of the human eye, as measured clinically using static perimetry, and to develop tools for quantification and analysis of the visual field for clinical diagnosis, characterization of the nature and extent of the field defects, and for monitoring the natural history of field loss in health and disease. Creation of this platform also enabled the development and evaluation of mathematical signatures based on Radial Fourier, Bessel, and Fractal analyses of the surface of the Hill of Vision to further characterize the field loss.

Threshold values from the perimeter are imported into the application and converted to differential luminous sensitivity (DLS) values. The Main Window of the application has a rectangular window spanning its width superiorly where the files of data for each eye are displayed, along with the Title of the file, the test date, session, patient number, Eye, Mean Sensitivity (total), Mean Sensitivity $\leq 30^\circ$, Mean Defect, MSV (Minimal Sensitivity Value), OCV (Octopus Conversion Value), and Pupil Size. Selecting one or more of the test data sets allows the operation of other functions that display the test points for the entire grid (with the points equal size or scaled by sensitivity), the 3D model of the DLS values fit with an infinitely differentiable elastic thin plate spline, the normal 3D model of the HOV for the age of the subject, and the Defect Surface, which is the 3D model of the normal HOV from which the patient's data has been subtracted. Another menu allows display of the HOV by a Mercator Projection map and signature graphs of the Mercator function and Radial Fourier spectrum (by column and row), the Bessel function, and the Fractal analysis.

Other menus allow creation of selection templates for measuring the volume in decibel-steradians (dB-steradians) of the entire HOV as assessed by the grid pattern and distribution, selections based on circles of specific diameters (which can be moved eccentrically on the nonlinear-Mercator polar projection to measure a circular representation of the HOV other than centered on the pole), and for selections of scotomas that can be anywhere in the field and different for each eye. All of these selection can be stored for future use for measurements of other field data.

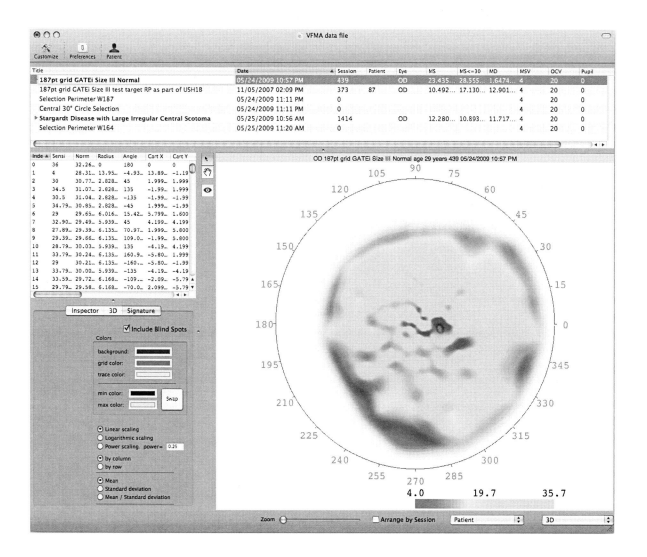

Figure 11: Typical appearance of the Hill of Vision within the VFMA application.

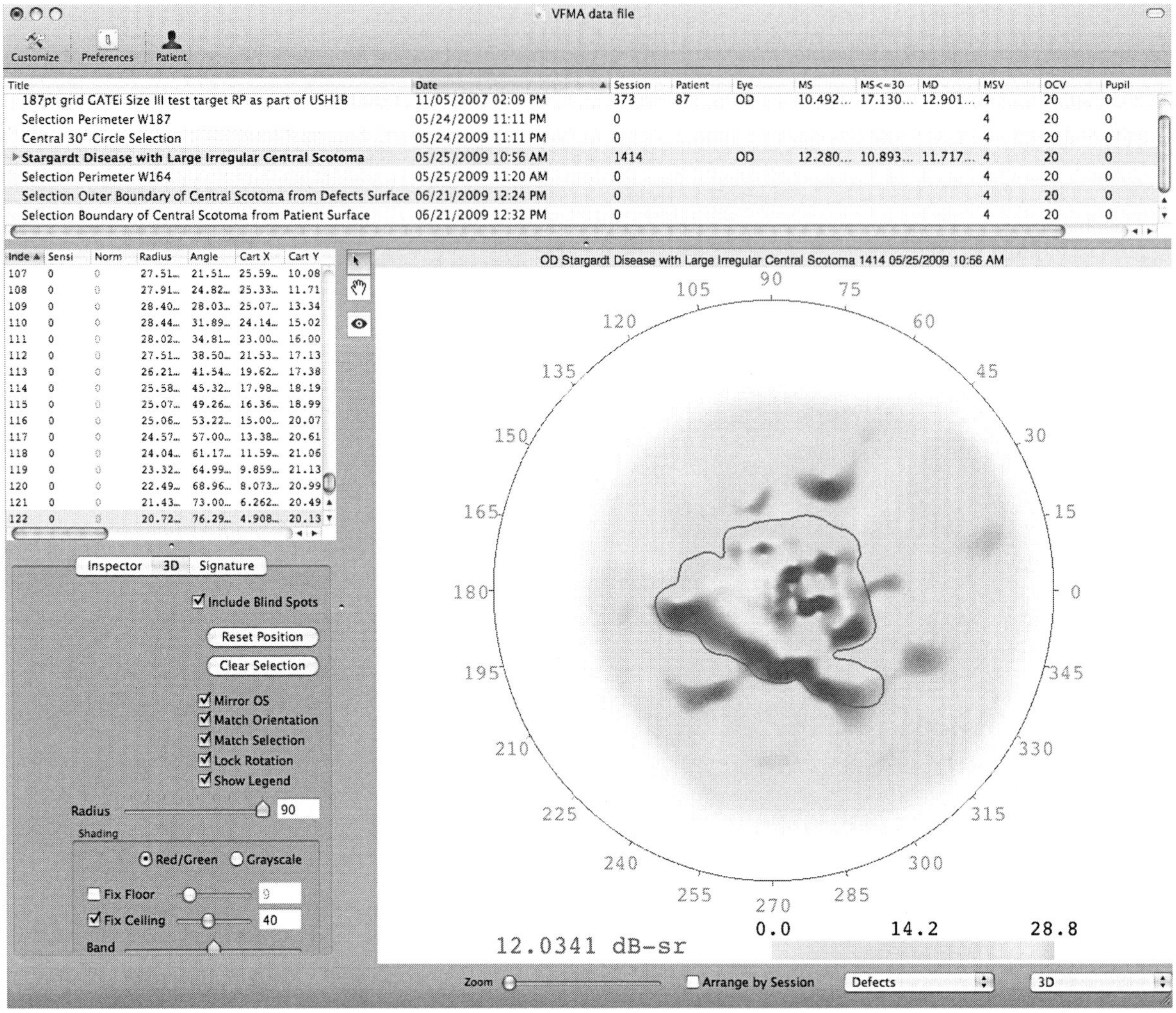

Figure 12: The VFMA application also allows "scribing" of specific areas of the Hill of Vision either as remaining sensitivity or as a scotoma or defect within the Hill of Vision. In this instance, a "Defects" model was created by subtraction of the Hill of Vision for patient MD from Figure 2 from the Hill of Vision of an age-adjusted normal subject. Outlining the base of the inverted scotoma in this "Defects" space allows the measurement of the volume of the scotoma, which is reported as 12.03 dB-steradians.

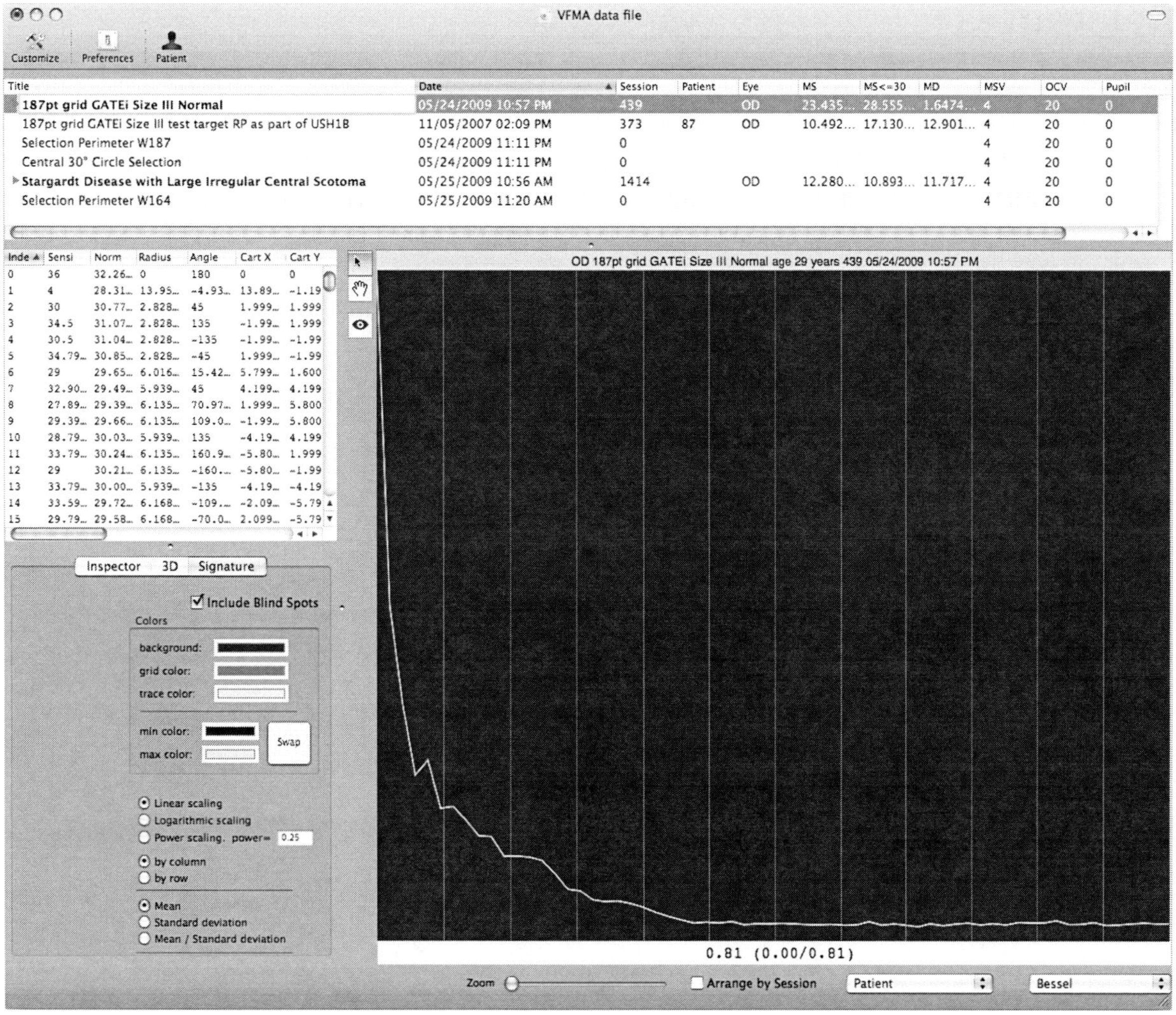

Figure 13: Typical appearance of a signature; here, a Bessel signature for a normal (NO) patient.

References

[1] Elonen, J., `http://elonen.iki.fi/code/tpsdemo/`, 2008

[2] Schiefer U, Paetzold J, Dannheim F., *Konventionelle Perimetrie*, Teil 1 Einfhrung–Grundbegriffe. [Conventional techniques of visual field examination. Part I: Introduction–basics], *Ophthalmologe*,102:627–646, 2005.

[3] Schiefer U, Pascual JP, Edmunds B, Feudner E, Hoffmann EM, Johnson CA, Lagrze WA, Pfeiffer N, Sample PA, Staubach F, Weleber RG, Vonthein R, Krapp E, Paetzold J., "Comparison of the new perimetric GATE strategy with conventional full-threshold and SITA Standard strategies," *Invest Ophthalmol Vis Sci*, 50:488-494, 2009.

[4] Traquair HM , *An Introduction to Clinical Perimetry*, 1st edn. London: Henry Kimpton, 1927.

4. NLA system for medical-data classification

Discussion

This collaborative work at the Center for Advanced Computation was in response to a request from Dr. M. Quarum to classify medical-bill adjudication types. This task may seem lackluster—essentially involving highly sophisticated accounting for medical bills—but as often happens with algorithm problems, the rewards can be enormous. In this case, the reward is a quantitative start to medical-fraud detection, as explained in the paper.

It is refreshing that a neural-network approach can handle human-generated paperwork, and not just digitized images, quantitative trends, and so on.

Challenges

Of course the ultimate goal in such problems is 100 per cent accuracy—a marvelous result if one can attain it, especially marvelous because machine detection is perforce immune to human bias, which bias being an aspect of the fraud problem in the first place.

Source

R. Crandall, K. Lynagh, T. Mehoke, and N. Pepper, "NLA system for medical-data classification," (preprint April 2010).

Subsequent publication: R. Crandall, K. Lynagh, T. Mehoke, and N. Pepper, "Automated Classification of Medical-Billing Data," in ARTIFICIAL INTELLIGENCE APPLICATIONS AND INNOVATIONS; IFIP Advances in Information and Communication Technology, Volume 364/2011, 389-399, (2011).

NLA system for medical-data classification

R. Crandall, K. Lynagh, T. Mehoke, and N. Pepper

April 20, 2010

Abstract. We describe a neural-layer algorithm (NLA) system for medical data classification. A canonical embodiment is a medical-bill classifier with a view to medical-fraud detection. Such a system employs an artificial neural network (ANN) with one hidden layer, the key invention being transformed-histogram inputs, in a sense we clarify mathematically. In this way the NLA system typically outperforms, say, Bayesian counterparts. Our results suggest that the NLA algorithm can go beyond classification per se; namely, other channels of fraud detection may well be possible, especially when the NLA is "ensembled" with other classifiers.

1

1   Brief history of and introduction to ANNs

The history of the modern artificial neural network (ANN) goes back to the seminal work of McCulloch and Pitts, c. 1943, then through the works of Minsky, Rosenblatt, along with many other researchers, eventually culminating in modern software packages of various sophistication and utility. A typical reference on modern ANN implementations is [1].[1]

ANNs are constructed based on the model of their biological counterparts. The primitives of ANN construction are nodes, called *neurons*, and the connections between neurons which form the network. Neurons can be organized into *layers*. The network receives input vectors into the input layer, makes computations and may pass the result on to subsequent 'hidden' layers, until the last layer, the output layer, produces a final output vector. There are two general types of NN topology. The first is a *feedforward* network in which there are no loops and connections within same layer – all connections only go forward to the next layer. The other kind of network topology is *recurrent*, allowing possible feedback loops [1].

Rather than executing direct commands, a neural network contains instructions for adaptively adjusting its own parameters in response to the data. This means that an ANN can 'learn' to self-organize and detect statistical, even non-linear, patterns from the data. Further, an ANN is able to generalize from training data to make predictions for previously unseen data. Therefore, ANNs are well-suited to problems that lack an accepted mathematical model, predetermined step-by-step instructions, or for which there is incomplete understanding of the data [5].

Execution of an ANN requires two initial phases: a training phase and a testing phase. Conventionally, 2/3 of the data are used for training while the final 1/3 is reserved for testing. Training and testing are generally conducted in one of two modes: *supervised* or *unsupervised learning*. In a *supervised learning* mode, an auditor, usually a human, gives feedback to indicate whether the ANN output is correct (a supervisor must have access to the 'right' answers for comparison).

Since the computations of each neural layer depend on what that layer received as the input propagated through the network, error correction cannot be handled only by the output layer. The network must use error *back-propagation* to adjust the behavior of all layers leading up to the output layer as well [2].

For our present research, the particular type of ANN we chose is an open source application called the Fast Artificial Neural Network (FANN), which is a multilayer, feed-forward, fast ANN library with support for fixed-point arithmetic. FANN is designed for situations requiring fast execution, where training is not as time-critical [2].

---

[1] For the present research we have employed *FANN*, a fast artificial neural network package (see [2]).

2 Our basic NLA paradigm

Figure 1 shows the style of neural layer connection we wish to employ. Note that

- There are T data types.

- Vector inputs from the data, generally represented by I, could be either analog (A) vector inputs or digital (D) vector inputs to the neural network.

- For analog data, F data fields assumed. There are T data types, and so, in the case of analog information, $F \cdot T$ input nodes may be thought of as a $T \times F$ matrix of input signals $\{A_{tf}\}$;

- For digital data, G, data fields are assumed; There are T data types, and so, in the case of digital information, $G \cdot T$ input nodes may be thought of as a $T \times G$ matrix of input signals $\{D_{tg}\}$;

- There are H perceptrons—labelled P_0 through P_{H-1} in the hidden layer; The label H may be considered the *height* of the neurons 'stacked' in the layer.

- There are T outputs, labelled $O_0, \ldots, O_{T-1}$.

3 Canonical embodiment: Medical bill analysis

A specific embodiment of the general scheme in Figure 1 runs as follows. In prevailing medical-bill databases, there are at least 7 different *adjudication types* with which providers can classify and report services on medical bills. These adjudication types describe what general category of service was provided which is directly relevant for assessing how much compensation the provider can expect to receive from an insurer, such as Medicare. The standard 7 specific adjudication types, in alphabetical order, are

- t=0 Ambulance (AMB),

- t=1 Ambulatory service center (ASC),

- t=2 Durable medical equipment (DME),

- t=3 Emergency room (ER),

- t=4 Inpatient hospital (IPH),

- t=5 Outpatient hospital (OPH),

- t=6 (Professional services) (PRO).

3

t

Note that we have here given the t value we use, consistent with the basic scheme of Figure 1; and, of course, for this specific embodiment the total number of types is $T = 7$.

Why does modern medical-bill analysis require accurate assessment of these adjudication types? Traditionally, in medical bill review human auditors will determine the adjudication type of a bill manually and then apply the appropriate rules or fee schedules. There are two problems with this approach which an automated solution can help solve. First, humans are much slower than an optimized machine learning engine. While the engine described in this paper can classify around 14,000 bills per second a human takes anywhere from 10 seconds to 45 seconds to classify a single normal bill. Not only does the algorithm classify bills many orders of magnitude faster than a human, it also saves a large amount of worker time which translates to huge savings and increased efficiency. Additionally, if one were to build a fully automated system for applying fee schedules and rules to medical bills this would eliminate the human bottleneck entirely.

Second, the algorithm can be carefully tuned to reduce errors. Humans tend to make errors when trying to do many detail-oriented tasks rapidly. While many bills may be classified using an extensive rules-based automated system, such a methodology is inferior to a machine learning approach. Rules- based systems cannot reliably handle edge cases and are labor intensive to modify or update. The genesis of this project was a desire to solve a software engineering problem: accurately route bills through a fraud detection and cost containment pipeline. Initial attempts at solving this problem using manually coded rule sets created error rates that were unacceptable. This lead to much of the classification being redone manually. The machine learning approach presented here was developed for two purposes. First, the system allows for improvement of classification accuracy of manual rule sets and manual auditors by examining the output of the machine learning algorithm. Second, this system can be integrated as a part of this larger information system.

3.1 Analog data

By "analog data" we mean data that is in some appropriate sense ordered in analog fashion; i.e., ordered as to physical meaning. Examples are:

- The total medical bill C, in cash dollars (so, a float),

- The number of sub-bills S (an integer),

- The calendar duration of sub-bills D (an integer),

- The sub-bill entropy E, in bits (a float).

The first three (C, S, D) field definitions are self-explanatory. We define the entropy E as follows:

$$E = -\sum_{i=0}^{S-1} \frac{C_i}{C} \log_2 \frac{C_i}{C},$$

4

where C_i are the sub-bill costs, and of course $\sum_i C_i = C$.

Interestingly, we have found that histograms of the above analog parameters—generally speaking—are nicely dispersed *provided that the horizontal histogram axis is logarithmic*. This notion is quantified in Section 4.

On the other hand digital (equivalently, categorical) data are represented by strings indicating a medical procedure (such as with icd9 codes). These data have no magnitude or natural distance metric.

4   A new idea: Histogram transformation and inversion

The most powerful histogram transformation we have found in the canonical context of medical-data input is as follows. Fix the data field $f \in [0, F-1]$ and assume the drawing of N sample data $\{x_{it} : i \in [0, N-1]\}$ from a training set, for each type $t \in [0, T-1]$. A standard histogram can be constructed, with (say) equal-spaced bins decimating the horizontal axis from $x_{\min,t}$ to $x_{\max,t}$, with total bin sum equal to N. Unfortunately, it often occurs that such a linear-bin histogram is too heavily skewed, as in Figure 2.

The key to proper transformation of histograms is to establish first the global extrema, defined

$$ m_f := \min_t x_{\min,t}, $$

$$ M_f := \max_t x_{\max,t}, $$

so that *every* datum x from the total of NT sample data lies in the interval $[m_f, M_f]$.

See Figure 3, a pictorial of transformed histograms under $W(x) := \log_{10} x$. One convenient way to visualize the new, transformed histograms is to append to every fixed-t set of N samples the isolated values $W(m_f), W(M_f)$ and then create the nonlinear histograms having horizontal axis $W(x)$.

See Figure 4 for a pictorial of the general NLA input scheme. The nonlinearly transformed histograms are used to provide "probabilities" π_t for any given data-field entry x outside the training set.

5   Confusion matrices

Our primary vehicle for pictorializing the results of NLA training is the confusion matrix. A (hopefully large) set of bills is used, typically with 2/3 of them for NLA training, and the remaining 1/3 for testing. The top labels (left-to-right) of the confusion matrix is the true type under test. To create the matrix, the "guessed" type is tallied according to the type label at matrix-left (top-to-bottom labels). The ideal confusion matrix will have 100% on the diagonal, with 0% everywhere off-diagonal.

In Figure 4 the log-histogram heights for a new "mystery" bill under test are normalized and fed to the NLA. Denote for a given data-field index f the trained-histogram heights by π_i, where $i \in [0, T-1]$ runs through the adjudication types. Then for frequency mode set to 0, we define

$$I_{i,f} := \frac{\pi_i}{\sum \pi_i},$$

So that "probabilities" summing to unity are being fed directly to the NLA, as in Figure 4. If, however, we want frequency mode to be 1, we renormalize by:

$$\pi_i := \pi_i \cdot P_i,$$

where P_i is the overall probability of type i occurring in the training set. In these cases with frequency mode = 1, we again feed the NLA with inputs $I_{i,f} := \pi_i / \sum \pi_i$. Thus, in either frequency mode = 0,1 cases, the NLA inputs loom as probabilities that sum to unity. We allow this frequency-mode degree of freedom, because we do not currently have a rigorous theory of when the "natural probability" of a type's occurrence should be used as a weight.

In order to be rigorous in our reporting, and allow future investigators to fairly reproduce our kinds of results, we provide with each confusion matrix figure a table of parameters. The general parameters table reads as follows (we admit that figures thus need rather tedious notation/definition; such is often the case with neural-network results(:

bills, train, epochs, test: (# bills, #training bills, #NLA training epochs, # test bills)
hidden, neurons: (list of (#layers, #neurons/layer))
polarity, offset: (mono/bi, value of extra offset line)
histogram bins, mode: (list of (parameter, # bins in log-histogram, frequency mode))
digital fields: (list of (medical codestrings), occurrence threshold)

In the last row here, the occurrence threshold is the minimum number of times a medical code must appear before being flagged—such a threshold serves to minimize random/chaotic misidentification.

To summarize our NLA reporting via pictorials:

- Figure 7 shows confusion matrices for representative, 2-by-2-type competition,

- Figure 8 shows a matrix using analog data-fields only,

- Figure 9 shows a result for digital data-fields only,

- Figure 10 shows the "grand confusion matrix" with both analog and digital data fields included.

References

[1] K. Cios, W. Pedrycz, and R. Swiniarski,"Data mining methods for knowledge and discovery," Kluwer Academic Publishers, 1998.

[2] S. Nissen, "Implementation of a Fast Artificial Neural Network Library (FANN)," Department of Computer Science University of Copenhagen (DIKU), 31 Oct 2003. `http://leenissen.dk/fann/index.php`

[3] A. Chariatis, "Very Fast Online Learning of Highly Non Linear Problems," Journal of Machine Learning Research, 2007.

[4] S. Haykin, "Neural networks and learning machines," Prentice Hall, 2009.

[5] A. Tettamanzi and M. Tomassini, "Soft computing: integrating evolutionary, neural, and fuzzy systems," Springer, 2001.

[6] C.M. Bishop, "Pattern Recognition and Machine Learning," Springer Science + Business Media, LLC, 2006.

[7] C.M. Bishop, "Neural Networks for Pattern Recognition," Oxford University Press Inc., New York, 2007.

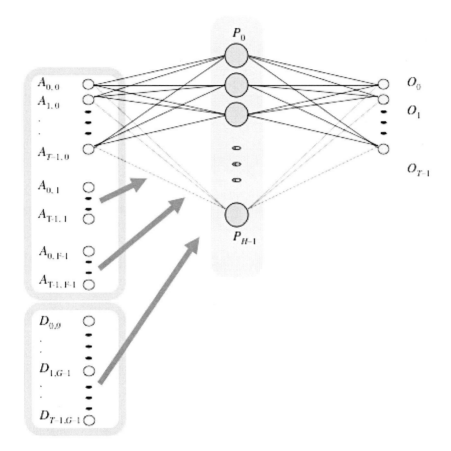

Figure 1: Basic neural-layer algorithm (NLA) configuration. There are T data types to be classified, for input type A (analog) with F data fields and input type D (digital) with G data fields, H hidden-layer perceptrons, and T output nodes. The input data is a matrix of values $\{A_{tf}\}$ or $\{D_{tg}\}$ where $t \in [0, T-1]$ is the type index, and $f \in [0, F-1]$ or $g \in [0, G-1]$ is the data-field index.

Figure 2: Typical standard histogram; here the horizontal axis is in dollars, for the cost C data field. The left-skewness is not only obvious, but prevents sharp analysis across types.

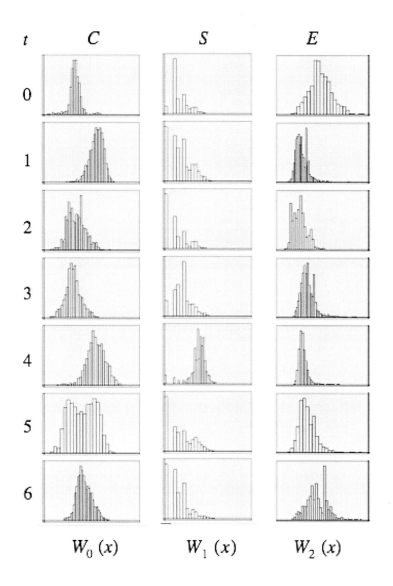

Figure 3: Collection of 21 histograms; 7 per type $t \in [0,6]$ and 3 per data field $f \in [0,2]$. The data fields are C (total cost), S (number of sub-bills), E (sub-bill entropy). The horizontal data axes are $W_f(x)$ for nonlinear transformation functions W that serve to disperse the histograms.

10

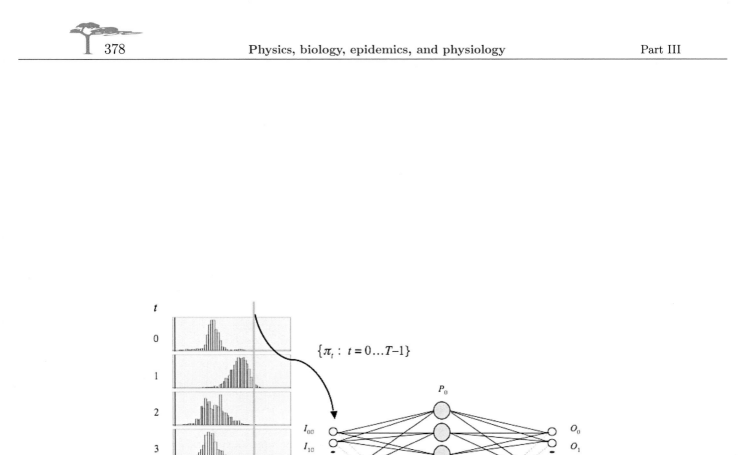

Figure 4: The new idea of histogram transformation prior to NLA input. A data-field-dependent transformation W_f is used to transform histograms of the T types in a dispersive manner, to isolate statistical features. For a given data value x, the relative probabilities (along blue vertical line) are accumulated, normalized (see text), then become the T NLA inputs.

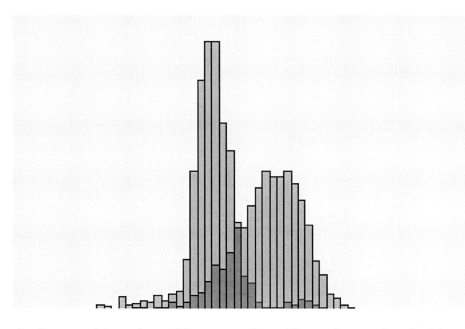

Figure 5: Superposition of two histograms from Figure 4; namely, the histograms for $t = 1, 2$ (AMB and ASC types). The important phenomenon is the near-orthogonality of such examples: This allows filtering/pruning as an aid to training.

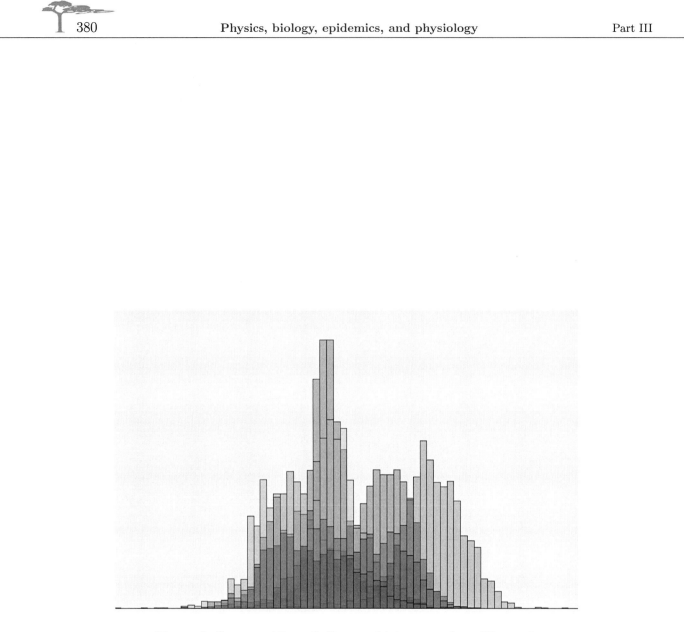

Figure 6: Superposition of all seven histograms from Figure 4.

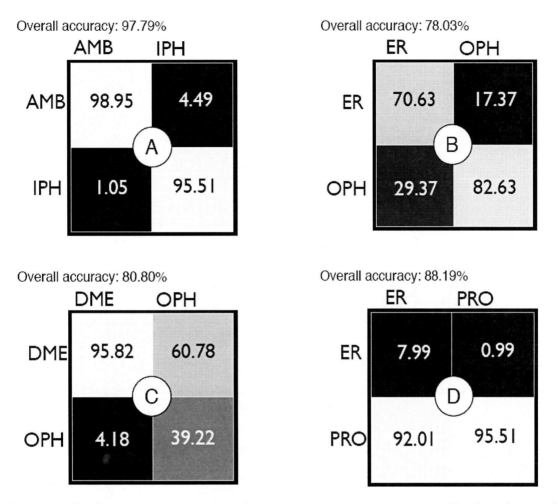

Figure 5: Confusion matrices arising from two-histogram inputs; namely, the scheme of Figure 4 is used with $T = 2$ input types. The four example matrices labelled A, B, C, D show various levels of accuracy/inaccuracy. For example, confusion matrix A is quite successful, mirroring the fact that histograms for types AMB, IPH tend to be disjoint (of minimal overlap). On the other hand, matrix D exemplifies the scenario in which a type—PRO—is simply so predominant in the bill population that the ER type is difficult to detect, at least via analog histograms alone.

bills, train, epochs, test: (30000, 20000, 600, 10000)
hidden, neurons: ((1, 4))
polarity, offset: (mono, +1)
histogram bins, mode: ((C, 512,0),(S, 64,0), (D,32,0), (E, 256,0))
digital fields: N/A

Overall accuracy: 58.10%

	AMB	ASC	DME	ER	IPH	OPH	PRO
AMB	47.75	1.37	4.10	4.97	0.00	1.37	1.75
ASC	8.13	76.05	6.30	21.47	36.57	37.28	8.44
DME	13.47	2.10	41.27	4.80	1.28	6.81	8.85
ER	2.00	0.56	0.35	4.28	0.00	0.50	0.67
IPH	0.00	1.90	0.05	0.46	59.08	2.38	0.08
OPH	0.14	0.09	0.20	0.23	0.26	1.80	0.20
PRO	28.51	17.94	47.72	63.79	2.81	49.86	80.01

Figure 8: Confusion matrix using analog columns only. Some adjudication types—such as ASC and PRO—do well as evidenced by their respective diagonal elements.

bills, train, epochs, test: (100000, 66666, 600, 33334)
hidden, neurons: ((1, 11))
polarity, offset: (mono, +1)
histogram bins, mode: ((C, 512,0),(S, 64,0), (D,32,0), (E, 256,0))
digital fields: N/A

15

Overall accuracy: 86.28%

	AMB	ASC	DME	ER	IPH	OPH	PRO
AMB	76.12	0.55	1.34	0.06	0.00	0.00	0.10
ASC	1.43	77.3	0.10	12.34	7.80	18.70	1.32
DME	5.63	1.28	93.53	0.06	0.26	0.22	2.67
ER	0.57	4.71	0.00	70.93	2.30	4.65	0.43
IPH	0.07	1.23	0.04	1.77	72.12	2.34	0.10
OPH	0.50	8.46	0.52	10.74	13.68	68.91	1.67
PRO	15.68	6.47	4.47	4.11	3.84	5.19	93.71

Figure 9: Confusion matrix using only digital columns; thus, the histogram methods are not relevant. The accuracies are good, but can be improved by combining with the analog algorithms.

bills, train, epochs, test: (100000, 66666, 600, 33334)
hidden, neurons: ((1, 11))
polarity, offset: (mono, +1)
histogram bins, mode: N/A
digital fields: ((TIN, ICD9, SVCS, MS, HRCS, NCDS), 5)

16

Overall accuracy: 88.52%

	AMB	ASC	DME	ER	IPH	OPH	PRO
AMB	87.88	0.63	1.51	0.17	0.00	0.00	0.13
ASC	1.21	82.37	0.08	13.59	7.93	23.78	1.34
DME	5.84	1.40	94.90	0.06	0.64	0.36	1.83
ER	0.14	4.71	0.00	75.56	0.51	3.46	0.46
IPH	0.00	0.91	0.03	0.69	82.74	1.44	0.04
OPH	0.29	5.65	0.13	6.97	4.60	66.43	1.84
PRO	4.63	4.32	3.35	2.97	3.58	4.54	94.37

Figure 10: Grand confusion matrix resulting from analog-and-digital data fields combined. The overall accuracy is nearly 90%. It is evident that, while digital fields are more acute as adjudication predictors, analog fields do contribute to overall accuracy, especially for specific matrix elements such as the ASC-ASC diagonal element here (compare with previous figure's such element).

bills, train, epochs, test: (100000, 66666, 600, 33334)
hidden, neurons: ((1, 11))
polarity, offset: (mono, +1)
histogram bins, mode: ((C, 512,0),(S, 64,0), (D,32,0), (E, 256,0))
digital fields: ((TIN, ICD9, SVCS, MS, HRCS, NCDS), 5)

5. On the fractal distribution of brain synapses

Discussion

It was in early 2008 that the eminent Dr. S. Smith of Stanford Medical School asked me if one could carry out mathematical analyses of their new, 3-dimensional mouse-brain data. The website URLs and other references in this paper are wonderful to behold—and I understand some such brain videos—involving motion down into brain tissue—will be exhibited at the Smithsonian in the near future.

The paper here is a mathematical one, as this author is neither a biologist nor could even hope to pass for one. Moreover, these results were first presented at a 2011 conference dedicated to the work of J. Borwein. Indeed, the present author and Borwein had worked together for a decade on so-called "box integrals," being statistical moments within an ideal box, and this was the takeoff point for the mathematics herein. (Note that the reference given under Source below will be the Conference Proceedings from that initial event—including many other papers on other mathematical subjects.)

It must be claimed that the primary result of this work is the successful detection of actual neural layers. Instead of layer detection via human-microscope symbiosis, now we have purely mathematical means for isolating layers.

Challenges

One scintillating—and still tough— goal is to use fractal statistics to diagnose brain pathology. The Alzheimer's example is given as a footnote herein. Another goal is to deduce features of brain function—not just the critical layers, say, but deeper structural patterns.

Source

R. Crandall, "On the fractal distribution of brain synapses," *Computational and Analytical Mathematics*, Springer, (to appear 2012).

On the fractal distribution of brain synapses

Richard Crandall

Abstract Herein we present mathematical ideas for assessing the fractal character of distributions of brain synapses. Remarkably, laboratory data are now available in the form of actual 3-dimensional coordinates for millions of mouse-brain synapses (courtesy of Smithlab at Stanford Medical School). We analyze synapse datasets in regard to statistical moments and fractal measures. It is found that moments do not behave as if the distributions are uniformly random, and this observation can be quantified. Accordingly, we also find that the measured fractal dimension of each of two synapse datasets is 2.8 ± 0.05. Moreover, we are able to detect actual neural layers by generating what we call probagrams, paramegrams, and fractagrams— these are surfaces one of whose support axes is the y-depth (into the brain sample). Even the measured fractal dimension is evidently neural-layer dependent.

1 Motivation

Those who study or delight in fractals know full well that often the fractal nature is underscored by structural rules. When the author was informed by colleagues[1] that 3D synapse data is now available in numerical form, it loomed natural that mathematical methods should be brought to bear.

Thus we open the discussion with the following disclaimer: The present paper is not a neurobiological treatise of any kind. It is a mathematical treatise. Moreover, there is no medical implication here, other than the possibility of using such measures as we investigate for creation of diagnostic tools.[2]

Richard Crandall
Center for Advanced Computation, Reed College, e-mail: crandall@reed.edu

[1] From Smithlab, of Stanford Medical School [15].

[2] Indeed, one motivation for high-level brain science in neurobiology laboratories is the under-standing of such conditions as Alzheimer's syndrome. One should not rule out the possibility of

2   Richard Crandall

There is some precedent for this kind of mathematical approach. Several of many
fractal studies on neurological structures and signals include [10] [8] [9]. Studies on
random-point clouds per se have even been suggested for the stringent testing of
random-number generators [7]. Some researchers have attempted to attribute notions of context-dependent processing, or even competition to the activity within
neural layers [1]. Indeed, it is known that dendrites—upon which synapses subsist—
travel through layers. Some good rendition graphics are found in [16]. Again, our
input datasets do not convey any information about dendritic structure; although, it
could be that deeper analysis will ultimately be able to suggest dendritic presence.

2 Synapse data for mathematical analysis

Our source data is in the section Appendix: Synapse datasets. It is important to note
that said data consists exclusively of triples (x,y,z) of integers, each triple locating
a single brain synapse, and we rescale to nanometers to yield physically realistic
point-clouds. There is no neurological structure per se embedded in the data. This
lack of structural information actually allows straightforward comparison to random
point-clouds.

To be clear, each synapse dataset has the form

$$x_0 \ \ y_0 \ \ z_0$$

$$x_1 \ \ y_1 \ \ z_1$$

$$\cdots$$

$$x_j \ \ y_j \ \ z_j \qquad (=\mathbf{r})$$

$$\cdots$$

$$x_k \ \ y_k \ \ z_k \qquad (=\mathbf{q})$$

$$\cdots$$

$$x_{N-1} \ \ y_{N-1} \ \ z_{N-1}$$

where each x,y,z is an integer (Appendix 1 gives the nanometer quantization). There
are N points, and we have indicated symbolically here that we envision some row
as point $\mathbf{r}$ and some other row as point $\mathbf{q}$, for the purposes of statistical analysis.
(A point $\mathbf{r}$ may or may not precede a $\mathbf{q}$ on the list, although in our calculations we
generally enforce $\mathbf{r} \neq \mathbf{q}$ to avoid singularities in some moments.)

"statistical" detection of some brain states and conditions—at least, that is our primary motive for
bringing mathematics into play.

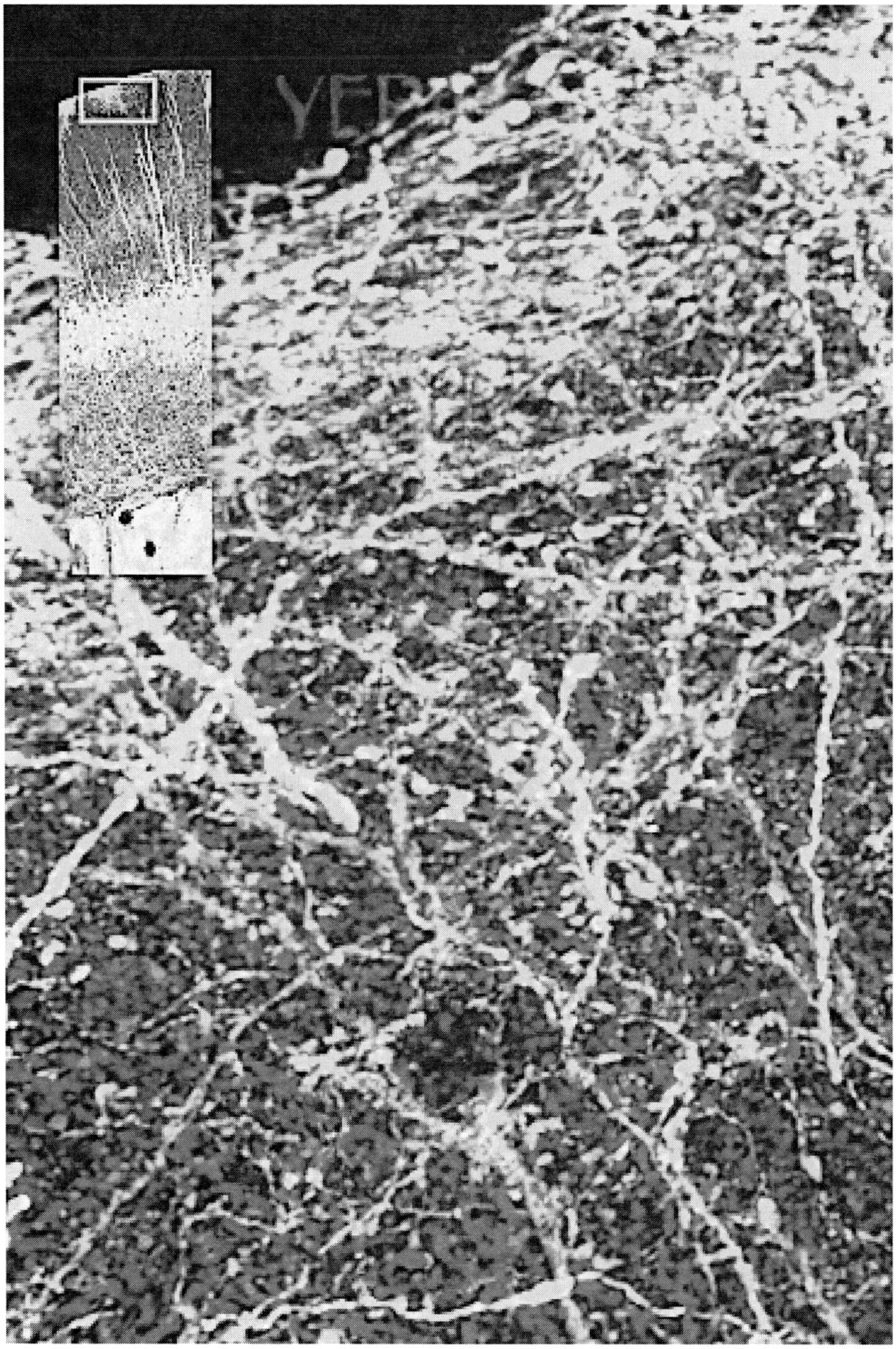

On the fractal distribution of brain synapses   3

Fig. 1 Frame from video: The beginning (top layer, $y \sim 0$) of a mouse-brain section. Synapses (our present data of interest) are red points. The vertical strip at upper left represents the complete section—the small light-pink rectangle indicates the region we are currently seeing in the video (courtesy of Smithlab, Stanford Medical School [15]).

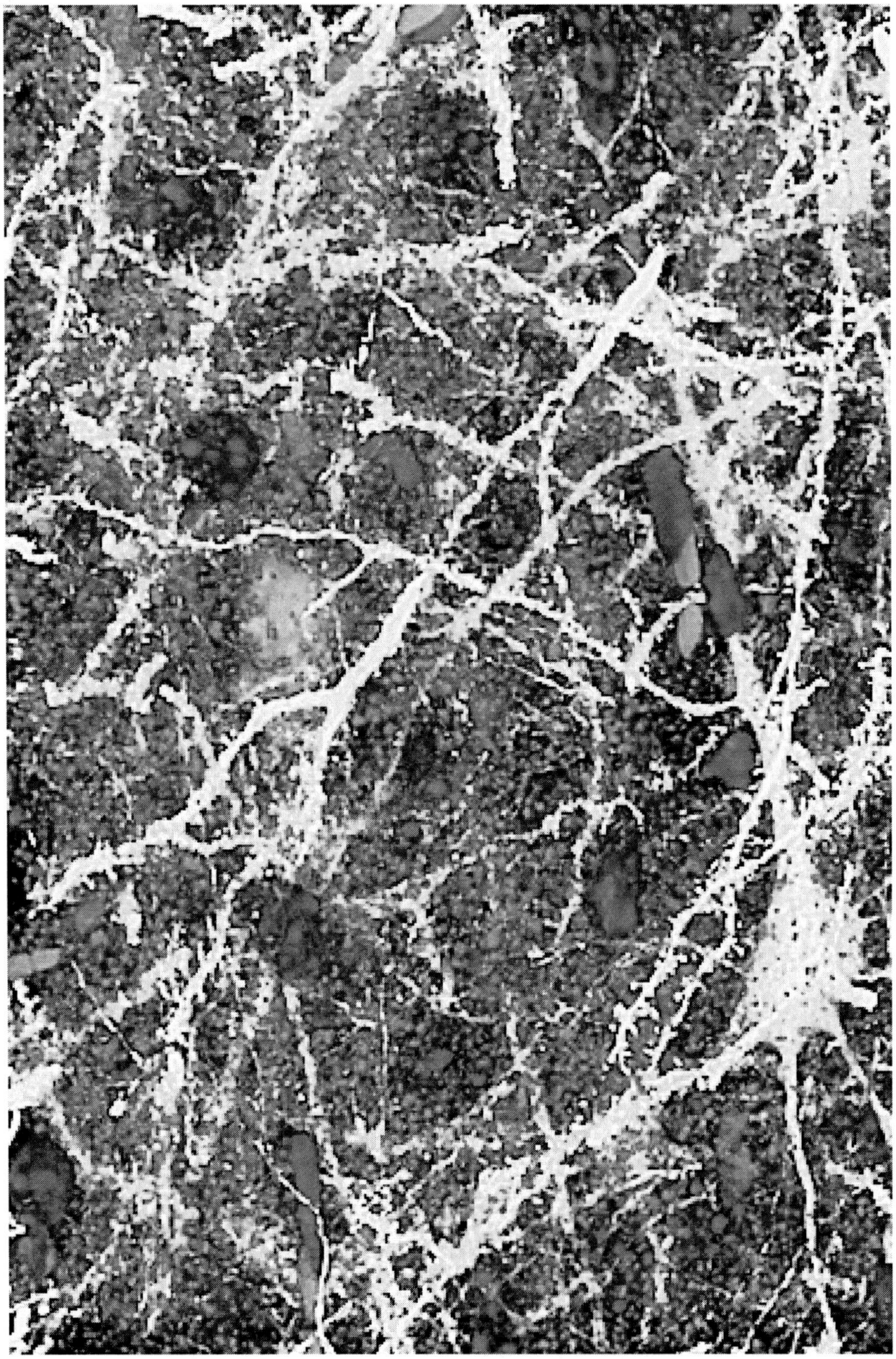

4 Richard Crandall

Fig. 2 A subsection in neural layer 5b. The chemical color-coding is as follows. Green: Thy1-H-YFP (Layer 5B Neuron Subset); Red: Synapsin I (Synapses); Blue: DAPI (DNA in all nuclei). All of our present analyses involve only the Synapsin-detected synapses.

On the fractal distribution of brain synapses   5

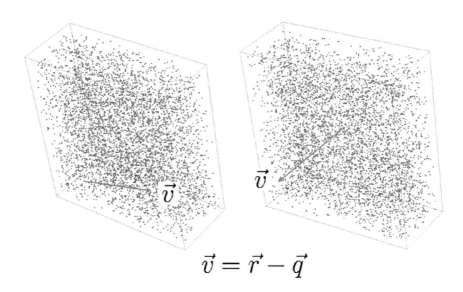

$$\vec{v} = \vec{r} - \vec{q}$$

Fig. 3 Views of 5000 random points (left) and 5000 actual synapses (right) in a cuboid of given sides as follows (all in nanometers); $a = \Delta x \sim 103300; b = \Delta y \sim 78200; c = \Delta z \sim 11400$, for horizontal, vertical, and transverse (angled into page) respectively. To convey an idea of scale: A millimeter is about 10x the horizontal span of either point-cloud. It is hard to see visual differences between the random points at left and the actual brain points at right. Nevertheless, sufficiently delicate statistical measures such as moments $\langle |\mathbf{v}|^s \rangle$ as well as fractal measurement do reveal systematic, quantifiable differences.

3 The modern theory of box integrals

Box integrals—essentially statistical expectations, also called moments, over a unit box rather than over all of space—have a rich, decades-long history (see [3, 4, 5] and historical references therein). The most modern results involve such functions as

$$\Delta_n(s) := \langle |\mathbf{r} - \mathbf{q}| \rangle |_{\mathbf{r},\mathbf{q} \in [0,1]^n}$$

$$= \int_0^1 \cdots \int_0^1 \left(\sum_{k=1}^{n} (r_k - q_k)^2 \right)^{s/2} dr_1 dq_1 dr_2 dq_2 \cdots dr_n dq_n.$$

This can be interpreted physically as the expected value of v^s, where separation $v = |\mathbf{v}|$, $\mathbf{v} := \mathbf{r} - \mathbf{q}$ is the distance between two uniformly random points each lying in the unit n-cube.

It is of theoretical interest that $\Delta_n(s)$ can be given a closed form for every integer s, in the cases $n = 1, 2, 3, 4, 5$ [5]. For example, the expected distance between two points in the unit 3-cube is given exactly by

6   Richard Crandall

$$\Delta_3(1) \;=\; -\frac{118}{21} - \frac{2}{3}\pi + \frac{34}{21}\sqrt{2} - \frac{4}{7}\sqrt{3} + 2\log\left(1+\sqrt{2}\right) + 8\log\left(\frac{1+\sqrt{3}}{\sqrt{2}}\right)$$

$$= 0.6617071822671762351558311332484135817464001357 9095\ldots.$$

The exact formula allows a comparison between a given point cloud and a random cloud: One may calculate the empirical expectation $\langle|\mathbf{r}-\mathbf{q}|\rangle$, where $\mathbf{r},\mathbf{q}$ each runs over the point cloud, and compares with the exact expression $\Delta_3(1) \approx \ldots$. Similarly it is known that the expected inverse separation in the 3-cube is

$$\Delta_3(-1) := \left\langle \frac{1}{|\mathbf{r}-\mathbf{q}|} \right\rangle =$$

$$\frac{2}{5} - \frac{2}{3}\pi + \frac{2}{5}\sqrt{2} - \frac{4}{5}\sqrt{3} + 2\log\left(1+\sqrt{2}\right) + 12\log\left(\frac{1+\sqrt{3}}{\sqrt{2}}\right) - 4\log\left(2+\sqrt{3}\right)$$

$$= 1.8823126443896601601056008388683675878524628803 1070\ldots.$$

Such exact forms do not directly apply in our analysis of the brain data, because we need volume sections that are not necessarily cubical. For this reason, we next investigate a generalization of box integrals to cuboid volumes.

4 Toward a theory of cuboid integrals

In the present study we shall require a more general 3-dimensional box integral involving a cuboid of sides (a,b,c).[3] Consider therefore an expectation for two points $\mathbf{r},\mathbf{q}$ lying in the same cuboid:

$$\Delta_3(s;a,b,c) := \langle|\mathbf{r}-\mathbf{q}|\rangle|_{\mathbf{r},\mathbf{q}\in[0,a]\times[0,b]\times[0,c]}$$

$$= \frac{1}{a^2 b^2 c^2} \int_0^a \int_0^a \int_0^b \int_0^b \int_0^c \int_0^c |\mathbf{r}-\mathbf{q}|^s \, dr_1 \, dq_1 \, dr_2 \, dq_2 \, dr_3 \, dq_3.$$

This agrees with the standard box integral $\Delta_3(s)$ when $(a,b,c)=(1,1,1)$.

Figure 6 shows the result of empirical assessment of cuboid expectations for dataset I.

We introduce a generalized box integral, as depending on fixed parameters k, a_1, a_2, a_3 (we use a_i here rather than a,b,c just for economy of notation):

$$G_3(k;a_1,a_2,a_3) := \langle e^{-k|\mathbf{p}-\mathbf{q}|^2}\rangle$$

$$= \frac{1}{\prod a_i^2} \int_0^{a_1} \int_0^{a_1} \cdots \int_0^{a_3} \int_0^{a_3} e^{-k|\mathbf{r}-\mathbf{q}|^2} \, dr_1 \, dr_2 \, dr_3 \, dq_1 \, dq_2 \, dq_3,$$

---

[3] A cuboid being a parallelepiped with all faces rectangular—essentially a "right parallelepiped."

On the fractal distribution of brain synapses   7

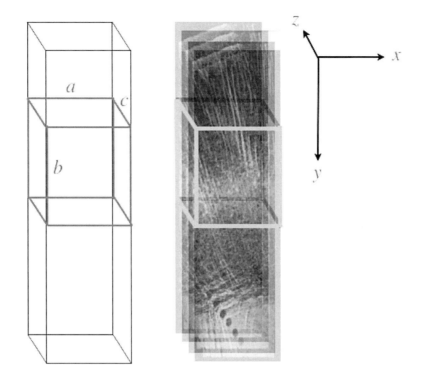

Fig. 4 Pictorial of the role of cuboid calculus in our analysis scenario. The right-hand entity pictorializes an array-tomography section of mouse brain (see Appendix: Synapse datasets for details). At the left is an idealized, long cuboid representing the full brain-sample, inside of which is a chosen subsection as an (a,b,c)-cuboid. The idea is to statistically compare the synapse distribution within an (a,b,c)-cuboid against a random distribution having the same cuboid population. By moving the (a,b,c) cuboid downward, along the y-axis, one can actually detect neural layers.

which, happily, can be given a closed form:

$$G_3(k;a_1,a_2,a_3) = \frac{1}{k^3}\prod_i \frac{e^{-a_i^2 k} + a_i\sqrt{\pi k}\,\mathrm{erf}\left(a_i\sqrt{k}\right) - 1}{a_i^2},$$

where $\mathrm{erf}(z) := 2/\sqrt{\pi}\int_0^z e^{-t^2}\,dt$ denotes the error function. The closed form here is quite useful, and by expanding the erf() in a standard series, we obtain for example a 3-dimensional summation for G_3. The question is, can one write a summation that is of lower dimension? One possible approach is to expand the Gaussian in even powers of $|\mathbf{p}-\mathbf{q}|$ and leverage known results in regard to box integrals Δ_n of Bailey, Borwein and Crandall [3, 4]. Such dimensionality reduction remains an open problem.

Yet another expectation that holds promise for point-cloud analysis is what one might call a Yukawa expectation:

8   Richard Crandall

$$Y_3(k;a_1,a_2,a_3) := \left\langle \frac{e^{-k|\mathbf{r}-\mathbf{q}|}}{|\mathbf{r}-\mathbf{q}|} \right\rangle.$$

This is the expected Yukawa potential—of nuclear physics lore—between two points within the cuboid. The reason such potentials are of interest is that being "short-range" (just like nuclear forces) means that effects of closely clustered points will be amplified. Put another way: The boundary effects due to finitude of a cuboid can be rejected to some degree in this way.

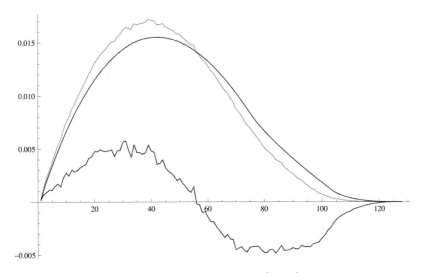

Fig. 5 Probability density curves for the separation $v = |\mathbf{r}-\mathbf{q}|$ (horizontal axis), taken over a cuboid of data, in the spirit of Figure 4. The green curve (with highest peak) is extracted from subsegment 2 of dataset I, under the segmentation paradigm $\{12,1,128,\{1,128\}\}$. The red curve (with rightmost peak) is theoretical—calculated from the Philip formula for $F_3(v;146700,107900,2730)$. The blue "excess curve" is the point-wise curve difference (amplified 3×), and can be used in our "probagram" plots to show excess as a function of section depth y. The expected separations within this cuboid turn out to be $\langle v \rangle = 62018,66789$ for brain, random, respectively.

4.1 Cuboid statistics

Not just the exact expectation $\Delta_3(1;a,b,c)$ but the very probability density $F_3(v;a,b,c)$ has been worked out by J. Philip [11]. Both exact expressions in terms of a,b,c are quite formidable—see Appendix: Exact-density code for a programmatic way to envision the complexity. By probability density, we mean

$$\text{Prob}\{|\mathbf{r}-\mathbf{q}| \in (v,v+dv)\} = F_3(v;a,b,c)\,dv,$$

On the fractal distribution of brain synapses   9

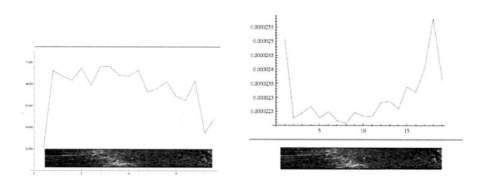

Fig. 6 Results for cuboid expectations of separation v and $1/v$ for cuboids of the type in Figure 4, running over all y-depth. (The dark green horizontal strip represents the full sample, oriented left-right for these plots.) In both left- and right-hand plots, the horizontal red line is the calculated from the exact formula for $\Delta_3(1;a,b,c)$. The segmentation paradigm here is $\{12,2,80,\{1,32\}\}$, dataset I.

hence we have a normalization integral with upper limit being the long cuboid diagonal:

$$\int_0^{\sqrt{a^2+b^2+c^2}} F_3(v;a,b,c)\,dv = 1.$$

More generally we can represent the moment Δ_3 in the form

$$\Delta_3(s;a,b,c) = \int_0^{\sqrt{a^2+b^2+c^2}} v^s F_3(v;a,b,c)\,dv.$$

The Philip density for separation v can also be used directly to obtain the density for a power of v, so

$$f_3(X := v^s;a,b,c) = \frac{1}{|s|}X^{\frac{1}{s}-1}F_3(X^{\frac{1}{s}};a,b,c).$$

For example, if we wish to plot the density of inverse separation $X := 1/v$ for a random point-cloud, we simply plot $X^{-2}F_3(1/X;a,b,c)$ for X running from $1/\sqrt{a^2+b^2+c^2}$ up to infinity; the area under this density will be 1.

5 Fractal dimension

For the present research we used two fractal-measurement methods: The classical box-counting method, and a new, space-fill method. For a survey of various fractal-dimension definitions, including estimates for point-cloud data, see [6].

10   Richard Crandall

As for box-counting, we define a box dimension

$$\delta := \lim_{\varepsilon \to 0} \frac{\log \#(\varepsilon)}{-\log \varepsilon} \, ,$$

where for a given side ε of a microbox, $\#(\varepsilon)$ is the number of microboxes that minimally, collectively contain all the points of the point-cloud. Of course, our clouds are always finite, so the limit does not exist. But it has become customary to develop a $\#$-vs.ε curve, such as the two curves atop Figure 7, and report in some sense "best slope" as the measured box dimension.

There are two highly important caveats at this juncture: We choose to *redefine* the box-count number, as

$$\# \to \# \cdot \frac{1}{1 - e^{-N\varepsilon^3}} \, ,$$

when the cloud has N total points. This statistical warp factor attempts to handle the scenario in which microboxes are so small that the finitude of points causes many empty microboxes. Put another way: The top curve of the top part of Figure 7—which curve should have slope 3 for N random points—stays straight and near slope 3 for a longer dynamic range because of the warp factor.

The second caveat is that we actually use not ε-microboxes, but microcoboids. When the segment being measured is originally of sides (a, b, c), we simply rescale the cuboid to be in a unit box, which is equivalent to using a "microbrick" whose aspect ratios are that of the cuboid, and transforming that microbrick to a cube of side $\varepsilon := (abc)^{1/3}$.

5.1 Space-fill method for fractal measurement

During this research, we observed that a conveniently stable fractal-measurement scheme exists for point-cloud datasets. We call this method the "space-fill" algorithm, which runs like so:[4]

1. Assume a 3-dimensional unit cube containing a point-cloud, and construct a Hilbert space-filling curve, consisting of discrete visitation points $\mathbf{H}(t)$, where t runs over the integers in $[0, 2^{3b} - 1]$. (The resolution of this curve will be b binary bits per coordinate, therefore.)
2. Create a list of "pullback" rationals $t_k/2^{3b}$, corresponding to the points r_k of the point-cloud data.
3. Perform a 1-dimensional sort on the set of pullbacks, and measure the fractal dimension δ_1 using a simple interval counter.
4. Report fractal dimension of the point-cloud data as $\delta = 3 \cdot \delta_1$.

---

[4] The present author devised this method in 1997, in an attempt to create "$1/f$" noise by digital means, which attempt begat the realization that fractal dimension could be measured with a Hilbert space-fill.

On the fractal distribution of brain synapses   11

We do not report space-fill measurements herein—all of the results and figures employ the box-counting method—except to say, a) the space-fill method appears to be quite stable, with the fractagram surfaces being less noisy; and b) the dimensions obtained in preliminary research with the space-fill approach are in good agreement with the box-counting method. Figure 8 pictorializes the space-fill algorithm.

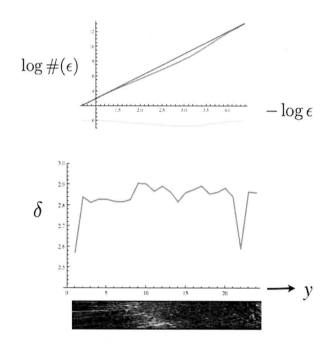

Fig. 7 Fractal-dimension measurement. Within a given cuboid we use the standard box-counting method; namely, in the upper figure is plotted $\log\#$ vs. $\log(1/\varepsilon)$ for random points (upper, blue curve), then for the actual synapse points (lower, red curve), with the excess as the green (lowest) plot. In the bottom figure, we use the excess to estimate fractal dimension for each cuboid in a segmentation paradigm $\{12, 2, 80, \{2, 80\}\}$. Evidently, the fractal dimension fluctuates depending on layer characteristics at depths y, with an average fractal dimension of ~ 2.8 for the whole of dataset I.

6 Probagrams, paramegrams, and fractagrams

Our 'grams we have so coined to indicate their 3-dimensional-embedding character.[5] Each 'gram is a surface, one of whose support dimensions is the section depth y. In our 'grams, as in the original synapse datasets, $y = 0$ is the outside (pial) surface, while y increases into the brain sample. We typically have, in our 'grams, *downward increasing* y, so that the top of a 'gram pictorial is the outside surface.

---

[5] As in "sonogram"—which these days can be a medical ultrasound image, but originally was a moving spectrum, like a fingerprint of sound that would fill an entire sheet of strip-chart.

12   Richard Crandall

$$0 \underline{\hspace{8cm}} 1$$

Fig. 8 The "space-fill" method for measuring point-cloud dimension. This algorithm as described in the text yields similar results to the more standard box-counting method, yet preliminary research reveals the space-fill method to be rather more stable with respect to graph noise. The basic idea is to create a set of pullbacks on the line $[0,1)$, then use a quick sort and a simple 1-dimensional fractal assessment.

Precise definitions are:

- Probagram: Surface whose height is probability density of a given variable within a cuboid, horizontal axis is the variable, and the vertical axis is the y-depth into the sample.
- Paramegram: Surface whose height is a parameterized expectation (such as our function $G_3(k; a, b, c)$, horizontal axis is the parameter (such as k), and the vertical axis is y-depth.
- Fractagram: Surface whose height is the excess between the fractal-slope curve for a random cloud in a cuboid and the actual data cloud's fractal-slope curve, horizontal axis is $-\log \varepsilon$, and vertical axis is as before the y-depth.

In general, we display these 'grams looking down onto the surface, or possible at a small tilt to be able to understand the surface visually.

What we shall call a segmentation paradigm is a set P of parameters that determine the precise manner in which we carve (a, b, c)-cuboids out of a full synapse dataset. Symbolically,

$$P := \{M, G, H, \{b, e\}\},$$

where

- M is the "magnification" factor—the y-thickness of a cuboid divided into the full y-span of the dataset.
- G is the "grain"—which determines the oversampling; $1/G$ is the number of successively overlapping cuboids in one cuboid.

On the fractal distribution of brain synapses   13

- H is the number of histogram bins in a 'gram plot, and we plot from bin b to bin e.

We generally use $g < 1$ to avoid possible alias effects at cuboid boundaries. The total number of cuboids analyzed in a 'gram thus turns out to be

$$S = 1 + G(M-1).$$

For example, with grain $G = 3$ and $M = 10$, we calculate over a total of 28 cuboids. This is because there are generally $G = 3$ cuboids overlapping a given cuboid. In any case, one may take cuboid dimensions a, b, c as

$$a = x_{max} - x_{min}; \; b = \frac{y_{max} - y_{min}}{M} ; c = z_{max} - z_{min},$$

where min, max coordinates are deduced from the data. (In our 'grams, we continually recompute the min, max for every cuboid to guard against such as corner holes in the data.)

14   Richard Crandall

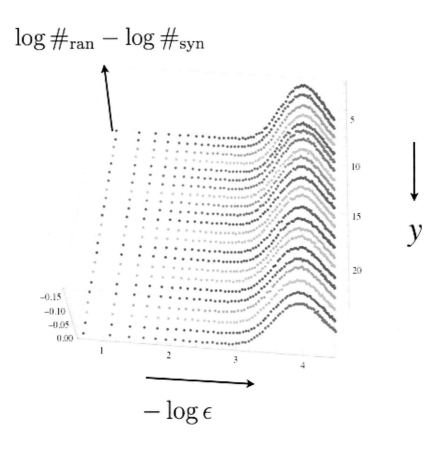

Fig. 9 The "fractagram" concept—which is similar for probagrams and paramegrams. For each cuboid in a given segmentation paradigm (here, paradigm $\{12, 2, 80, \{2, 80\}\}$) we generate the fractal-slope excess as in Figure 7. The resulting "strands" of fractal data vs. y-depth in the dataset (here, dataset I) are much easier to interpret if plotted as a surface, which surface we then call a fractagram as pictured in Figure 11.

On the fractal distribution of brain synapses   15

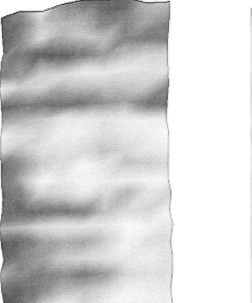

Fig. 10 A baseline experiment. At left is the probagram for dataset I and density $f_3(X := 1/v^2; a, b, c)$ under segmentation paradigm $\{12, 2, 16, \{2, 10\}\}$. At the right is the result of using the same number of points ($N = 1119299$) randomly placed within the full sample cuboid. This kind of experiment shows that the brain synapses are certainly *not* randomly distributed.

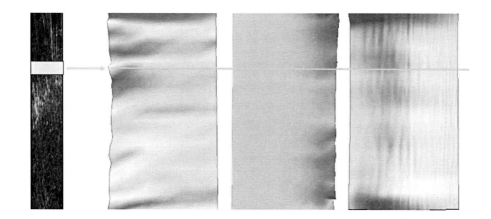

Fig. 11 Typical set of three 'grams: At far left is a pictorialized a full section-sample, with a small box indicating a cuboid subsection. As said section is moved downward (increasing y), we obtain, left-to-right, and for separation $v := |\mathbf{r} - \mathbf{q}|$, the probagram for v^{-1}, then the paramegram for $\langle \exp(-kv^2) \rangle$, then the fractalgram. The phenomenon of neural layering is evident, and qualitatively consistent (either correlated or anticorrelated) across all three 'grams for this sample (dataset I, detailed in Appendix: Synapse datasets).

16   Richard Crandall

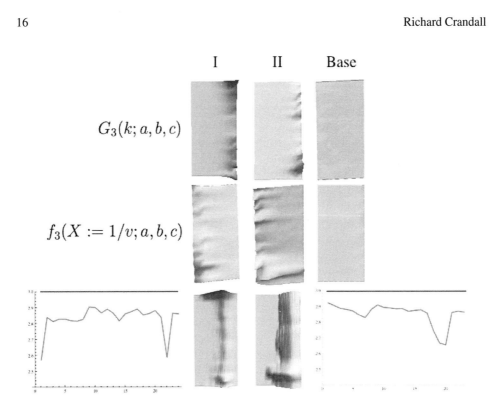

Fig. 12 The 'grams for the synapse-location datasets I, II. The top row shows G_3 paramegrams and baseline test for segmentation paradigm $\{12, 2, 32, \{1, 10\}\}$. The second row shows probagrams for inverse separation $1/v$, in the same segmentation paradigm. The two 3D plots at bottom are the fractagrams. At far-left and far-right bottom are graphical displays of the per-cuboid fractal-dimension estimate. Note that the baseline test here is for a randomly filled cuboid; the horizontal lines at dimension 3.0 really are less noisy than one pixel width. Thus the datasets I, II can be said both to have overall fractal dimension 2.8 ± 0.5, although the dimension is evidently neural-layer dependent.

7 How do we explain the observed fractal dimension?

Let us give an heuristic argument for the interaction of cuboid expectations and fractal dimension estimates. Whereas the radial volume element in 3-space is $4\pi r^2 dr$, imagine a point-cloud having the property that the number of points a distance r from a given point scales as $r^{\delta-1}$ where $\delta < 3$, say. Then, if the characteristic size of a point sample is R (here we are being rough, avoiding discussion of the nature of the region boundaries), we might estimate an expectation for point-separation v to the s-th power as

$$\langle v^s \rangle \sim \frac{\int_0^R u^s u^{\delta-1}\, du}{\int_0^R u^{\delta-1}\, du}.$$

Note that we can avoid calculation of a normalization constant by dividing this way, to enforce $\langle v^0 \rangle = 1$. This prescription gives the estimate

On the fractal distribution of brain synapses   17

$$\langle v^s \rangle \sim \frac{\delta}{s+\delta} R^s,$$

showing a simple dependence on the fractal dimension δ. In fact, taking the left-hand plot of Figure 6 we can right off estimate the fractal dimension of the whole dataset as

$$\delta \sim 2.6,$$

not too off the mark from our more precise fractal measurements that we report as 2.8 ± 0.05.

So one way to explain our discovered fractal dimension $\sim 2.8 < 3$ for both datasets is to surmise that the distance metric is weighted in some nonuniform fashion.

7.1 Generalized Cantor fractals

One aspect undertaken during the present research was to attempt to fit the observed fractal properties of the datasets to some form of Cantor fractal. There is a way to define a generalized Cantor fractal in n dimensions so that virtually any desired fractal dimension in the interval $[n\frac{\log 2}{\log 3}, n]$ (see [2]).[6] Such generalized Cantor fractals were used to fine-tune our fractal measurement machinery.

Interestingly, the cuboid expectations for dataset II seem qualitatively resonant with the corresponding expectations for a certain generalized Cantor set called $C_3(\overline{33111111})$ having dimension $\delta = 2.795\ldots$. However, dataset I does *not* have similar expectations on typical cuboids. For one thing, the highest-peak curve in Figure 5—which is from a cuboid within dataset I—shows $\langle v \rangle$ for the laboratory data being less than the same expectation for random data; yet, a Cantor fractal tends to have such expectation *larger* than random data.

We shall soon turn to a different fractal model that appears to encompass the features of both datasets. But first, a word is appropriate here as to the meaning of "holes" in a dataset. Clearly, holes in the laboratory point clouds will be caused by the simple fact of synapses not subsisting within large bodies.[7] So, too, Cantor fractals can be created by successive removal of holes that scale appropriately. But here is the rub: The existence of holes *does not in itself necessarily alter fractal dimension*.[8] For example, take a random cloud and remove large regions, to create essentially a swiss-cheese structure in between whose holes are equidistributed points. The key is, fractal-measurement machinery will still give a dimension very close to $\delta = 3$.

---

[6] Mathematically, the available fractal dimensions for the generalized Cantor fractals are dense in said interval.

[7] Synapses live on dendrites, exterior to actual neurons.

[8] Of course, the situation is different if hole existence is connected with microscopic synapse distribution; e.g. if synapses were to concentrate near surfaces of large bodies, say.

18   Richard Crandall

7.2 "Bouquet" fractal as a possible synapse-distribution model

We did find a kind of fractal that appears to lend itself well to comparison with synapse distributions.[9] We shall call such artificial constructs "bouquet" fractals. A generating algorithm to create a bouquet point-cloud having N points runs as follows:

1. In a unit 3-cube, generate N_0 random points (N_0 and other parameters can be used to "tune" the statistics of a bouquet fractal). Thus the point-cloud starts with population N_0.
2. Choose an initial radius $r = r_0$, a multiplicity number m, and a scale factor $c < 1$.
3. For each point in the point-cloud, generate m new points a mean distance r away (using, say, a normal distribution with deviation r away from a given point). At this juncture the point-cloud population will be $N_0 \cdot m^k$ for k being the number of times this step (3) has been executed. If this population is $\geq N$, goto step (5).
4. Reduce r by $r = c \cdot r$ and goto step (3).
5. Prune the point-cloud population so that the exact population is achieved.

The bouquet fractal will have fractal dimension on the order of

$$\delta \sim \frac{\log m}{-\log c},$$

but this is an asymptotic heuristic; in practice, one should simply tune all parameters to obtain experimental equivalencies.[10] For example, our dataset I corresponds interestingly to bouquet parameters

$$\{N_0, r_0, m, c\} = \{1000, N_0^{-1/3}, 23, 1/3\}.$$

The measured fractal dimension of the resulting bouquet for population $N = 1119299$ is $\delta \sim 2.85$ and statistical moments also show some similarity.

Once again: Something like a bouquet fractal may not convey any neurophysiological understanding of synapses locations, but there could be a diagnostic parameter set—namely that set for which chosen statistical measures come out quantitatively similar.

---

[9] Again, we are not constructing here a neurophysiological model; rather, a phenomenological model whose statistical measures have qualitative commonality with the given synapse data.

[10] The heuristic form of dimension δ here may not be met if there are not enough total points. This is because the fractal-slope paradigm has low-resolution box counts that depend also on parameters N_0, r.

On the fractal distribution of brain synapses   19

7.3 Nearest-neighbor calculus

Another idea that begs for further research is to perform nearest-neighbor calculus on synapse cuboids. This is yet a different way to detect departure from randomness.

In an n-dimensional unit volume, the asymptotic behavior of the nearest-pair distance for N uniformly-randomly-placed points, namely

$$\mu_1 := \langle \min |\mathbf{r} - \mathbf{q}| \rangle_{\mathbf{r}, \mathbf{q} \in V}$$

is given—in its first asymptotic term—by

$$\mu_1 \sim \Gamma\left(1 + \frac{1}{n}\right) \frac{2^{1/n}}{\sqrt{\pi}} \Gamma^{1/n}\left(1 + \frac{n}{2}\right) \frac{1}{N^{2/d}} + \cdots$$

In our $(n = 3)$-dimensional scenarios, we thus expect the nearest-pair separation to be

$$\mu_1 \sim \frac{\Gamma(4/3)}{\pi^{1/3}} \frac{1}{N^{2/3}} \approx \frac{0.6097}{N^{2/3}}.$$

It is interesting that this expression can be empirically verified with perhaps less inherent noise than one might expect.

Presumably a nearest-pair calculation on the synapse distributions will reveal once again significant departures from randomness. What we expect is a behavior like so:

$$\mu_1 \sim \frac{\text{constant}}{N^{2/\delta}}$$

for fractal dimension δ. Probably the best research avenue, though, is to calculate so-called k-nearest-pairs, meaning ordered k-tuples of successively more separate pairs, starting with the minimal pair, thus giving a list of expected ordered distances $\mu_1, \mu_2, \ldots, \mu_k$.

8 Acknowledgements

The author is grateful to S. Arch of Reed College, as well as N. Weiler, S. Smith and colleagues of Smithlab at Stanford Medical School, for their conceptual and algorithmic contributions to this project. T. Mehoke aided this research via statistical algorithms and preprocessing of synapse files. Mathematical colleague T. Wieting supported this research by being a selfless, invaluable resource for the more abstract fractal concepts. D. Bailey, J, Borwein, and M. Rose aided the author in regard to experimental mathematics on fractal sets. This author benefitted from productive discussions with the Advanced Computation Group at Apple, Inc.; in particular, D. Mitchell provided clutch statistical tools for various of the moment analyses herein.

20   Richard Crandall

Appendix 1: Synapse datasets

File, voxel nm×nm×nm	N	(x_{min}, x_{max})	(y_{min}, y_{max})	(z_{min}, z_{max})
I KDM-100824B 100x100x70	1119299	$(2800, 151300)$	$(2300, 1298000)$	$(105, 2835)$
II mMos3_Syn 100x100x200	1732051	$(100, 103400)$	$(100, 1252600)$	$(105, 4095)$

Table 1 Synapse dataset characteristics. The point-cloud population N exceeds 10^6 for each dataset. The min, max parameters have been converted here to nm.

Referring to Table 1: Both datasets I, II are from adult-mouse "barrel cortex" which is a region of the somatosensory neocortex involved in processing sensation from the facial whiskers (one of the mouse's primary sensory modalities). The long y-axis of the volumes crosses all 6 layers of the neocortex (these are layers parallel to the cortical surface, and the long axis is perpendicular to the surface).

Neurophysiological considerations including array-tomography technology are discussed in references [13, 14, 12] and web URL [15]; we give a brief synopsis:

Array tomography (AT) is a new high-throughput proteomic imaging method offering unprecedented capabilities for high-resolution imaging of tissue molecular architectures. AT is based on (1) automated physical, ultrathin sectioning of tissue specimens embedded in a hydrophilic resin, (2) construction of planar arrays of these serial sections on optical coverslips, (3) staining and imaging of these two-dimensional arrays, and (4) computational reconstruction into three dimensions, followed by (5) volumetric image analysis. The proteomic scope of AT is enhanced enormously by its unique amenability to high-dimensional immunofluorescence multiplexing via iterative cycles of antibody staining, imaging and antibody elution.

Appendix 2: Exact-density code

```
(* Evaluation of the exact Philip density F3[v,a,b,c]
 for an (a,b,c)-cuboid. *)

h11[u_, a_, b_, c_] := 1/(3 a^2 b^2 c^2) *
      If[u <= b^2, -3 Pi b c u + 4 b u^(3/2),
         If[u <= c^2,
              4 b^4 + 6 b^2 c Sqrt[u - b^2] -
              6 b c u ArcSin[b/Sqrt[u]],
              If[u <= b^2 + c^2, 4 b^4 + 6 b^2 c *
```

```
                        Sqrt[u - b^2] +
                        6 b c u (ArcCos[c/Sqrt[u]] -
                        ArcSin[b/Sqrt[u]]) -
                        2 b (2 u + c^2) Sqrt[u - c^2],
                        0
                    ]
                ]
        ];

    h12[u_, a_, b_, c_] := 1/(6 a^2 b^2 c^2) *
            If[u <= a^2,
                12 Pi a b c Sqrt[u] - 6 Pi a (b + c) u +
                8 (a + c) u^(3/2) - 3 u^2,
                If[u <= c^2,
                    5 a^4 - 6 Pi a^3 b +
                    12 Pi a b c Sqrt[u] +
                    8 c u^(3/2) - 12 Pi a b c *
                    Sqrt[u - a^2] - 8 c *
                    (u - a^2)^(3/2) -
                    12 a c u ArcSin[a/Sqrt[u]],
                    If[u <= a^2 + c^2,
                        5 a^4 - 6 Pi a^3 b +
                        6 Pi a b c^2 -
                        c^4 + 6 (Pi a b + c^2) u +
                        3 u^2 - 12 Pi a b c *
                        Sqrt[u - a^2] -
                        8 c (u - a^2)^(3/2) -
                        4 a (2 u + c^2)*
                        Sqrt[u - c^2] +
                        12 a c u *
                    (ArcCos[c/Sqrt[u]]-ArcSin[a/Sqrt[u]]),
                        0
                    ]
                ]
            ];

    h22[u_, a_, b_, c_] := 1/(3 a^2 b^2 c^2) *
            If[u <= a^2, 0,
                If[u <= a^2 + b^2,
                    3 Pi a^2 b (a + c) - 3 a^4 -
                    6 Pi a b c Sqrt[u] +
                    3 (a^2 + Pi b c) u +
                    (6 Pi a b c - 2 (b + 3 c) a^2-4 b u)*
                    Sqrt[u - a^2] -
                    6 a b u ArcSin[a/Sqrt[u]],
```

22   Richard Crandall

```
              If[u <= a^2 + c^2,
                   3 a^2 b (Pi a - b) - 4 b^4-
                   12 a b c Sqrt[u]*
ArcSin[b Sqrt[u]/(Sqrt[a^2 + b^2] * Sqrt[u - a^2])]]-
                   6 a c (a - Pi b) Sqrt[u - a^2]-
                   6 c (b^2 - a^2 +
                   2 a b ArcSin[a/Sqrt[a^2 + b^2]])*
                   Sqrt[u - a^2 - b^2] -
                   6 a b (a^2 + b^2)*
                   ArcSin[a/Sqrt[a^2 + b^2]] +
                   6 b c (a^2 + u) *
                   ArcSin[b/Sqrt[u - a^2]],
                   3 a^2 (a^2 - b^2 - c^2)-4 b^4-
                   3 a^2 u - 12 a b c  Sqrt[u]  *
(ArcSin[b Sqrt[u]/(Sqrt[a^2 + b^2] * Sqrt[u-a^2])]-
ArcCos[a c/(Sqrt[u - c^2] Sqrt[u - a^2])]) +
                   2 b (a^2 + c^2 + 2 u) *
                   Sqrt[u - a^2 - c^2] -
                   6 c *
(b^2 - a^2 + 2 a b ArcSin[a/Sqrt[a^2 + b^2]]) *
                   Sqrt[u - a^2 - b^2] -
                   6 a b (a^2 + b^2) *
                   ArcSin[a/Sqrt[a^2 + b^2]] +
                   6 b c (a^2 + u) *
(ArcSin[b/Sqrt[u - a^2]] - ArcCos[c/Sqrt[u - a^2]])+
                   6 a b (c^2 + u) *
                   ArcSin[a/Sqrt[u - c^2]]
                 ]
                ]
              ];

h32[u_, a_, b_, c_] := h22[u, b, a, c];

h33[u_, a_, b_, c_] := 1/(6 a^2 b^2 c^2) *
        If[u <= b^2, 0,
            If[u <= a^2 + b^2,
                3 (2 Pi a b + b^2 + u) (u - b^2) -
                4 c (b^2 + 3 Pi a b + 2 u) *
                Sqrt[u - b^2],
                If[u <= b^2 + c^2, 3 (a^2 + b^2)^2 -
                    3 b^4 + 6 Pi a^3 b -
                    4 c (b^2 + 3 Pi a b + 2 u) *
                    Sqrt[u - b^2] +
                    4 c (a^2 + b^2 + 3 Pi a b + 2 u)*
                    Sqrt[u - a^2 - b^2],
```

On the fractal distribution of brain synapses   23

```
      3 (a^2 + b^2)^2 + c^4 +
            6 Pi a b (a^2 + b^2 - c^2) -
            6 (Pi a b + c^2) u - 3 u^2 +
            4 c (a^2 + b^2 + 3 Pi a b + 2 u)*
            Sqrt[u - a^2 - b^2]
      ]
      ]
    ];


(* Next, the Philip density function for separation v.
   It must be arranged that a <= b <= c. *)

F3[v_, a_, b_, c_] :=
  2 v (h11[v^2, a, b, c] + h12[v^2, a, b, c] +
  h22[v^2, a, b, c] + h32[v^2, a, b, c] +
  h33[v^2, a, b, c]);
```

24   Richard Crandall

References

1. H. Adesnik and M. Scanziani, "Lateral competition for cortical space by layer-specific horizontal circuits," *Nature*, Vol 464:22 (April 2010).
2. D. Bailey, J. Borwein, R. Crandall, and M. Rose, "Box integrals on fractal sets," preprint (2012).
3. D. Bailey, J. Borwein, and R. Crandall, "Box integrals," *Journal of Computational and Applied Mathematics* 206, 196-208 (2007).
4. D. Bailey, J. Borwein, R. Crandall, "Advances in the theory of box integrals," *Math. Comp.* 79, 1839-1866 (2010).
5. J. Borwein, O-Y Chan, and R. Crandall, "Higher-dimensional box integrals," *Exp. Math.*, Vol. 19, No. 3, pp. 431–445 (2010).
6. K. J. Clarkson, "Nearest-Neighbor Searching and Metric Space Dimensions," in *Nearest-Neighbor Methods in Learning and Vision: Theory and Practice*, eds G. Shakhnarovich et al., MIT Press (2005).
7. M. Fischler, "Distribution of Minimum Distance Among N Random Points in d dimensions," Technical Report FERMILAB-TM-2170 (May 2002).
8. S. Georgiev et al. "EEG Fractal Dimension Measurement before and after Human Auditory Stimulation ," *BIOAUTOMATION*, 12, 70-81 (2009).
9. A. Grski and J. Skrzat, "Error estimation of the fractal dimension measurements of cranial sutures," *J Anat.*, 208(3): 353359 (March 2006).
10. N. Lynnerup and J. C. Jacobsen, "Brief communication: age and fractal dimensions of human sagittal and coronal sutures," *Am J Phys Anthropol.*, Aug;121(4):332-6 (2003).
11. J. Philip, "The Probability Distribution of the Distance between Two Random Points in a Box," TRITA MAT 07 MA 10 (Dec 2007).
12. K. D. Micheva and S. J. Smith, "Array tomography: A new tool for imaging the molecular architecture and ultrastructure of neural circuits," *Neuron* 55:25-36 (2007).
13. K. D. Micheva, B. Busse, N. C. Weiler, N. O'Rourke, and S. J. Smith, "Single-synapse analysis of a diverse synapse population: proteomic imaging methods and markers," *Neuron*, Nov 18;68(4):639-53 (2010).
14. K. D. Micheva, N. O'Rourke, B. Busse, and S. J. Smith, "Array Tomography: High-Resolution Three-Dimensional Immunofluorescence," In: *Imaging: A Laboratory Manual*, 3rd Ed. Cold Spring Harbor Press, Ch. 45, pp. 697-719 (2010).
15. Smithlab: Array tomography,
http://smithlab.stanford.edu/Smithlab/Array_Tomography.html
16. N. Spruston, "Pyramidal neurons: dendritic structure and synaptic integration," *Nature Reviews Neuroscience*, 9, 206-221 (March 2008).
17. M. Teich et al., "Fractal character of the neural spike train in the visual system of the cat," *J. Opt. Soc. Am. A* , Vol. 14, No. 3 (March 1997).